Naked Science

Naked Science

Anthropological Inquiry into Boundaries, Power, and Knowledge

edited by
Laura Nader

Routledge
New York & London

Published in 1996 by
Routledge
29 West 35th Street
New York, NY 10001

Published in Great Britain by
Routledge
11 New Fetter Lane
London EC4P 4EE

Copyright © 1996 by Routledge

All rights reserved. No part of this book may be reprinted or reproduced or utilized in any form or by any electronic, mechanical, or other means, now known or hereafter invented, including photocopying and recording or in any information storage or retrieval system, without permission in writing from the publishers.

Library of Congress Cataloging-in-Publication Data
Naked science : anthropological inquiry into boundaries, power, and knowledge / Laura Nader, editor.
     p. cm.
   Includes bibliographical references.
   ISBN 0-415-91464-7. — ISBN 0-415-91465-5 (pbk.)
    1. Anthropology—Philosophy. 2. Science—Philosophy. 3. Science—Social aspects. 4. Knowledge, Sociology of. 5. Power (Social sciences) I. Nader, Laura.
GN33.N35 1996
301'.01—dc20   95-23650
                                                                                      CIP

In recognition of the U.S. taxpayer for generous support of science

List of Figures and Tables

Figures

1   Carolinian Navigator's World.
2   The "Star Structure" (Sidereal Compass).
3   "Island Looking" Exercise.
4   "Drags" on Course from Puluwat to Truk.
5   The "Great Trigger Fish."
6   Carolinian's Schematic Representation of Linked "Trigger Fish."
7   Location of the Tzeltal and Tzotzil Maya.
8   Ethnoepidemiological Reports of Health Problems Grouped by Major Ethnomedical Category among the Highland Maya.
9   Reports of Gastrointestinal Conditions in Ethnoepidemiological Surveys

Tables

1   Major Gastrointestinal Conditions and Their Key Characteristics
2A   Key Signs and Symptoms of the Most Important Diarrheas and Worms
2B   Key Signs and Symptoms of Gastrointestinal Pains
3   Five Most Frequently Mentioned Species Used in the Treatment of Bloody Diarrhea, Tzeltal Traveling Herbarium
4   Three Most Frequently Mentioned Health Conditions for the Five Top-Ranked Species Used in the Treatment of Bloody Diarrhea, Tzeltal Traveling Herbarium
5   Principal Species Employed in the Treatment of Gastrointestinal Disease among the Tzeltal and Tzotzil
6   Proportional Distribution of Medical Use Reports for the Principal Species Employed in the Treatment of Gastrointestinal Disease by the Tzeltal Maya

7 Proportional Distribution of Medical Use Reports for the Principal Species Employed in the Gastrointestinal Disease by the Tzotzil Maya

8 Results of Pharmacological Screening and Literature Search on Principal Species Employed in the Treatment of Gastrointestinal Conditions

9 Common Diarrheal Disease Agents by Class and Stool Characteristics

10 Organs and Probable Manifestations in Cases of Abdominal Pain

11 Association of Pharmacological Effects with Types of Gastrointestinal Pain

12 Summary of Authors' Credentials

13 Past and Present Directors, Los Alamos and Livermore Laboratories

14 Assistant to the Secretary of Defense for Atomic Energy and Chair of the Military Liaison Committee

Preface

This volume is the result of a desire to join two bodies of anthropological research that rarely meet. Consideration of both ethnoscience and technoscience research within a common framework is long overdue as a means for raising questions about the deeply held beliefs and assumptions we all carry about scientific knowledge. We need a perspective on how to regard different science traditions because public controversies over science should not be reduced to polarization and polemics asserting a glorified science or a despicable science. The development of science and technology in the West considerably overlaps with beliefs in material and social progress. But in the course of this century there is a growing awareness of the ideological nature of such belief, for domination of nature has double-edged consequences, and science and technology are not necessarily synonymous with social progress.

Our book is about science and power, particularly the power of Western science over the other "sciences" around the globe. It is thus about the contested domains of science. The chapters cover a variety of research on science and include the study of several different science cultures: physics, molecular biology, primatology, immunology, and ecology, as well as environmental, medical, mathematical, and navigational domains. Everything in this volume rests on the key assumption that science is not autonomous, that it draws its form from its social and cultural roots, a linkage that scholars have made many times. But the book is distinguished by its perspective across cultures—the study of different sciences in a global context. Examining empirical knowledge systems within a planetary frame forces us to think about boundaries erected by Western scientists (including anthropologists) that silence or otherwise affect knowledge building. In the insistent dressing of all science in Western garb we limit answers to the philosophical question: what is science, or more important, what is Western science? By examining the processes, some hidden, some disguised precisely because they are taken for granted, we can see a more naked science; stripped of our garb scientists look more alike across cultures.

The recent controversy over the Smithsonian exhibit "Science in American Life" indicates that even at the highest levels of educational achievement a cultural framework with which to untangle the complex threads of debates over science is absent (Vackimes 1995). There is a parochial nature to the anti-antiscience position found in popular books like *Higher Superstition* (Gross and Levitt 1994), one that points to the need for a more inclusive approach. Indeed, a global scope is a requirement for a nonethnocentric, broad-gauged understanding of science, in the plural. In this sense, the anthropological study of science is an old tradition, one that covers the laboratories, the streets, the suites, the steppes—any place that is touched by human affairs. The role of anthropology has been grounded in an awareness of the implications of traditional knowledge for scientific communities. Modern science is in part a continuation of the human effort to understand and give meaning to our existence.

Higher superstitions have always been part of culture and certainly part of our academic writings, whether to make distinctions between religion and science, or between clear thinking and the nonsense of those who still think that Western rationalist traditions can be contrasted with presumably non-Western, nonrationalist traditions (see Gutmann 1992). Another story needs to be moved to center stage, one less dogmatic and, I dare say, more reasoned. The important questions about science ideology, practice, and consequence encounter walls of resistance to reasoned thinking when issues are phrased in binary modes—science and antiscience. Such bites are all too simplistic when what has opened up is an opportunity to rethink the issues in this post–cold war period: the political events of the past fifty years through which large segments of science became heavily militarized; the enormous funding of science that often resulted in wondrous discoveries that changed the human landscape forever; the boundaries and degree to which science is an isolated activity. The isolation and insulation of science activity does not result in absence of impact, however, and academics have not been mute on the role of science in the modern world. If the chapters in this book are less marred by pure dogma, it is because the authors are not in lockstep agreement. Nor should they be.

Most of the authors in this book are anthropologists or have had anthropological training. Most have worked in cultures other than their own. One author could be described as a "native speaker"—of physics. Most agree as to the importance of examining Western science in context and in contrast to knowledge traditions elsewhere, and in relation to science boundaries that force consideration of power dimensions. Agreement, however, is not the purpose. Unlike Gross and Levitt's (1994) characterization, there is no American academic left that has concocted a single story line about the abuse and irrationality of science. Indeed if our work results in treating science as scientists treat their subjects, the chapters that range from technoscience to ethnoscience merge, and the science of culture—as anthropology used to be called—takes on the culture of science.

Although science has interested me since student days, I more or less stumbled into the anthropology of science. My interest focused in the seventies, sparked by participation in public policy debates on energy and sustainability. I was inspired to understand the workings of science and scientists, and the barriers to new thinking about energy, at the National Academy of Sciences (Nader 1980). At the Department of Energy; the National Laboratories at Berkeley, Livermore, Los Alamos, and Oak Ridge; at the United Nations; and since then at local, regional, national, and international sites, I learned that scientists were a heterogeneous lot, not at all the self-satisfied drones they are sometimes made out to be. On the other hand, as a group they do often seem unaware of other possibilities, other worlds, such as that of the Mexican Zapotec, whose traditional knowledge often left me thinking about other possibilities.

The understanding of "science" across cultures has had a revolutionary effect on the way we think about knowledge more generally. Science is not free of culture; rather, it is full of it. Militarization has certainly had an effect on American science. It is also likely that conversion or the demilitarization of science cultures and institutions will stimulate different directions and content in our knowledge systems. Today we are witnessing the faltering, sometimes valiant attempts at conversion of nuclear weaponry in our national laboratories—attempts to stimulate different science directions. The transition has also fired the pervasive commercialization of scientific effort, and throughout this time there are still voices that call for a "purer" science; others call for a fuller science, one that does not so easily brush aside knowledge that has been acquired and tested for hundreds of years even though individual inventors of such knowledge are rarely known.

The growing recognition of the relevance of "traditional knowledge" to modern society requires us to eschew provincialism. In this sense the anthropological study of science, at times led by anthropologists such as Bronislaw Malinowski and Franz Boas (who were trained in physics and mathematics), has a long history. They understood that a global scope is a requirement for understanding the worldwide varieties of knowledge production and use. The assumption that Western science functions autonomously from other parts of the world is informed by findings in archaeology and ethnology. The observation that Western science does not develop independent of its laity is also informed by ethnography. The findings of anthropology and the social sciences contradict the ubiquitous images of Western science as pure, independent from politics.

In this regard, the 1991 American Association for the Advancement of Science meeting in Washington, DC was instructive.[1] The opening session was particularly dramatic. The invited keynote speaker, President George Bush, did not appear in person, but by means of high-tech television. The enlarged image of Bush that played on three screens for about twenty minutes prior to his appearance numbed the audience. There was little applause (though the telecast had its own applause), and there was no opportunity for dialogue. The speaker observed that modern weapons were making it possible to face down aggression in the Gulf War, while also noting that science and technology have brought mobility, prosperity, health, and security. The meetings that followed were dominated by strategy talk and exhibits that further enhanced the commodification of science while bemoaning the state of scientific literacy.[2]

It is no wonder that the recent ferment in debates about science involves discussion of the worth of science, its function, its cultural ascendency, its ethnocentricity, and its universality.[3] These issues are being discussed by scholars in history, philosophy, medicine, economics, science education, sociology; by scientists; by those who call themselves deprofessionalized scholars; and by the general public.

The anthropological contribution to these debates is critical to a relocation and a rethinking of the future of Western science traditions at a time when the Western myth of total superiority is shrinking. We need not be defensive, but we do need to move in new directions. As I explain later on, this volume is not a conversation about the academic and interdisciplinary aspects of the anthropology of science or technology and science studies in the United States. Nor is the book a summary of current work in the anthropology of science. It is about the need to chart a more inclusive paradigm.

The authors in this book iterate the importance of empirical appraisal to avoid distortions of the value or lack of value of Western science, and to avoid the problem of undervaluing how lasting knowledge is produced in the first place. While there is tension and provocation in this collection, a common concern is clarified—to wit, boundaries, demarcations, and margins as important aspects of present debates over the official science stories that distinguish the West from the rest. A naked science turns out to be more like the worlds in which it is embedded than something of a different order.

Laura Nader
Berkeley, California
May 1995

Notes

1. With colleagues from sociology, political science, geography, and physics, anthropologists presented a four-part symposium at the 1991 AAAS meetings under the rubric of the Anthropology of Science and Scientists: 1) Is Science Universal? 2) the Study of Knowledge Formation and Its Use, 3) the Behavior of Scientists, and 4) Science Traditions across Cultures. The participants were mainly self-selected, having responded to a widely circulated letter of invitation that I sent out as Head of Section H for Anthropology. Some of the papers have developed into this volume. Other papers by participants not included here were seminal to the discussion: D. Irvine, Cultural Survival; U. Fleising, University of Calgary; A. Young, McGill University; L. Lieberman, Central Michigan University; W. Chambliss, George Washington University.

    A series of contemporary developments moved me to organize this symposium: the intolerance for dissent in our culture, an intolerance that reverberates into the scientific community thereby truncating the creative impulse; the increasingly ideological nature of thinking about science whether pro or anti (related to the first), and the nature and direction of creative research in the sciences; the absence of mutual respect and the presence of mutual ignorance between scientists and laypersons; and the relatively undemocratic nature of decision making in and about science in our country with effects around the globe.

2. During that same period the editor of *Science,* journal of the AAAS, wrote that science advances "can remove some of war's barbarities," and bring about more "civilized" conduct (Koshland 1991). The self-deception was extraordinary, ignoring the whole-

sale destruction endured by the Iraqi populace during the Gulf War because of assertions that smart weapons spare civilians.
3. Mary Midgley (1992b) attributes the critical examination of science as a reaction to exaggerated claims of salvation through science to solve problems of poverty, pollution, sickness, and overpopulation. Others point to the stance of the nonnegotiability (that is, dismissal of dissent from the outside) of scientific assumptions, methods, and knowledge, and the depreciation of other forms of knowledge as of marginal value.

Acknowledgments

For encouragement and criticism, for helping me formulate my ideas about science, for comments on draft materials, for ideas about how best to articulate this endeavor, for titles, for helping to pull the book together, for typing and retyping, and for all that one dreams for in an editor I recognize the following people. They know what their part in this effort was, and I know that without them, the contributors, and many more, I might not have persisted: Di Beach, Nancy Chaisson, Elizabeth Colson, Troy Duster, Roberto Gonzalez, Hugh Gusterson, Ellen Hertz, Carol MacLennan, JoAnn Martin, Norman Milleron, Rania Milleron, Tarek Milleron, Ralph Nader, C. Jay Ou, Lori Powell, Anna Roosevelt, Sharon Traweek, Vivian Wang Tung, Sherwood Washburn, Marlie Wasserman, Beatrice Whiting, and John Whiting.

A sabbatical leave in Spring 1995 at the University of California, Berkeley made possible the completion of the final phase of this work.

Introduction

Anthropological Inquiry into Boundaries, Power, and Knowledge

Laura Nader

Science may refer to a body of knowledge distinguishable from other knowledge by specific methods of validation. It may define a self-conscious attitude toward knowledge and knowing that embodies curiosity with empiricism. In Western society, science also connotes an institutional setting, a set of concerns ruled by the notion of ordered rationality, a group of people united by a common competence. Science is systematized knowledge, a mode of inquiry, a habit of thought that is privileged and idealized. Much about science is taken for granted—its bounded and autonomous nature, its homogeneity, its Westernism, its messianic spirit.

In Western society, science is contrasted with superstition and with practices of the occult. Scientists are contrasted with "primitive" peoples or with laypeoples in their own society. Contemporary Western science is contrasted with sciences of other civilizations—such as that of China, India, Islam, or the Old and New World earlier civilizations—as if our science is unrelated to theirs or as if their science existed only in the past. Science is often characterized as detached from and above ordinary daily life, although science defines everyday life in innumerable ways, apart from social, political, or economic contexts, and even apart from technology. Debates about science are embroiled in definitions. These debates are found not only in the academy—they are held in backyards, in museums, on reservations, among the lay public, in and out of government agencies, and among indigenous peoples. Their purposes are manifold, and from them different worldviews emerge.

A book by A. Oscar Kwagley (1995), an Alaskan Yupiaq, illustrates this point. Kwagley is an indigenous educator dedicated to teaching science in a way that respects both traditional worldviews and the sciences of the modern world, a romantic notion that takes him the long way home. Some of the ground he walks is time-honored—for example, that science and technology destroy traditional ways of life—but he repeatedly reminds us that without a critical stance, one that recognizes complexities, new forms of knowledge are not born and old ones are silenced. Kwagley's great grandparents forbade his grandmother from attending school, saying that she would become dumb. They were alluding to the loss of knowledge through schooling, knowledge that was their means of survival:

> Once a person gets out of the village on the river or into the tundra, the homogenization and standardization brought on by man-made things slowly fades away to the heterogeneity of landscape, flora and fauna, weather, and ever-changing conditions. One becomes part of the ecological system, as if assuming the "cosmic" consciousness, a sense of oneness, a synchronicity with the universe. (1995:84)

He continues:

> Down through the millennia, the Yupiaq people produced and maintained a science and technology to support a sustainable social and economic system.... At the advent of the White society, the Yupiaq ways were pronounced primitive and savage.... (ibid)

Throughout Kwagley's book the contrasts appear: primitive/civilized, modern/traditional, native knowledge/science. These contrasts are commonly repeated in debates about science. Although the process of marking science can be traced back to the seventeenth century, anthropologists (using secondary sources from travelers, government workers, missionaries, and the like) also played a part in marking difference. In demarcating science from other systems of knowledge, they used a style formed by the very contrasts that bedevil Kwagley—science/religion, rational/magical, universal/particular, theoretical/practical, developed/underdeveloped—one that fixed hegemonic categories in the popular imagination. These dichotomous categories are powerful mind organizers, and remain a source of contention in small communities. One could argue that the demarcation of science is part of a general tendency to establish formal structures through which we think about the world, whether it be in terms of science, medicine, or art. But the demarcation of science—a keystone of modernity—is of particular interest in arguments about boundaries and power.

What is it about boundaries that makes them important to power relations? A style favored by contrast includes some things, excludes others, and creates hierarchies privileging one form of knowledge over another (Gutmann 1992). Contrast also tends to fix a positional superiority in the mind of the categorizer—the notion that one is superior by virtue of being in a position to create the categories, or to draw the lines. As one author notes: "Boundary-work describes an ideological style found in scientists' attempt to create a public image for science by contrasting it favorably..." (Gieryn 1983:78). The theme of boundaries and power is central to the conception of this volume, as is the awareness that the boundaries of science are drawn and redrawn. Borders are contentious, and as any scientist knows, science is not a revealed and unambiguous truth—today's science may be tomorrow's pseudoscience or vice versa.

What is at issue is whether a narrowly demarcated science—one restricted to contemporary Western ways of knowing—provides us with the greatest source of truth. The idea that it might is a recent cultural fact. In this regard a history of the term "science" as it evolved in the English language is instructive. It appeared in the 1300s and remained a "very general . . . term for knowledge as such" (Williams 1985:277). It was not until the late 1700s that a distinction was made between theoretical and practical knowledge, and the notion that scientific knowledge means technological power over nature can "scarcely be dated before about 1850, save in the chemical industries" (White 1967). Science became associated with theoretical knowledge and experiment, and soon came to include a particular method as well. During most of human history, there have been different ways of knowing and an enormous range of ways of understanding the world. Because the modern Western idea of science is a contemporary fact, the process of its demarcation, its construction, its dress, and how it fits as part of the dynamics of power are crucial to a critical understanding of the basis for modern knowledge about the world. For science is not only a means of categorizing the world, but of categorizing science itself in relation to other knowledge systems that are excluded.

To demarcate science from the problems it confronts, from the solutions it yields, and from the technology it engenders is to remove it from its context. Western science is generally considered an autonomous activity—separate from social, political, or economic contexts, and even, separate from technology. Science decontextualized becomes privileged, dressed up, understood by its ideology rather than its practice, lacking in reflexivity. Because ethnocentrism runs deep, creative ideas in the practices of indigenous people and our own folk are routinely overlooked, as are ways in which modern science responds to financial and intellectual reward initiatives. Furthermore, we overlook the consequences of science work—in cost, in health, and in resources. Modern Western science is a habit of mind that mirrors the compartmentalized societies in which it is embedded (Young 1972). In our collection of essays the outsider is inside and what has been previously compartmentalized is refashioned. The resultant picture departs radically from ideologies of science, and allows us to re-view science afresh, plain, and undecorated.

Demarcating Science: The Place of Anthropologists

> The idea of society is a powerful image. It is potent in its own right to control or to stir men to action. This image has form; it has external boundaries, margins, internal structure. Its outlines contain power to reward conformity and repulse attack. There is energy in its margins and unstructured areas. For symbols of society any human experience of structures, margins or boundaries is ready to hand. (Douglas 1970:114)

Anthropologists have been in the middle of debates about science, today as in the past. Some of the most distinguished early anthropologists were effective statesmen for science in arguments over what distinguishes science from other forms of knowledge and belief. Their popular writings constructed a science for the public that mirrored scientists' views of the place of science. More recently the anthropological posture included a more reflexive, critical approach, a consideration of the culture that is in and around science. But from whatever vantage point, anthropologists have been key in drawing on a long intellectual tradition to demarcate science and separate it from other forms of knowledge, thereby blurring markers of difference, and in generating new methods and questions in descriptions of the actual content of knowledge systems. As in any anthropological encounter, we are keenly aware that boundary battles about what to include and what to exclude are often arbitrary, rarely neutral, and always powerful.

Anthropology's early success in drawing lines or in recognizing multiple traditions of science has been due to the popular manner in which anthropologists wrote, and to the uses they made of other cultures, rather than to any intellectual break from Western tradition. They were in an ideal position to exploit differences in human cultures as a way of privileging one over another. After all, their subject matter included the cultures and societies of the entire world. Anthropologists were well-positioned to write about universals and to explore the diffusion of ideas and material culture worldwide, and we did both. However, because nineteenth-century anthropologists (in the spirit of the Enlightenment) delineated science for the modern age by contrast, hierarchy became endemic in analyses; science was conceptualized as separate from pseudoscience and from religion, and other cultures were used to justify the position that science could best answer fundamental questions in life.

Anthropologists bounded science by comparison with magic and religion, probably because both magic and religion were potent and competitive with the science of their times. Edward B. Tylor, Sir James Frazer, and Bronislaw Malinowski were among the important figures who formulated these contrasts. Their reading publics were diverse and their ideas and influence, I wish to underscore, remain intact among educated and influential people (including scientists) into this last decade of the twentieth century. A critical analysis of these authors and the intellectual history that framed their writings was the objective of Stanley Tambiah in his pioneering book, *Magic, Science, Religion and the Scope of Rationality* (1990), and although intellectual history is not my primary aim here, it is worth noting the bare bones of the positions of these three anthropologists because of continued and widespread acceptance of their ideas.

Edward B. Tylor (1871) wrote in the context of the entrenched nineteenth-century belief that science would ultimately destroy religion by showing people the irrationality of their myths and rituals and by improving their lives: religion would diminish in importance as a result of the increasing rationality of science. Science would gradually encroach upon the domains of magic and religion. Tylor's view of science as a replacement for religion is still widely accepted in spite of the fact that the opposite tendency is observed all around us in the resurgence of both Christian and Islamic fundamentalist religions exhibiting both antiscience and proscience positions. Today some scholars argue that religion fulfills functions not satisfied by science, or alternatively, that science itself is religion (Feyerabend 1978). The ongoing debates between biologists and creationists over questions of biological evolution and creationism provide examples of both positions.

Tylor's views were modified and popularized by Sir James Frazer. In his book *The Golden Bough* (1911–1915) Frazer used a straightforward, little-nuanced social evolutionary scheme of cognitive systems. Magic corresponded to the "lower cultures" while religion and science corresponded to the "higher cultures." Societies that were technologically developed (such as those of the English theorists who were developing these ideas) were placed on the higher end of the spectrum. Those not as technologically developed (such as the Australian aborigines) were classed on the lower end. Again comparison was key to illuminating popular concepts of progression.

Frazer's scheme, arranging magic, religion, and science as a linear development, leads the reader from the age of magic to the age of religion and, finally, to the age of science. Belief in the contemporary validity of Frazer's scheme is used as justification of development projects as a way to shrink the gap between primitive and civilized, poor and rich, and magic and science in favor of the latter, although even Frazer had an interest in the origins of scientific thinking in habits of observation found among practitioners of magic.

It is interesting that in contemporary textbook discussions of Tylor and Frazer, anthropologists concentrate on what the authors had to say about magic and religion rather than on what they wrote about science. They also ignore the manner in which science was privileged by the juxtaposition of it with magic and religion on a linear scale. In spite of the recent upsurge of interest in the anthropology of science, contemporary anthropology textbooks ignore anthropological studies of science and treat modern science as if it were beyond analysis. The same applies to considerations of Malinowski's work on the Trobriand Islanders of the Western Pacific, *Magic, Science, and Religion* (1925).

Unlike Tylor and Frazer, Malinowski (who held a doctorate in physics and mathematics) was not an armchair theorist. He was an exemplary field researcher who wrote vivid ethnography about the Trobrianders for general audiences, and an influential scholar widely read and appreciated. In his work, Malinowski made a further demarcation. Science, he argued, was a secular activity, while magic and religion belonged in the sacred domain. Malinowski's contribution was to demonstrate how magic, science, and religion interact in a single society, using examples from Trobriand gardening and canoe building. But even so, his work conceptualized the three domains as separate. To this end he demonstrated that there was a language of technology and science, as distinct from the language of magic, and that these two languages intermingled in the same Trobriand coral gardens.

The break that Malinowski did make with his predecessors was to reason that "primitive humanity was aware of the scientific laws of natural process, that all people operate within the domains of magic, science, and religion" (1925:196). Needless to say, his view touched some nerves. The study of the other thereby becomes the study of us. Even today, those who take the position that science is a universal property of humanity are roundly criticized regularly both outside and inside the discipline of anthropology. Sigfried Nadel referred to Malinowski's stance as "simply untrue" (Nadel 1957: 198), while more recently Stanley Tambiah (1990:167) stated that Malinowski's position is "simplistic."

In his classic monograph on the Azande of Africa, E. E. Evans-Pritchard (1937) reflected on the limitations of science. His focus is on questions that our science does not address, such as why a piece of fruit falls on one man's head rather than another's. Evans-Pritchard illustrates that from an internal perspective, Azande beliefs are coherent and rational, an idea developed further by Robin Horton (1967). In *The Savage Mind* (1966), Claude Levi-Strauss distinguished knowledge systems of "savages" from those of the "civilized"—the primitive, improvising "bricoleur" was contrasted with the modern, the "engineer." Meyer Fortes is quoted by Tambiah (1990:54) as noting that the Tallensi of West Africa never perform rain magic in the dry season. Elizabeth Colson (1973) describes how people use divination when they have to cope with the consequences of decision making about an unknown future where risks cannot be calculated: the British Coal Board consults with their statisticians, while the Gwembe Tonga of Zambia appeal to traditional diviners. The conclusion that anthropology draws from diverse ethnographic settings is that decision making is emotion-laden, and that knowledge is invented to create or plan for solutions to problems. These and other field ethnographies cloud the theories of the early armchair anthropologists who, unencumbered by data, made self-serving distinctions and false comparisons that have lived on long after field anthropologists discarded them in favor of different interpretations, sometimes no less self-serving. By now it is clear that Western rationality is not the benchmark criterion by which other cultural knowledge should be evaluated.

Two main research directions emerged. The first continues Malinowski's earlier work by describing knowledge in traditional society, as for example, Harold Conklin's 1954 work "The Relation of Hanuoo Culture to the Plant World," or later the work of David Hyndman (1979, 1994) and Eugene Anderson (1967), or that of Brent Berlin (1974) on botanical classification among the Tzeltal of Chiapas, Mexico. A second and more recent direction involves the ethnographic study of the sociocultural context of Western science as a category to be critically examined. In some instances, there is comparison across cultures, as in the work of Sharon Traweek (1988) on high energy physics in the United States and Japan. Other anthropologists have examined the world of biomedical science (Martin 1987; Patton 1990; Lock 1993), nuclear weapons testing sites (Gusterson 1996), high-technology factories (Dubinskas 1988), and development ideologies (Escobar 1995).

A third direction constitutes a strategy. It links studies of technoscience and other knowledge traditions, focusing on both context and content. Linking the West and the rest erases boundaries or at least makes them less formidable, enabling ethnographers to lay bare Western science practices; linkage encourages mutual interrogation.[1] I am not trying to say that the emperor has no clothes. Rather, rapid globalization renders the search for a more balanced, indeed more scientific, treatment of disparate knowledge systems inevitable as notions of intermingling idea systems themselves become objects of study and manipulation. The late twentieth century is special in that increasing numbers of modern scientists are crossing paths with indigenous peoples who have recovered from the shock of first contact and now deal directly with outsiders. The ecological knowledge of northern Canada as developed by indigenous people and that produced by biologists studying in the Arctic are placed side by side, thereby bridging the chasm which has traditionally separated pre- and postliterate scientific traditions (Bielawski, this volume). There is power in juxtaposition, and this, at the very least, is bound to stimulate our curiosity about how "scientific" knowledge is produced in two very different cultures. Implied in such a juxtaposition is the Malinowskian thesis that a scientific habit of mind is universal, which, to use Paul Rabinow's phrase (1992), serves to "lower case" the abstraction of science.

Science versus Knowledge

The question of a universally present science is an interesting one. If knowledge is born of experience and reason, as Malinowski argues, and if science is a phenomenon universally characterized (after the insight) by rationality, then are not indigenous systems of knowledge part of the scientific knowledge of mankind? And, if Homo sapiens dispersed from one or more locations, then couldn't one argue that contemporary sciences are in part the cumulative result of the ideas and material progress that diffused along with Homo sapiens? We speak of Archimedes as a classic source for Western science, and his work was influenced by ideas diffusing from the Middle and Far East—an old process for Timewalkers (Gamble 1994), and the movement of human beings into every corner of the globe.

In many parts of the world, ethnoscientists, geographers, and ethnobotanists have found a spectrum of knowledge in the domains of agricultural techniques, resource management, pharmaceutical and navigational knowledges, irrigation systems, the provision of food, and transportation. These same ethnoscientists report on collective or individual discoveries and experiments that are the products of centuries of refinement, like lines in the sand. There is a sense of urgency. Reports include a variety of societies worldwide, some already extinct, almost any sampling of which indicates the difficulties of demarcating modern science as a purely Western phenomenon (Weatherford 1988). Scholars who study the Maya of Mexico and Central America recognize that science was an integral part of Mayan civilization at its peak, and the contemporary Maya benefit from the medical knowledge they retain. Recent documentaries record the rediscovery of navigational science among the Polynesian peoples, as lost knowledge recaptured. These are the same islanders who taught modern marine biologists about the biology of fish populations, a knowledge that entails a complex understanding of ocean currents and the life within the seas (Johannes 1981). Elsewhere, sophisticated pharmaceutical knowledge has been recorded for years for all people on inhabited continents, a knowledge that has either been diffused, lost, or borrowed by peaceful means or by piracy, by equals, or by those who enter with a "civilizing mission."

In numerous societies, collective strategies for managing the ecosystem are represented or sanctioned in ritual rather than in government. New World Northwest Coast Indians consciously managed fish populations in this manner prior to disruption by white settlers. To preserve such strategies, the Canadian government has mandated the creation of an intellectual (written) tradition to integrate both traditions of knowledge construction; meanwhile knowledge gathered over hundreds of years continues to slip away. Peoples of the Amazon use methods to conserve resources, to maintain ecological balances for stabilizing their food supplies. Often when scientists are trying to understand something about the tropics, it is already understood by indigenous peoples. Indeed, our understanding of resource conservation among native peoples is either hindered by misinterpretations and the idealization of indigenous systems or an assumed convergence of a Western conceptual system with an indigenous one.

The traditional environmental knowledge field (sometimes known as TEK) is a burgeoning one and includes dozens of researchers. Others are working in regions where indigenous peoples are collecting their own TEK, as noted in the example of Oscar Kwagley (1995). The many observations of rationalities other than Western rationality point to the conclusion that scientific attitudes and methods of validation are not unique to a cognoscente of the West. Controversy in "traditional" societies over how to make sense of their world is complex, sophisticated, and innovative. In human-plant interaction, indigenous people carry a knowledge from which Western scholars have much to gain (Etkin 1994).

In sum, the dominant Western tradition of science is one among many traditions. Historians of science who describe science as a tradition originating from Europe are incorrect and ignorant of the remarkably diverse science traditions internal to Europe itself, traditions that have vied for control over knowledge production. It is likely true that a scientific attitude common to all humans is what has made it possible for people to exploit their biological advantage, to survive.

Still, Malinowski's question about the scientific validity of native knowledge remains unanswered. The question itself may be the problem. In the anthropological literature there are hints of defensiveness, nostalgia, respect, practicality, and the need to correct the record. Yet, the sum total of the work could be read as a challenge to the use of science with a capital S as a means of asserting absolute positional superiority.

Science versus Society or Science in Society

If the first research strategy focuses on documenting systematic knowledge produced by people without literate traditions, the second reflects an interest in the study of Western science as an antidote to the idealization of science as autonomous and separate from society (Hess and Layne 1992; Traweek 1993; Downey et al., in press). *Science* magazine occasionally publishes commentaries that point to the inadequacies of narratives of the scientific past. The historical narrative found in science textbooks, according to reviews, often depicts science as a technorational activity. As a result, the operations of science in society are so poorly understood as to become controversial.

As I mentioned earlier, the notion of science in society is at the core of the controversy over the 1994 opening of the Smithsonian exhibit "Science in American Life" (Molella and Stephens 1995). The exhibit documents such themes as "The establishment of science in American universities in the late 19th century, the growing cultural authority of science after the First World War, and the public questioning of scientific authority after the 1960s." The exhibit's most vociferous critics were those who object to broad definitions of science that include technology at the cost of an image of science essentially nonutilitarian, neutral, and separate from society. The focus was not on discovery. Science was there, unpretentious in its nakedness.

Recent ethnographic writing uses participant observation among scientists to render clearer histories of science in process. Histories that result from examining how scientists create knowledge and how that knowledge is received and used dismantle stereotypic images of science and society. Similar observations are sometimes shared by members of the scientific community, who themselves are able to separate image from practice. Physicists who observe themselves with some candor may refer to ideas and behaviors that are irrational, workplaces that are undemocratic, dissent that is not tolerated, and cultural practices that are in conflict (Nader 1980). While others idealize physicists' behavior as more rational, less subjective, and more advanced than the lay public, physicists interested in moving beyond the meaning of idealized versions of science explore how science and scientists have been affected by military and corporate interests, the primary sources of funding, and the users of the discoveries of physics. Introspection by physical and biological scientists provides anthropologists with useful commentary for understanding the way cultures of science are formed and institutionalized.

In the underlying politics of science, disciplines develop and are shaped by tension and power struggles, the dynamics of which are rarely chronicled. Though most scientists and engineers would (at least publicly) deny that they are engaged in or touched by political maneuvering, their behavior is affected by those who control funding and who often determine research questions! It is foolish not to recognize that the behavior of biotechnologists and physicists is affected by funding patterns, as in other fields as well. Nuclear weapons politics is linked to national and international hierarchies of power, just as biotech is to commerce. How does this play out in the way science is done? Such "naked" observations are rarely if ever discussed in biochemistry or physics classes.

Denial of a contextualized science, or the assertion that science is autonomous, strikes at the scientific endeavor, defined as a process of free inquiry. Yet the politicization of science is unavoidable, not only because politicians, corporations, and governments try to use what scientists know, but because virtually all science has social and political implications. When the notion of an elegant, pure science defines as external the context in which science is practiced, a wider dialogue is considered irrelevant. Purity in this case is the pursuit and the myth. The threat inherent in a contextualized science is that it incorporates science publics. When laity has a window stripped of obfuscation, science practice becomes of participatory interest to wider publics.

Over the past three centuries there has been the most extraordinary expansion of Western science.[2] We have also witnessed an increase in human and environmental problems in which the use and abuse of Western science has made major contributions. In the light of what we know about non-Western, nonliterate peoples, Western science in Western and non-Western places is increasingly revealed as a self-conscious expansive science, self-consciously detached from its practical effect, detached from other scientific traditions and detached from the lay public, a situation that hardly bodes well for a renaissance in science on its comparative merits rather than "because the show has been rigged in its favour" (Feyerabend 1978:42).

In order to avoid false or misleading comparisons between a model of science identified with reason and the domination of nature, and "native" uses of knowledge (sometimes entwined with magical and religious practices) of non-Western and "scientifically illiterate" Western peoples, it is imperative to document the process of knowledge formation and its use. When we do, we sometimes find that popular beliefs have played an important role in developing knowledge. The work of Ludwik Fleck (1937) connects the history of the development of a blood test for syphilis (see Löwry 1988) with ideas circulating between scientists, practitioners, and the lay public. Interestingly, Fleck's work, which was inspired by anthropology (Gonzalez, Nader, and Ou 1995), was ignored for decades and was only currently rediscovered. Perhaps one way of shifting paradigms is to frame the discovery of knowledge as occurring within different systems of production, consumption, and power.

As anthropologists examine the content and context of knowledge systems, the contrasts erupting from observing different traditions blur scholarly markers and reveal processes of privileging or suppressing knowledge. Robert Johannes' study of fishing in the Palau district of Micronesia turns the human-in-nature query "how does this environment influence you?" to "what can we learn about this environment from you?" (1981:ix). He quotes observers who note that the Pacific Islanders' knowledge of fish behavior is of such precision that the corresponding poverty of the interviewer makes inquiry difficult. Johannes made a serious attempt to collect and record this knowledge. "The native fisherman searches with his eyes and ears and he is ... more in touch with his prey and their surroundings than his modern, mechanized counterpart" (vii). Although the natural history data he collected would be blended with more sophisticated forms of biological research, Johannes admitted that he had "gained more new (to marine science) information during sixteen months of fieldwork ... than ... during the previous fifteen years using more conventional research techniques." He explains, "This is because of my access to a store of unrecorded knowledge gathered by highly motivated observers over a period of centuries." When it comes to understanding fish behavior so as to manage its exploitation efficiently, full-time fishermen may know more than marine biologists.

Science is supposedly culture-free, but studies of science practice suggest it is not. Scientists trained in high energy physics but originating from two different cultures—American and Japanese—exhibit differences in the doing of science and in beliefs about risks, leadership, and conflict that are shaped by class, region, school, and gender, as well as by national cultures, and these differences color their respective science cultures (Traweek 1988). As any ecologist might expect, science, a particular way of looking at the world, is influenced by its surroundings. Yet in spite of differences, there is the common theme of human societies doing science or accumulating knowledge by verifying observation, or by borrowing from others knowledge that works in some way. Thus one way of looking at modern science is as the ongoing result, though not the cumulative result, of the discoveries, inventions, and collective sciences of others. We have been munching on each other for millennia.

Some years ago philosopher Alfred North Whitehead wrote that contemporary scientists have become the chief promulgators of a delusion that our knowledge is infallible and final. Humanity has suffered from such delusions. In development work worldwide failed attempts to eradicate malaria in Southeast Asia and failed results from projects such as the Green Revolution, as in the Punjab (see Shiva 1991), are salient. It is still assumed that Westerners who go to nonliterate cultures will find knowledge blanks, tabula rasa that need to be filled in with the products of our science and technology. But cultures by definition are not a blank, and in agriculture and pharmacology, the knowledge of "under developed" peoples may surpass that of Western scientists (Sachs, ed. 1992). This is not news. People are impressively entrepreneurial when it comes to learning what other cultures know; we have been exploiting indigenous knowledge for centuries. When there is advantage to be gained, we are neither Eurocentric nor ethnocentric. Among the best entrepreneurial examples are the pharmaceutical companies who for decades have been exploring Amazonia, Mexico, Polynesia, and other parts of the world to find products that work. They reproduce these products in laboratories and eventually sell them back to native peoples who are described as being ignorant of modern medicine. While some contemporary scientists may boast that "We solve puzzles not problems" (referring to a disinterest in real world problems), the peoples of the world have been geared for millennia toward practical problem solving to enhance biological advantage or human longevity. The assertion of superiority based on narrow conceptions of what is scientific is of short-term value. We are dismantling a knowledge base built from centuries of trial and error and observation (Brokensha et al. 1980; Warren 1991). The rigor and disambiguity of "hard" science is based upon the selection of convenient approximations that some scientists term puzzles. Real world problems are apt to be of such complexity that this sort of "rigor" cannot be demonstrated because fundamentally, rigor is illusive.

It is preposterous to think that we live at a time when science proponents consider it outrageous to allow that there are different science traditions. Over centuries the principles that sustain traditional agricultural systems have been intercropping, agroforestry, and shifting cultivation based on an understanding of the interactions between vegetation and soils, animals, and climate. Diverse plant usage is part of the nutrient cycle, insect pest management, weed control, and soil conservation methods—as we are beginning to rediscover in the wake of disillusionment with modern agricultural science (Altieri 1983). Practicality does not replace inquisitiveness and not all native peoples are ecological sages, but as we see in the collection of papers that follow, solving practical problems tests the genuine flexibility of conceptual systems and their ability to recognize the limitations in drawing lines, dressing up or down, and other such activities associated with cultures of science.

Charting a New Direction

The significance of a naked science—an open science stripped of its ideologized vestments—is worth serious thought. Even before the industrial revolution, achievements in science and technology influenced the shaping of European perceptions of non-Western peoples. From the first decades of European expansion in the fifteenth century, ships, tools, weapons, and engineering technologies of others were compared by missionaries and travelers to their own (Adas 1989). Science and technology became measures of human worth. They were key components of "the civilizing mission," and were used to justify European political hegemony. It has been the tragedies of war that have shaken the complacency of unquestioned belief in Western science and technology, as during World War I when men were massacred in the trenches, or during World War II when the bombs were dropped on the Japanese cities of Nagasaki and Hiroshima.

The silencing effect of modern Western science led geographer Paul Richards (1985) to note the often devastating effects of the preference for "universal" explanation. As he observes, science as universally applicable knowledge is supposed to override ecological particularism and site-specific knowledge. Science derives its power precisely because it is not confined to particularities. But principles in an ecological rather than physics model may not be true for all times and places. Unfortunately, inflexible commitment to universalist assumptions can misdirect efforts for generations. As one critic put it, "Modern science makes knowledge scarce because it asserts unrivalled hegemony" (Alvares 1988).

If science anchors power, as Foucault (1980a) and his followers would have it, then the battle between scientific instruments becomes important. Lynn White, Jr., aptly notes:

> Mathematics plays a part in ecological research, but conclusions rest on an almost aesthetic perception of the counterpoint among a vast array of qualitatively different qualities ... the laboratory methods power lies precisely in its isolation of the phenomenon to be studied ... ecological science is on principle anti-scientific, as science at present is usually conceived and practiced.... (1980:76)

In *Science as Salvation* (1992a) Mary Midgley makes similar distinctions and remarks: "Science education is now so narrowly scientistic that many scientists simply do not know that there is any systematic way of thinking besides their own" (Midgley 1992a:25). More important, perhaps, is the observation that scientists are always thinking systematically. Raymond Williams (1980) observes that "while the utopian transformation is social and moral, the science fiction transformation is not social and moral, but natural," inviting a science that is divorced from consideration of people. These are all items that need to be addressed by studies of what it is to practice science.

There are costs to demarcating technoscience in a manner that excludes, because if there is nothing to which big science can be compared, it is not a subject for critical judgment. There are also costs to analyzing science activity without studying how people actually do science and think about science. Modern scientists, like any other group of people, are apt to turn to religion, emotion, or denial when confronted with the unknown. In this respect, there is nothing special about scientists qua scientists. However, when juxtaposed to other traditions, big science is unlike small science, which by definition is less ensnared in bureaucratic turf protecting, institution building, and financial and intellectual reward incentives. Because of these contextual differences, traditional knowledge is more capable of being appraised than a nonnegotiable technoscience.

In the contributions that follow, the stage is set for new collaborations between those who study the workings of technoscience and ethnoscience, those producing knowledge, and those consuming science. The book is divided into three parts: "Discovering Science," "Culture, Power, and Context," and "Conflicting Knowledge Systems." As the collection of papers indicate, there is a good deal at stake in our efforts. It would be easier to move toward a world renaissance if we had a space in which to discuss questions that did not mirror the world's dominant political and economic structures.

The Collection

Part I provides rich and startling descriptions of local science and knowledge that do not take for granted the status and credibility of indigenous knowledge systems. From the Micronesian islands to highland Mexico, from U.S. grocery stores to cities we learn about systematic knowledge gathering, testing, and verifying. While their findings are not news to anthropologists and many social scientists, the authors raise questions that are often unspoken. They use science practice from peoples that Westerners commonly exclude from the scientific mainstream to argue for the inclusion of these peoples and their practices in considering the question—What exactly does constitute science? Why do we tend to exclude certain "others" from our notions of legitimacy, and what are the implications of exclusion? All five essays in this section are joined by the notion that there is science outside of the "expert science" paradigm, both within and outside the West.

The first example deals with Micronesian navigators. In classic work on Micronesian navigators, Thomas Gladwin (1970) likened their knowledge of the island and star courses to a map, just part of a network of social, economic, and political ties. Ward H. Goodenough (1953) has illustrated that native astronomy is not merely practical, it is almost entirely abstract. In this volume, Goodenough asks the reader to consider if Micronesian navigation rests on mental operations of a kind with which we are largely unfamiliar, or if it represents ingenuities similar to those exemplified in products of Western thought. In his paper about the knowledge and practices associated with indigenous navigation, Goodenough deals with conceptual tools such as the "star structure," and the "trigger fish" schematic map. He then describes the practice of navigation, which requires a more contingent set of techniques: reading "ocean swells," making course adjustments, and using "dragging" information about relative position. The first set of techniques is based on static information, while the second is more concerned with dynamic situations. Goodenough then moves the discussion to the sociology of navigators, exploring their role in society and the way they transmit knowledge across the atolls and down through the generations. Although there are parallels between this practical science and Western practical science, Goodenough notes the navigators are more "sensible": that is, Melanesians rely more on sensory information and their own bodies in the absence of our instruments of observation. From what ethnography we have gathered, one could argue that Melanesian navigators are sophisticated scientists.

There are striking parallels between the star structure and our own magnetic compass. Equally striking is their complex knowledge regarding wave nodes, involving the interaction of ocean currents, swells, and winds. One might also be struck by these traditional scientists having their own specialized technical jargon, licensing procedures, schools of training, and professional conferences.

As interest in alternative medicine grows in the United States, so does our knowledge of indigenous medicine, especially in the area of human-plant interaction (Etkin 1994). In the second essay, the Berlins and their coauthors learn that the Highland Maya system of traditional medicine is a complex one, based on an understanding of disease symptoms that could only have been elaborated after empirical experimentation with herbal remedies. The authors arrive at their conclusion after analyzing herbal remedies for gastrointestinal disorders in fourteen municipalities in Chiapas, Mexico. In addition, they conduct laboratory tests on plant species which suggest that the plants induce effects that, according to our own biomolecular medical models, would likely improve the condition of an ill person.

What is interesting about the Maya system is the extremely high percentage of plant species that showed antimicrobial activity in lab tests. The authors indicate that the Maya system of medicine is like ours—both rely upon the identification of symptoms to dictate a remedy. This approach to "naturalistic" illness is by no means random, nor is it unsubstantiated folklore; it is scientific, they argue, because it is based on a process of observation and experimentation.

In a similar vein, Colin Scott addresses the question of whether the Canadian James Bay Cree practice science. Anthropology has tended to highlight the mythics—ritual framing of knowledge in non-Western societies—in opposition to Western scientific notions of literal-empirical rationality. However, Scott argues that while hunting the Cree rely on practical, empirical knowledge to draw deductive inferences, and that these inferences are deliberately and systematically verified. Scott reports that the Cree practice a science different from ours because it builds on root metaphors of panspecies personhood, communication, and reciprocity—metaphors not typically associated with Western science. He uses the example of goose hunting to demonstrate how this system is put to practical use in a scientific way. Central to Scott's argument is the idea that all sciences—including contemporary Western science—are constructed upon certain "root metaphors." In the West, the metaphor of nature as something apart from culture and human beings undergirds science.

Two scientific understandings emerge from the Cree case; one in their terms and one in our terms of cultural ecology. An even more powerful provocation is Scott's observation that positive reciprocity in Cree society enhances reciprocity with geese, while the conventional social context of Western society is one of hierarchy and control that is projected in our own relationship with nature and which encourages the disqualification and subjugation of indigenous knowledge. The reader is left to decide whether Cree hunters practice science, pseudoscience, protoscience, or whether they are unscientific in the light of Scott's depiction of how mythico-ritual categories are implicated in hunters' literal modeling of social-environmental practice.

The last two papers in this section deal with problems inherent in viewing laypersons as ignorant "savages." Jean Lave's article is a commentary on the tendency in scholarship to create an inferior "other" (whether non-Western "primitives" or Western nonscientists) by means of a mental-model approach. She challenges this notion by studying everyday mathematics and finds that people adapt their mathematical reasoning as they move from one practical context to another, and that they are remarkably accurate in their calculations, even by standards of school math practices. What Lave is talking about here is the degree to which the myth of the "inferior other" is deeply incorporated into our cultural practices such that "just plain folks" collude in their self-identification as incompetent "others." Science characterizes everyday practice in ways that inevitably help to reinforce its own hegemonic role. The everyday then contrasts with science, and scientists and "just plain folks" are juxtaposed.

A striking point made by Lave is that in their own kitchens, people invent literally hundreds of units of measurement and procedures for generating accurate portions of food. Of interest to those who care about scientific literacy is the observation of extremely high accuracy—nearly 100 percent in some cases—of everyday mathematics. Lave questions why it is so difficult to address everyday mathematical activity in terms other than its perjorative contrast to "real" math to dispel far-reaching rationales and defenses of Western culture in their analyses of schooling, ethnicity, literacy, intelligence, expert knowing, gender, and more. As she points out, such rationales rest upon programatic idealizations of mathematics and scientific thinking as opposed to common sense and concrete thinking. Studies of everyday mathematical practices and studies of the practice of science converge to reveal what each has in common with the other.

Like Lave, Emily Martin and her students examine who holds what knowledge within our society. Worldviews are inscribed in scientific images. Earlier, Martin (1987) argued that working-class women reject dominant medical views of menstruation but that this rejection is actively constructed, a response to the gendered configurations of science rather than the result of being "ignorant of science." In this volume she and her students question who it is that is scientifically literate, suggesting that literacy should be broadened to include many kinds of knowledge.

Rapidly expanding knowledge in the science of immunology is being widely disseminated in the United States by means of a variety of media and technologies. But the way this knowledge is taken up—what it means to have an immune system—varies dramatically. To illustrate, the authors consider the immune system from four different points of view: 1) a professor of microbiology who teaches an undergraduate class on cancer and AIDS; 2) a gay couple; 3) a heterosexual man, married and living in a racially integrated neighborhood in Baltimore; and 4) a married woman who has had many close friends die of AIDS. In each case, a distinct picture emerges about what the immune system is and how it works. Each interpretation carries its own metaphors linked to broader knowledges and understandings about contemporary society. The complex interactions between science and culture are differentiated by gender, identity, race, and class.

What might be surprising to Western scientists and laypeople alike is the way in which facts are interpreted and located in the knowledge systems of individuals. For example, it is the microbiology professor who connects the immune system with a "scientific version of . . . an angel or a protector or something." Also we all might think about the idea that Western scientists are scientifically illiterate because they have an extremely narrow interpretation of what facts are and what they might mean. The authors call our attention to the illiteracy that accompanies a narrowly technocratic conception of knowledge, thereby implying the need for a genuine scientific literacy, one that is indispensable as a tool for shifting attention from supposed scientific ignorance and altering unreflective assumptions about what is relevant to laypeople who may correctly sense the unfathomable complexity of the forces that surround them.

In Part II, the papers are about expert scientists and how politics and sociocultural assumptions affect their behaviors. They cover research that is somewhat representative of the field of science studies.[3] Each paper focuses on one science and opens up the cultural and political world within the particular scientific establishment. The first three papers raise important questions of the cultural authority of science in society as well as how power politics operate in science. They contribute to ideas of how authority and hence legitimacy is established within a particular science, which is directly related to issues of the silencing effect of

Western science as illustrated by examples in Parts I and III. The last two papers deal with the cultural settings of science work and demonstrate how cultural details direct the science itself.

This section represents the well-clad sector of the technoscience continuum. High-energy physicists, nuclear scientists, and molecular biologists exist in heavily dressed institutional structures that span the world's hundreds of nation-states and even more businesses and industrial interests. The papers deal with communities of scientists and their social contexts, and with how such contexts affect the science itself. Contrary to social analysts who help construct the framework for science, the movement to unpack cultural and social baggage insists that any assessment of the role of modern science in everyday affairs must begin with a rigorous understanding of the very nature of the scientific enterprise itself.

Scientists exist in social and economic structures. Inasmuch as they may be constrained or legitimated by such structures, they themselves have visible impacts on social polity, and international relations, and may effectively rework such structures. This dynamic role of scientists in Western society has been the subject of much recent acclaim in the field of science studies, leading to debates on the nature of modern life. Scientific communities were never autonomous; ideas always moved between communities. This means that there must be a careful balance between an analysis of what is within science, and what is influenced from the outside. The approaches taken by the papers are diverse in method, but they all push toward the goal of understanding the meaning of received opinions and beliefs about science.

Troy Duster uses perspectives from the sociology of knowledge to analyze the increasing frequency of genetic explanations for human behaviors and conditions (crime, mental illness, alcoholism, gender relations, and intelligence). By the phrase "prism of heritability," he is referring to the genetic lens through which we peer at behavior. Duster points out that in his review of scientific journals of human genetics, only about one-fourth of the articles were written by authors with credentials in human genetics or cytogenetics. Even when taking into account the possible disciplinary backgrounds that cross over into genetics, well over two-thirds of the authors were far removed from molecular genetics.

Science cannot be understood apart from the decade of the 1990s against which it now appears. Duster is suggesting an appropriation of the imprimatur molecular genetics to the explanation of these behaviors. He then asks why it is that those outside of genetics are engaged in genetic explanations. Specifically, he discusses the relation between genetic explanations for criminal behavior and the construction of race and stratification in American society. The issue of biological versus social factors (nature/nurture) has had a long-standing contention, and it is the current drift toward genetic explanations of an increasing variety of human behaviors by those outside the field of genetics that interests Duster: How do differing interests shape the questions that then become building blocks of knowledge structures? Consumers use "science" to get where they want to go, indicating that the cultural space of science is not stable.

Charles Schwartz would agree that science is not value free. His document is a physicist's evaluation of physics from the perspective of social responsibility. He recognizes the enterprise of physics in the United States since World War II as the preeminent model of success in science, one that experienced rapid growth in membership, journals, schools, and funding, one that achieved the highest level of prestige relating to its production of new knowledge about the "fundamental understanding of the universe." But what, he asks, are the institutional structures directing physics? How are the elites of physics related to national elites, to the military and large corporate interests that have been sources of funding and users of the products and trained personnel?

Schwartz's approach is one common to power structure research. Network analysis shows how the small world elites move in, how methods of training and competition serve to keep members of the profession in line, as well as students and the public at large. One could also argue that elite scientists are using external power to push their cultural authority into spaces previously occupied by others, to draw science close enough to a politics congruent with their interests and programs. Schwartz raises the question of social responsibility and advocacy directly, arguing that if young scientists are not politically committed, they should at least not be blind to politics.

The advocacy Schwartz pursues is directly related to the nuclear problem. Using ethnographic techniques, Hugh Gusterson moves beyond networks into the community of nuclear weapons scientists to examine scientific experiment as ritual. He asks how it is that a world comes to be constructed such that deterrence depends on reliability not unreliability of nuclear weapons, on certainty rather than uncertainty about the behavior of weapons. After outlining the procedures of designing and testing weapons, Gusterson is drawn to describe the ritual and culture of nuclear weapons scientists. His focus is on the rituals that induct individuals into the culture. If nuclear weapons scientists are the products of a scientific rationality that subordinates nature and which disciplines human anarchic impulses, then each successful nuclear test reaffirms their faith in the redemptive power of the system of rationality.

Testing is not only a rite of passage, but successful testing means greater social status for the designer. Gusterson points out that by assuming that a select group of experienced scientists must test out different weapons over and over again (and thereby be in control of that knowledge), there is assurance that continued testing is necessary. One begins to understand why battles rose around critiques that suggested testing be carried out through different means; different means implies an entirely different sociotechnical system for weapons testing. Gusterson indicates that mastering the design and testing of nuclear weapons leads to a feeling of security in deterrence and freedom from fear of the weapon itself. In the end, he is arguing that our most expensive scientific experiments are saturated with elements of myth and ritual, and the tests themselves serve to legitimate the broader social organization of weapons labs and arguably the nature of international politics, especially in relation to deterrence and global security.

Dozens of researchers are now drawn to study the social meaning of the Human Genome Initiative, and some are documenting the move of biology from experimental to theoretical description as biology becomes a technological science. Joan H. Fujimura and Michael Fortun are interested in the construction of knowledge through the growing division of labor in molecular biology, and the possible dominance of representations of nature constructed by computers over other constructions. Molecular biologists standardize in order to do sequences and then practice the same routine to represent nature. Fujimura and Fortun examine the computerized databases, which store the information from genetics research in the Human Genome Initiative, to discuss the community of scientists, their work practices, and the implications for how scientists conceptualize and intervene in nature. How do heterogeneous elements such as people, diseases, cell lines, genes, DNA probes, sequences, and the entire human genome become connected? The answer is, as they point out, through a great deal of work—experimental, representational, organizational, and rhetorical. Important issues arise as databases are being used in genetic research. The most common operations are homology searches constructed between laboratories, representations, and social worlds. Instead of carrying out each project in a wet lab, through the database project, a homology search is faster and more efficient. There arises, then, the problem of ownership of matches, a problem being negotiated by companies such as GenBank. A related concern is the acceptance of this research by the larger scientific community, and the conflicts that arise.

Fujimura and Fortun point to a conflict between computational theoretical biology and laboratory molecular biology/biochemistry. For example, an article by scientists of the first group was rejected on grounds that the scientific research was not done, since they accessed their data through the computer. In conclusion, the authors express concerns with the uses to be made of the new sciences and the recent upsurge in genetic research and genetic reductionism and its concomitant effects on people's views of human societies and nature. Since they are social facts, biological representations are political. Linking such shifting representations of nature to changes in structure and thought for science and society is difficult work that remains unfinished.

In the last paper in this section Sharon Traweek tracks historical, cultural, and structural factors affecting the work of Japanese physicists. She wants to learn how different styles of research practices emerge and survive, how disputes and factions are formed and maintained—and how scientists are socialized within a given political economy. She foregrounds the strategies of the physicists who are in the laboratory as compared with the strategies of bureaucrats, politicians, and elders of the scientific community. As with science policy studies, the purpose is to document the process whereby high energy physics has developed in Japan since the post-war years of reconstruction as they correspond to career paths and generational differences. Traweek compares the Japanese high energy physics community with their U.S. counterparts according to the strategies used to make and maintain networks with businesses, politicians, bureaucrats, etc. Unlike the Americans, the

Japanese did not have close networks with the military. Rather, outside pressures, such as the energy crisis of the early seventies, were used to lobby for their laboratories.

In sum, Traweek provides an overview of how the Japanese high-energy physicists managed to build a community within the national and international science communities and larger political economies. She shows how the differences between nuclear and cosmic ray scientists on the one hand and high-energy physicists on the other are played out in national politics, and how the latter have had to negotiate their class, gender, and regional identities vis à vis the international community. In the end it appears as if *gaiatsu* (outside pressure) is the chief means by which Japanese high-energy physicists create and maintain their communities, while concurrently managing to make room for "big science."

Part III joins research on local knowledge with that on technoscience, cognizant of the context of science as well as its content. Each of the four chapters juxtaposes two sciences or scientific beliefs on a particular subject. These illustrate the boundaries with which I began my discussion and reflect a new direction in research, one in which the science(s) or knowledge systems are described in comparison with each other: fisheries, the Arctic environment, primatology, and surveillance technologies. We are impressed by the importance of creating bridges between the differing sciences in order to overcome their limitations and, in some instances, to comprehend their contradictory conclusions. More than any of the sections, these articles illustrate the importance of flattening the hierarchy of legitimate science which positions Western science at the apex. They urge us to consider other knowledge systems as legitimate in efforts to understand our world and to contribute to the need for broadly based problem solving. Urgency was evident in the United Nations University 1990s proposal titled *An Archive of Traditional Knowledge*:

> Traditional knowledge (or local knowledge more generally) is a record of human achievement in comprehending the complexities of life and survival in often unfriendly environments. Traditional knowledge, which may be technical, social, organizational, or cultural, was obtained as part of the great human experiment of survival and development.

M. Estellie Smith compares two knowledge systems held among groups involved in government policy-making for fishery stocks in New England. At the heart of her essay is a basic dilemma—the extent to which we should use or manage nature. In spite of the fact that fishing itself has become high-tech, commercial fishermen possess different knowledges with respect to fishing, and more generally about the "nature of Nature." The fishery managers (biologists, economists, and government administrators) see things in predictable, linear terms. The fishermen view nature in chaotic terms: fish stocks, the weather, the market, the actions of government are all seen as susceptible to disequilibrium. The chaotic and linear views represent apparently incommensurable "uncommon languages" that make communication between groups difficult. Surprisingly, "expert science" was being used in this case both to protect the resource and support the technology that exploits the very same resource more efficiently. Smith uses the fishing example to illustrate current trends in the formulation of public policy indicating the conflict between theory-driven "sciencing" and practicing. The trend favors multidisciplinary mingling of scientists and nonscientists, theoreticians and practitioners. Smith points to inherent problems in this trend.

Ellen Bielawski compares two distinct belief systems in use—one, Inuit indigenous knowledge, and the other, "Arctic science" as performed by Westerners. Arguments about rationality and relativism are philosophical issues for comparison in the Arctic. Inuit knowledge can be described in terms of scientific parameters such as measurement, prediction, experiment, and replication. According to Bielawski, the most significant difference between the two sets of belief is that the Inuit place humans in the sphere of nature and as inseparable from nature, while Arctic science does not. The Inuit group Bielawski studied were relocated between 1953 and 1956 from Inukjuak, a village on the east coast of the Hudson Bay, to a place far to the north. The move made it difficult to subsist as fewer plants and animals were available and kin groups were decimated; it also made it possible to study how the Inuit built knowledge when the Government of Canada resettled them in new and dramatically different hunting territories.

The most interesting point made about the Arctic scientists is that so few of them live in the Arctic. While Inuit knowledge is formed through "being told," "doing," "hearing about it," and "being there"—all interactive and personalized forms of knowledge transmission—Arctic science is shaped by external factors. The most shocking detail of this study is that this collision of Eskimo and Arctic science knowledges took place in relocation (of the Inuit). Inuit subsistence depends both on local, detailed environmental knowledge and on sharing practices dictated by kin relationships. The relocated Inuit lost the kin on whom they relied for local knowledge and with whom they hunted and shared. In other words, the Inuit were isolated from at least half of their reality: the natural world. A look at a region like the Arctic allows one to see how far apart modern science is from indigenous knowledge and from their concrete world, and forces one to think about the impact that such separation has for the construction of knowledge.

The creation of covert surveillance systems were part of a period (1947–1949) crucial in the formulation of U.S. science policies for the second half of the twentieth century. In examining their subject David Jacobson and Charles A. Ziegler make a finding that is counterintuitive. The beliefs of U.S. expert scientists are compared to those of nonscientists—government administrators and laypeople—regarding a surveillance system for detecting the detonation of atomic bombs in the post–World War II era. Based on their knowledge, most scientists believed that the Soviets would not have a bomb until the 1950s.[4] On the other hand, laypeople and administrators suffered from the "popular delusion" that the Soviets might detonate a bomb at any moment. As it turns out, the experts were wrong. In this case popular delusions were more true than scientific facts. The Soviets did in fact detonate an atomic bomb in 1949, shortly after the U.S. Air Force had implemented a surveillance system—against the advice of the scientists.

Their analysis of a selected incident, the U.S. detection of Russia's first atomic test, demonstrates why scientific beliefs should not be canonized and argues for the importance of nonexpert knowledge in key public policy decision making. The surprise here is that the best scientific opinion then available wound up failing, while the popular delusion was correct. Again the irony is striking: the apparent delusion was based loosely on the work of expert scientists and that expertise was promulgated and transformed by the mass media.

In concluding this section Pamela J. Asquith illustrates just how scientists bring intellectual, religious, and other cultural predispositions into their methods and theories.[5] Positions on the nature of nature are mainly unarticulated assumptions. Primatology is the behavioral and evolutionary study of our closest living relatives—apes, monkeys, and prosimia. Through a comparison of Japanese and Western primate studies she indicates how society of origin can delimit the possibilities of knowledge building. Asquith compares two primatologies that developed independently of each other. In Japanese primatology, methodology reflects an interest in the individual and the group, in intergroup and intragroup relations, and in ranking. There is greater commitment to long-term observation of a less objective nature—a cultural approach. The Western primatologists are by contrast neo-Darwinist, with more emphasis on sociobiology and evolutionary reproductive advantages of specific adaptive behaviors. Its field focus is intense and short-term. For the Japanese scientists, the histories of individual primates and of groups were considered equally important, while Western scientists tended to consider individual primate histories more important.

The implications are all the more significant because the Japanese approach to primatology has been fruitful and unique, uncovering information that Western primatology has been slow to discover. This includes early observation of male monkeys transferring between groups, early recognition of the significance of female kinship, and the discovery that male dominance need not correspond to improved sexual access. She explores the reasons for initial disregard of the Japanese work and the later vindication of their results.

Asquith's findings are unexpected, not only the extent to which the Japanese approach to primatology differs from our own, but how such findings reflect differences in our respective societies. As she notes, Japanese society is one where a person's identity is to a great extent congruent with one's group identity; thus it was natural to look for similar phenomena in the monkey groups. Like Donna Haraway, who writes about the union of the political and the physiological in *Primate Visions* (1989), Asquith's study is another striking example of how culturally embedded scientific worldviews and explanations are.

In the closing paper, U.S. scientists speak about energy. They demonstrate an understanding of science practice and perceive scientific activity independent from traditional science ideologies. The "coal-face of science" (Le Grande 1988) has been described in science studies and its features are by now familiar to social scientists. However, this familiarity is still not in the public view. The three-cornered constellation of magic, science, and religion, together with the comments of physicists and engineers, bring the reader back to earlier contributions to ask how the subservience of the scientist comes about. Through their own observations, the scientists indicate the intellectual power inherent in asking the question, "When is science scientific?" Like the Trobriand Islanders, a more adequate democratic science would include knowledge acquired in life experience, if knowledge for survival is at stake.

Concluding Comments

This book should be read as if it were written by one author although it is unlikely to have been. In this sense the book is not an edited volume. It is an effort to examine science through a distant mirror as well as through a magnifying glass. Moving from the first to the last chapters—from localized knowledge to what some call "cosmopolitan" science and then to arenas where the two are found in the same place—one gets a complex, multilayered perspective full of crosscurrents. The subject is a central issue of our time—the primacy of a heavily dressed Science with a capital S and the consequences of its global expansion based on power rather than greater rationality.

There are hazards in our enterprise. Depending on who is looking, the colors change. Is the "cosmopolitan" scientist the hero or is the "traditional" heroic? I hope the papers are not read from such vantage point. The point of the work is not to "put scientists in their place" (although one might want to), nor to romanticize the people with the small s. The point is to open up people's minds to other ways of looking and questioning to change attitudes about knowledge, to reframe the organization of science—to formulate ways of thinking globally about science traditions. The issue for technoscience is not a "failure of image," as some might argue. Rather image making is the failure. There is a sense of betrayal when a publicly funded science is tied to the welfare of a few.

Public discourse about science is still saturated with notions of science as autonomous, value free, and omnicompetent in spite of 25 years of science studies that have documented the links between science and society, and described science as first and foremost a human enterprise (Jasanoff et al. 1995, Harding 1994). The credibility of science and its cultural authority is still achieved though attempts to exclude values from the space of science. Indeed, there has been an expansion of these notions globally. Public discourses worldwide portray science as autonomous, self-generated, unique, absolutely superior, and progressive. Yet, I am persuaded by the Yupiak educator, the marine biologist, and our collection of essays that knowledges are rapidly being overshadowed, replaced, and pushed aside by the introduction of a science and technology that assumes primacy. The recognition of lost knowledge, all but ubiquitously lost, is made all the more real by failed development projects.

Anthropology has been part of the problem in creating demarcations that continue to be used, often inappropriately. Words like dichotomy, contrast, demarcation, and binary are all used. While firm boundaries make for greater efficiency in processing information and coordinating action, sometimes dichotomies get us into trouble, as in confusing construct with reality. But some theorists argue that boundaries are episodic not fixed (Gieryn 1995), implying that at any historical moment what is included and excluded may change. Maybe. The very creation of a program in alternative medicine at the U.S. National Institutes of Health might signal changing boundaries. This book, however, makes it clear that some boundaries have been fixed for a long time, and that the likelihood of change is directly related to the untying of a discourse that is currently isomorphic with the dominant world political and economic structures of multinational corporations and nation-states.

The belief in the omnicompetence of science has been steadily gaining ground throughout this century in this culture, and operating on a core-periphery model, in the world. We can recognize the distinguished accomplishment of human achievement everywhere without framing these accomplishments as progressive. We need not idealize non-Western science to make the point that there are different types of knowledge that provide valid truths of use to human kind. If a dominant science silences that knowledge, we all lose. Consider a view that includes the footprints of time: a view of knowledge in which imagination and vision can be openly checked against criticism; the myth of a single science can be seen as myth; the false separation between science and nonscience may be considered as a barrier to new thinking; and a whole range of vital and experimental thinking is possible. We might in such an atmosphere be willing to concede that technoscience is a peculiar effort if intended to make life more meaningful. Ironically, standardization, uniformity, and conformity may not provide the best possibilities for new kinds of science in the long run. At the end of the millennium, the perspective of a distant mirror and the correctives of a magnifying glass both provide food for thought and action.

Notes

For help in clarifying the ideas in this essay I am especially grateful to R. Gonzalez, E. Hertz, J. Martin, N. Milleron, R. Milleron, T. Milleron, J. Ou, and an incisive Routledge anonymous reviewer for their comments, arguments, and several readings.

1. It is ironic that philosopher Sandra Harding's paper "Is Science Multicultural?" utilizes a common anthropological framework while Eurocentric anthropologists commenting on her paper appear inadequately prepared to deal with the ethnology in Harding's work, a work that requires broad understanding of the human condition in addition to the particularities of localized and contemporary events.
2. See Richard Adams' *Energy and Structure* (1975) for theoretical attention to causative agents involved in this expansion.

3. It is important to recognize that crucial ideas coming out of early anthropological work, such as relativism, comparison, and ethnographic fieldwork, greatly benefited these works and more broadly the work of science and technology in history, philosophy, and sociology. Publications resulting from studies in both ethnoscience and technoscience were each carried on primarily in isolation, as if there were no intellectual connections in content and methodologies. Regardless, such studies have put into question assumptions about the nature of scientific and technical knowledge. Sociologists and anthropologists have entered scientific laboratories. Their studies show how facts are "discovered" and the shaping of such knowledge by social factors: how scientists decide on research agendas, how machine inscriptions convert into "facts," how boundaries draw between good and bad results. See Jasanoff et al. (1995) for a full review. Brian Pfaffenberger's (1992) review "Social Anthropology of Technology" provides a useful overview.
4. Interestingly, physicist Irving Langmuir predicted in 1945 that "the Russians may begin to produce atomic bombs in about three years" (Langmuir 1960–62).
5. In a related context archaeologist Fumiko Ikawa-Smith (1990) documents how the Japanese ideology of cultural homogeneity is reflected in the way archaeological and physical anthropological materials are interpreted and organized into narratives of national history.

Part 1

Discovering Science

CHAPTER 1

Navigation in the Western Carolines
A Traditional Science

Ward H. Goodenough

The atoll dwellers of Micronesia are ocean voyagers, unlike the inhabitants of Micronesia's few high islands. In the atolls, knowledge of how to build seaworthy sailing canoes and how to navigate from tiny place to tiny place over fairly long distances of open sea has been actively maintained for centuries as vital to successful life. Somewhat different systems of navigation are used in each of Micronesia's three atoll regions: Kiribati (formerly Gilbert Islands), the Marshall Islands, and the Western Caroline Islands (fig. 1).

Best known to students of Micronesia's cultures is the system used by navigators in the Western Caroline Islands—the great chain of atolls that lies between Pohnpei (formerly Ponape) in the east and Yap and Belau (formerly Palau) in the west, a distance of over a thousand miles. Throughout this region the same basic system has been in use, with local schools differing only in ways that are not crucial to successful voyaging. Navigators, moreover, all of whom are men, have to observe food taboos that result in their eating separately prepared food in their boathouses. Visiting navigators from other atolls dine with them there, so there is opportunity for exchanges of information and displays of knowledge. Such exchange is facilitated by there being a chain of closely related dialects, neighboring ones being mutually intelligible (or very nearly so) over the entire area. Navigation has been developed here, even in the absence of writing, to a high degree. Like any practical science, its application is an art, requiring both knowledge and skill.[1]

The Star Structure

Fundamental to the system of navigation is the "star structure" *(paafúú)*, as it is called. Seen near the equator, stars appear to rotate across the heavens from east to west on a north-south axis. Some rise and set farther to the north and others farther to the south, and they do so in succession at different times. The "star structure" divides the great circle of the horizon into thirty-two points. Polaris, just visible, marks north, and the Southern Cross in upright position marks south.

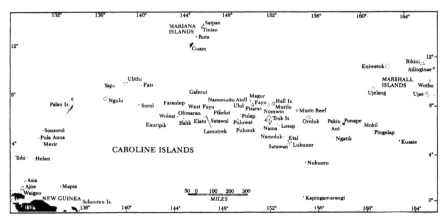

Figure 1   Carolinian navigator's world.

Other points, while conceptually equidistant, are named for the rising and setting of stars, whose actually azimuths of rising and setting are not equally spaced. Frake (1994) has observed, therefore, that the stars do not physically mark the points in the "star structure"; rather, they name them.

These thirty-two points, like the points on the European wind rose, form a conceptual compass and serve as the directional points of reference for organizing all directional information about winds, currents, ocean swells, and the relative positions of islands, shoals, reefs, and other seamarks. Every point has another that is conceptually diametrically opposite to it. These diametrical opposites are seen as passing through a point at the center of the compass, and a navigator thinks of himself or any place from which he is determining directions as at this central point, just as western navigators do when using a magnetic compass. Thus whatever point a navigator faces, there is a reciprocal point at his back.

Although navigators represent the thirty-two points of the sidereal compass as equidistant for instructional purposes, there seems to be recognition that the stars that name them are not evenly spaced at their points of rising and setting. Beta and Gamma Aquilae, which are very close to Altair, are omitted from some exercises, the compass being reduced in them to twenty-eight points. To use the stars effectively as directional guides requires either that sailing directions reflect empirical observation or that navigators take account of the difference between the observed position of a star and the position of the point on the compass to which it gives its name. On the basis of the evidence so far available, it seems that the former alternative is more probable. In the absence of writing and accurate maps, the discrepancies between the system as ideally represented and the sailing directions as empirically established are not likely to be a matter of concern. The "star structure" is represented in Figure 2 as a compromise between the ideal and the real, showing points as diametrical opposites but bringing them as close as allowed to their actual degrees of azimuth at rising and setting.

Figure 2 The "star structure" (sidereal compass).

A student first learns the compass points. Then he learns their reciprocals. For every pair of reciprocals he learns what other reciprocal pair lies at right angle to it. A compass star on one's beam can thus serve as a guide when the star on which one's course is set is not visible. Navigators develop a feel for the angular distances from one to another of all the points on the compass, just as we develop a feel for the angular distance of numbers on a clock face in describing directions. This feel enables them to maintain a course at the appropriate angle to any visible compass star or other star that "follows the same path" as a compass star. They can do the same thing with reference to any other phenomenon, such as an ocean swell or seasonally prevailing wind, whose compass direction is known.

Sailing Direction Exercises

All sailing directions are memorized in relation to the sidereal compass or "star structure." So are the relative locations of all places of interest, including such seamarks as reefs, shoals, and regions with distinctive marine life. After learning the "star structure," including its pattern of diametrical opposites, students begin

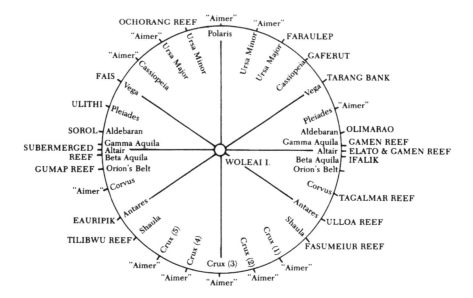

Figure 3 "Island looking" exercise, naming places and "aimers" (living sea marks) as one looks out from Woleai Island.

to learn all this information regarding the relative locations of islands and sea marks. Navigators have developed a variety of exercises as aids in memorizing and remembering this large body of information.

The most important set of exercises (fig. 3), called "Island Looking" (*woofanú*), takes all the known islands as points of reference and for each one the student goes around the compass naming the nearest places that lie along each radius (or very close to it). Navigators say that "Island Looking" is fundamental in their repertoire of knowledge.

Another exercise requires giving the names of all the "sea roads" (*yelán metaw*) or "sea directions" (*yitimetaw*) between various islands and reefs along with the reciprocal star directions on which they lie; and yet another requires naming the "sea brothers" (*pwiipwiimetaw*), the roads that lie on the same reciprocal star directions. An exercise called "Breadfruit Picker Lashing" (*fééyiyah*) uses as metaphor the breadfruit-picking pole (*yiyah*). In this exercise, one imagines reaching out with the pole along a given star course and picking off in succession all the islands and reefs that lie along it to the end of what is known, then turning and doing the same from that point along another course to its end, and so on until one has picked off all the places, real or imagined, in one's repertoire. Two other exercises, "Reef Hole Probing" (*yaaruwóów*) and "Sea-bass Groping" (*rééyaliy*), involve chasing a fish from island to island, each one cryptically identified by the name of a hole in one of its reefs to which the fish goes. These reef hole names provide a set of esoteric names, known only to navigators, with which they can discuss sailing among themselves without others present knowing what they are talking about.

In all such exercises the navigator follows a course from his home island to the one from which the exercise begins and then proceeds according to a set pattern from one place to another. The pattern may be to box the compass, as in "Island Looking"; to go in a series of zigzags, as in "Reef Hole Probing"; or to follow a main course northward, go east or west from it, and then back at a series of points along it, as in an exercise called "Sailing of the Red Snapper" (*herákinimahacca*).

Living Seamarks

"Aimers" or "aligners" (*yepar*), as the Carolinians call them, are living seamarks (*pwukof*) associated with particular areas in the vicinity of islands or midway between them. They consist of such things as a tan shark making lazy movements, a ray with a red spot behind the eyes, a lone noisy bird, a swimming swordfish, and so on. Each has its own name and is to be found in a particular drag on a particular star course from its associated island, often on a course along which no island lies. No one sails to find them; rather, one hopes to encounter one of them when one is lost. They serve as a last hope for the navigator who has missed his landfall or lost his bearing, enabling him, if he is lucky enough to encounter one, to align himself once more in the island world. When doing "Island Looking" exercises, advanced students include these "aimers" among the locations to be named in boxing the compass from a given island.

Keeping Track

A major problem for dead reckoning sailors is to estimate distance traveled and keep track of where they are in regard to their course. To do this, Carolinian navigators use what they call "dragging" or "drags" (*yeták*). It involves using a place other than one's destination as a point of reference. If you are traveling from Boston to New York, for example, Albany as point of reference lies to the west of Boston at the beginning of the journey and north of New York at the end of it. As you travel from Boston to New York, Albany moves in relation to where you are through several compass directions from west, to west northwest, to northwest, to north northwest, to north. The intervals between these changing compass directions divide the journey into four legs, as we would call them. In just this way, the Carolinians see the place of reference as being "dragged" through the intervening directions of their sidereal compass as a voyage progresses. The number of direction intervals through which the place of reference is "dragged" comprises the number of "drags" (legs) in the voyage.

Thus the course from Puluwat to Tol in western Truk, 160 miles away, is almost directly east on the "rising of Altair" (fig. 4). Pisaras lies 120 miles northeast of Puluwat on the rising of Vega and a like distance from Tol on the setting of Vega. During the voyage from Puluwat to Tol, Pisaras is "dragged" from the rising of Vega

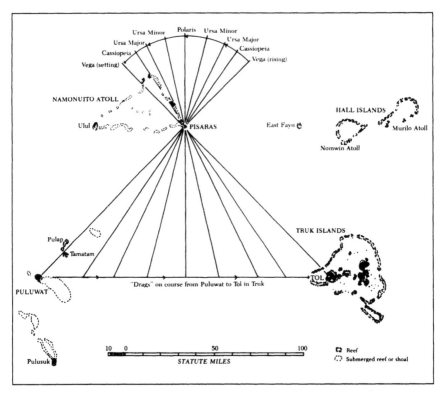

Figure 4 "Drags" on course from Puluwat to Truk.

through the rising of Cassopeia, the rising of the main star in Ursa Major, the rising of Kochab in Ursa Minor, Polaris, and on through the settings of Kochap, Alpha Ursa Major, and Cassopeia to the setting of Vega, dividing the journey into eight "drags" of roughly twenty miles each. Estimating the headway he is making, a navigator keeps track of his progress from drag to drag. As sailing conditions change, he adjusts his reckoning of progress from one drag to the next. Such reckoning greatly facilitates keeping track of overall progress and expectation of landfall.

Every course between two islands has an island or seamark of reference that serves to divide the journey into "drags." Ideally, the end of the first "drag," called the "drag of visibility" (*etákinikanna*), should come when the island of departure is no longer visible, and the next drag, the "drag of birds" (*etákini maan*), should end at the most distant point at which land-based birds feed at sea. Similarly the next to last begins where birds again appear and the last when the island of destination becomes visible. These correlations are understood to be rough, but are useful in that a navigator knows from his estimation of the number of "drags" traveled when he should soon be sighting land-based birds and when he should be able to see his destination. If these signs fail to appear when he has reason thus to expect them, a navigator knows that he is off course.

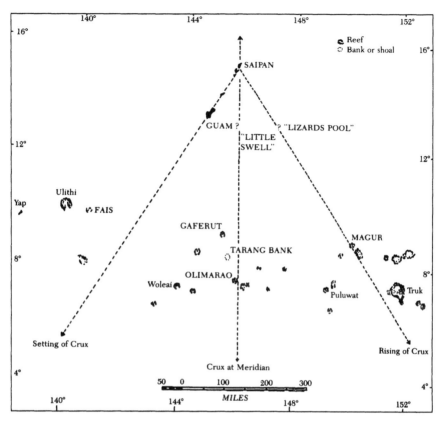

Figure 5 The "great trigger fish" (Fais ad Magus tail and head, Gaferut and Olimarao dorsal and ventral fins) with its northern flip (places in capital letters) as located on the map (cf. fig. 6).

Imaginary places can serve as points of reference for "dragging" as readily as real ones. All that is required is a convenient set of assumed compass directions to it from islands of departure and destination. For the voyage north from the Carolines to Guam and Saipan, there are no conveniently located islands. Here "ghost islands" and "aimers" are used as reference.

Schematic Mapping

Without maps or charts, navigators must devise ways of constructing mental equivalents. "Trigger Fish" (*pwuupw*) is the name for one such way of conceiving the geography of the navigator's world. It envisions the locations of five places. Four of them form a diamond to represent the head, tail, and dorsal and ventral fins of the trigger fish. The head is always the eastern point and the tail the western one. The dorsal and ventral fins can serve either as northern and southern or southern and northern points respectively. The fifth place, at the diamond's center,

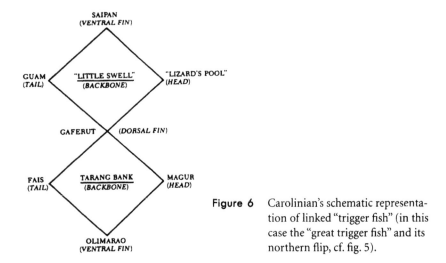

Figure 6 Carolinian's schematic representation of linked "trigger fish" (in this case the "great trigger fish" and its northern flip, cf. fig. 5).

is the fish's backbone. Any islands (real or imaginary), reefs, shoals, or living seamarks whose relative locations are suitable can be construed as a trigger fish. On a course between the dorsal and ventral fins, the head can serve as the reference island and the backbone marks midcourse. Trigger fish may be arranged so that the northern point of one is the southern point of another, or the backbone of one is the southern point of another.

"Great Trigger Fish" (*pwuupw lapalap*) are large-scale schematic maps. Of special importance is the one that has Magur and Fais (650 miles apart) as head and tail, and Gaferut and Olimarao as dorsal and ventral fins (fig. 5). As one looks south from Saipan, the rising of the Southern Cross, named "Trigger Fish," lies almost directly over Magur, the head, and sets a bit east of Fais, the tail. Sailing south from Saipan or Guam, if a navigator keeps his course within the rising and setting of the Southern Cross, he will end up in the heart of the Caroline Island chain, with its many reefs, shoals, and other seamarks of which he has knowledge.

In one scheme, the northern flip of this great trigger fish has Gaferut as dorsal fin, Saipan as ventral fin, Guam as tail, and the imaginary place "Lizard's Pool" as head (fig. 6). A set of lesser overlapping trigger fish involving a series of "ghost" places lying east and west of this north-south course between Saipan and Gaferut provide a series of reference points for dividing it into a convenient number of "drags."

Predicting the Weather

Prevailing weather conditions are equated with the "months" of a sidereal calendar. Although called "moons" (*maram*), they are not lunar months. In most calendars there are twelve or thirteen months of unequal length, each named for a star. A month begins when its star stands about 45 degrees above the eastern horizon just before dawn, when to look at it one must tilt one's head back to where one feels a roll of skin forming at the back of the neck. The month continues until the next month star reaches the same position.

A weather predicting system, called "storming of stars" (*mworenifúú*), involves the rising of storm stars and the first and last five days of the lunar cycle. Within the span of each month, one or two storm stars make their appearance in the east just before dawn. If there is one such star in the month, there will be stormy weather during the first five days of the next new moon in the west. If there are two such stars in the month, stormy weather will come during the last five days in the lunar cycle after the heliacal rising of the star.

More immediate weather conditions are forecast from the color of the sky at sunrise and sunset and from the shapes of clouds. Large cumulo-nimbus clouds, "house of wind" (*yimwániyang*), are believed to store up wind. If they are visible at dawn or dusk, a navigator expects the wind to come from their direction.

Putting the System to Work

It is one thing to learn the "star structure," sailing directions, sea-lane directions, and "aimers" and to become adept in the exercises and drills involving them. It is another thing to put it all to work in actual practice. The stars are not visible by day, the sky may be overcast at night, and the sailing directions in the exercises are, at best, only to the nearest compass point. Conditions vary with the seasons. To use the system, a navigator must rely on what he can actually see. He must also learn how to adjust the sailing directions in the light of his and his fellows' experience.

Ocean swells are a crucial guide in sailing. Navigators recognize up to eight different swells, one from each octant of the compass. Most dependable are those from the north, northeast, and east, associated with the tradewind season (our winter). During our summer, swells come from the southeast and south. The different swells have characteristic intervals. Navigators use opportunities to check the direction of swells against the stars. They then maintain course at the appropriate angle to the swells. When swell systems move across each other, they produce an effect somewhat like that of converging wakes of motorboats, forming an alignment of peaks or "wave nodes" (*pwukonó*) by which to steer. "Wave Nodes" is, indeed, the name used in reference to the whole body of knowledge relating to the interaction of currents, swells, and winds.

Currents reveal themselves by the shape of waves and ripple patterns. These patterns vary according to whether the current is going with the wind, across it, or against it, and according to their set in relation to the direction of swells. They make a significant difference in how one selects a course on the star compass. When setting out from an island, one uses an alignment of landmarks astern (*fótonomwir*) to set one's star course. A navigator should know the configuration of his home island as seen from every compass direction. On reaching the point, called "one tooth," where the island is just visible as a point on the horizon, one sights back along the bow-stern axis of one's vessel. If the island of departure lies directly astern on the axis, no compensation for drift from current is needed. The degree to which it is off, as measured by compass point intervals or their fractions, indicates to a navigator the degree of course adjustment he should make. He can measure the difference by holding out his hand at arm's length. The width of his hand corresponds rather well with the amount of arc on the horizon between adjacent points of a compass of thirty-two points.

In practice a navigator may begin with one star course and change to another at some "drag" point along the way. He adjusts for currents and changing wind conditions. Sailing against the wind may require planning a series of tacks from "drag" to "drag." In doing this, course settings from island of departure and island of destination to reference islands may help to structure the sequence of tacks. Using "drags" as a way to keep mental track of distances covered is crucial in such tacking, as when one plans to cross and recross the line of the direct course (*yallap*) from home to destination at the point where each new drag begins. Again, the set of current is critical in how far one tacks to right or left of one's true course.

Navigators must learn to make all these adjustments of course for the voyages they actually make. Years of sailing experience are necessary to develop skill as a practicing navigator. As with any other skill, not everyone is good at it.

Navigators as Ritual Specialists

Protective ritual is something else a navigator must learn. He is said to be the "father" of his crew, who depend on him for their safety and welfare. He must know how properly to invoke the patron spirits of navigation; he must carefully observe the associated taboos in regard to sex and food; he must know the spells that will prevent storms and repel sharks; and he must be able to provide his vessel with protective amulets. He and his crew must know the art of righting an overturned sailing canoe. Indeed, they may deliberately overturn it in order to ride out a bad storm without being blown far off course. In addition to all of this, it is useful for a navigator to know enough of the special rhetoric and spells associated with politics and diplomacy (*yitang*) to ensure hospitality and safe conduct for himself and his crew when visiting other islands. Where voyagers have kin or fellow members of their matrilineal clans, they can be relatively sure of hospitality, but otherwise they are liable to be treated with suspicion and hostility. Knowing how to interact properly with a community's official greeter of visiting canoes establishes a navigator as someone to be reckoned with and treated with respect.

Their ritual knowledge and their observance of food and sex taboos set navigators apart. They are perceived to be among the most learned and magically powerful members of Carolinian society. Having demonstrated ability on a test voyage, such as successfully making a 130-mile, direct trip from Puluwat to Satawal, a newly certified navigator is initiated (*pwpwo*) into the ranks of recognized senior colleagues (*palú*). Thereafter, he eats in the canoe house with fellow navigators, whose food must be prepared separately for them. He has achieved the equivalent of a Ph.D. degree or a professional license to practice medicine or law in Euroamerican society.

Keeping the Knowledge Alive

As should now be clear, sustaining the total body of knowledge in the absence of writing is accomplished by organizing it, making it systematic and schematic. It is taught and learned in this organized form. Indeed, it is overlearned with the use of drills and exercises that build in redundancy and are continually rehearsed.

For memory storage, some of the lore is embedded in chants, whose metric and tonal structures provide aids to recall. These chants are often cryptic in content, requiring commentary in order to understand them. A trainee will learn the words first. Only later will his teacher supply the necessary interpretation. If a teacher should die before passing the commentary on, his pupil must make the best sense he can of the chant. In time, he will develop his own interpretation in the light of his other knowledge and experience.

It is interesting to observe that the new interpretation may be quite different from the earlier one and still be workably consistent with reality. Interesting, too, is the evident elaboration of navigational lore beyond practical requirements, involving sailing directions to many named places that no one ever visits, that lie outside the known world or in the sky world. The living seamarks represent an elaboration beyond what is empirically known, also. Navigators seem to enjoy playing with the possibilities within their system, elaborating on them both for the fun of it and in order to show off superior knowledge to one another. Thus, the practical core of the system that is empirically tested in continual application remains much the same among the competing "schools" of navigation.[2] They differ in their living seamarks, in their chants, in the interpretation of specific chants, in their mythology, and in their magical rituals—in those respects, in short, where difference has little or no effect on successful voyaging and is in regard to what lies beyond the means of ready empirical testing. The navigators' separate mess in the boathouse, where visiting navigators also eat, appears to have played an important role in keeping the common system of navigation shared, at its practical level, over such a wide area.

We should note, finally, that the theoretical assumptions on which the system rests are that the sun and stars revolve around the earth, which remains stationary, and that the star Altair rises due east and sets due west. That these assumptions are false within the framework of modern Western scientific understanding does not deprive the system of practical utility for their purposes nor, indeed, for the purposes of any sailor without instruments and having to navigate by dead reckoning in Carolinian waters.

Carolinian Navigation as a Practical Science

Several things stand out about Carolinian navigational knowledge. It has all the features of a practical science.[3] It contains a massive amount of discrete information, which, in the absence of writing and reference books, has to be committed to memory. The information is highly organized in a systematic way; the different ways of organizing it provide much redundancy as an aid to recall. It involves highly abstract thinking: the compass as a set of imaginary points at equal intervals around the horizon, named for the stars and abstracted from their perceived motions, but not identical with them; the use of "drags" as imaginary divisions of one's course of travel; the use of imaginary places as points of reference to calculate "drags"; and schematic mapping in the form of "trigger fish." Gladwin (1970) has called attention to the fact that navigators tested low for abstract thinking, and used this discrepancy to question whether the psychological tests in fact were testing for concrete as against abstract thinking at all.

It as also clear that Carolinian lore is based on empirical observation in its practical aspects and becomes fanciful only beyond the bounds of readily verifiable experience or practical application. We should note, moreover, that navigators are ever ready to add to their knowledge. In these respects Carolinian navigational lore is quite similar to Western practical science. Indeed, demonstration of its fundamental soundness was made by Piailug of Satawal, who successfully navigated the Hokule'a, a replica of an ancient Hawai'ian double-hulled voyaging canoe, from Hawai'i to Tahiti by dead reckoning alone, using only his own knowledge of seamanship and navigation as a Carolinian *palú* (Finney 1979).

There are things about this body of knowledge that have mystified some observers. Gladwin found it strange to base the system of "drags" and its use in tacking on the "concept of a moving island" (1970:181), seeing the traveller as remaining at the center and the island world moving around him through the sidereal compass as he travels. He saw in this an example of a difference in the way human thinking may be culturally programmed; but I know of no other way to use a compass. If we speak of the direction of places changing in relation to us as we travel while Micronesians speak of places being "dragged" through different points of the compass, the figures of speech may be different, but the underlying understanding of a compass is not.

Lacking other instruments of observation, moreover, Micronesians train themselves to use their own bodies and senses far more than do Euroamericans. Piailug's ability accurately to gauge the direction and force of currents by observing faint patterns in surface ripples amazed Thomas (1987:32). It takes a lot of practice to be able to assess distance and time of travel accurately enough to use "drags" as reference in tacking upwind. It also requires training to learn to pick up such seamarks as the surface manifestations of the presence of a submerged reef. Yet, obviously, none of these skills are beyond human perceptual abilities.

Bodies of Knowledge and Cultural Anthropology

Describing the content and organization of the many diverse bodies of knowledge that comprise human understandings is one of the workaday tasks of cultural anthropology. These bodies of knowledge range from how to make fire and catch fish to how to build airplanes and computers, from how to conduct oneself acceptably in one's family relations to how to do so in negotiating a business deal or prosecuting a legal case. Ethnography, as it is called, aims, among other things, to describe what one needs to know in order to engage with a society's members in all their activities in a manner that meets their standards of acceptable performance. Such knowledge is what is technically meant by a people's "culture." Like a language or a game, a culture is something one has to learn in order to describe it. Ethnography is, therefore, an exercise in the systematic learning and presentation of what people are expected to know. Ethnography also tries to describe how people apply and use such knowledge in the affairs of life as that knowledge constructs those affairs, including the skills that application requires and the preferred performance styles. Ethnography is as applicable to what farmers have to know, aeronautical engineers have to know, elementary school educators have to know, or religious specialists have to know to perform acceptably by their respective standards, as it is to what Micronesian navigators have to know to perform acceptably by theirs.

Having said this, I must add that there are relatively few ethnographic accounts of bodies of knowledge that tell us what we need to know to be acceptable practitioners of that knowledge. As I have just said, the only way to record it is for ethnographers to learn it themselves, to pass the test of acceptability, and then, having made a record of what was being learned as it was being learned (the "field notes"), to use that record to describe what they think they now know. In this way they can produce an account whose accuracy can be checked by those already knowledgeable and evaluated by those who try to use it at as a guide to becoming knowledgeable themselves. Unfortunately, the customs governing dissertation research proposals in cultural anthropology are such as to steer students away from undertaking to learn and describe other systems of knowledge as a worthy end in itself. Ethnographies rarely describe activities in relation to all that one must know to perform them and all the decision points and criteria for making those decisions in the course of their conduct. But only by doing such things can we acquire good descriptions of the content of human cultures. Only by trying to do this can we confront and solve the methodological challenges it entails. Without such descriptions, however, we lack the information and the understanding that we need in order critically to examine propositions about the relation of culture to human cognition. With this last observation in mind, I ask the reader to consider if Micronesian navigation rests on mental operations of a kind with which we are largely unfamiliar or if, with better understanding of how it works, we find it represents ingenuities of much the same kind that are exemplified in the products of Western thought.

Notes:

1. This paper is adapted from a paper by Goodenough and Thomas (1987). Here, as in that paper, I try to integrate and summarize material from a number of sources. I have drawn heavily on the work of Gladwin (1970), who provides excellent information on the design of sailing canoes and on the use of "drags"; on that of Riesenberg (1972), who provides details on a number of star lore exercises and the workings of the "Great Trigger Fish"; and on that of Thomas (1987), who has greatly enriched earlier reports

with information on exercises, "drags," weather prediction, reading the sea's surface, and the practical application of the formal system of actual sailing. For an earlier account see Damm and Sarfert (1935). See also Lewis (1973) and Turnbull (1991). By and large the same technical terms are used through all of the Western Carolines, but dialects differ. I have chosen to render them here in the dialect spoken on Puluwat Atoll, where Riesenberg worked. It is very close to that of Satawal, where Thomas studied navigation.

2. Two schools are present in Puluwat; *Wáriyeng,*, "Wind Seeing," and *Fáánuur,* "Under the Banana Plant" (Gladwin 1970:132).

3. In calling this a practical science, I have in mind the kind of knowledge we have traditionally associated with engineering, knowledge that involves empirically tested principles and rules of thumb, organized into a coherent system of ideas, that works well in the achievement of practical objectives. Whether it is science, or craft, or art, or a mix of all three is a matter of how one chooses to fit it into Western intellectual categories about which we Western intellectuals are ourselves in some disagreement.

CHAPTER 2

The Scientific Basis of Gastrointestinal Herbal Medicine among the Highland Maya of Chiapas, Mexico

E. A. Berlin, B. Berlin, X. Lozoya, M. Meckes, J. Tortoriello, and M. L. Villarreal

in collaboration with D. E. Breedlove and M. G. Rodríguez

Introduction

Work over the last several decades in the emerging field of ethnobiology, and research in medical ethnobotany in particular, has demonstrated that the ethnobiological knowledge of traditional peoples conforms in many respects to basic scientific principles (Berlin 1973, 1978, 1992; Berlin, Breedlove, and Raven 1974; Raven, Berlin, and Breedlove 1971; Quiros et al. 1990; Toledo 1988) and that the curative properties of certain plants is not simply unsubstantiated folklore (Schultes 1984; Sofowara 1982; Elizabetsky 1986; Farnsworth and Pezzuto 1983).

In this chapter, we present the preliminary findings of medical ethnobiological research carried out among the highland Maya of Chiapas, Mexico that confirm the scientific basis of traditional medicine. Our data show that the highland Maya have developed a highly sophisticated system of herbal medication based on an astute understanding of the signs and symptoms of common disease conditions. The potential therapeutic efficacy of Maya herbal medicine is suggested by laboratory studies on the pharmacological properties affecting the agents associated with the ethnomedically identified clinical signs and symptoms that correlate closely with biomolecular medical disease categories.

Highland Maya Ethnomedicine

In the traditional medicine of the highland Maya, as with many other ethnomedical systems, the maintenance or reestablishment of a state of health is dependent on events and interactions in two separate realities: the natural, usually visible, reality that follows predictable physical norms and an alternate, frequently nonvisible, reality that relates to extranatural phenomena. We adopt Foster's (Foster and Anderson 1978) dual division of medical systems into *naturalistic* and *personalistic* cognitive frameworks; the naturalistic system being based on cause, response, and treatments that are explainable within the natural order of things; the personalistic system having relationship to all causes, responses, and treatments of illness that derive from extranatural phenomena and such sensate beings as souls, deities, demons, ancestors, and sorcerers. The participation of a health condition in the naturalistic system is empirically determined and is based primarily on immediately apparent *signs* and *symptoms*. For example, the naturalistic gastrointestinal condition *ch'ich' tza'nel* 'bloody diarrhea' (lit. 'blood' + 'feces') is recognized by the presence of frank blood in the stool. Self-treatment for these naturalistic conditions or consultation with individuals who are knowledgeable about medicinal plants is the norm.

On the other hand, diagnosis of the personalistic condition *jme'tik jtatik* (lit. 'our ancestral mothers and fathers'), which may demonstrate gastrointestinal symptoms including blood in the stool, is based on the *retrospective presumption of an etiologic agent*, such as an inadvertent encounter with these ancestral spirits. In this case, diagnosis and treatment frequently involve the intervention of healers with special powers. While personalistic conditions may at times also be treated with herbal medications, Maya curers normally employ remedies involving ceremonial healing rituals, special prayers, and shamanic divination.

Almost all earlier work dealing with highland Maya ethnomedicine has focused on its personalistic bases, and the general outlines of this aspect of Maya ethnomedicine are well known (for some of the more important examples, see Fabrega 1970a, 1970b; Fabrega and Silver 1970, 1973; Guiteras Holmes 1961; Harman 1974; Holland 1963; Holland and Tharp 1964; Metzger and Williams 1963; Silver 1966; Vogt 1970, 1976). Unfortunately, this preoccupation with the symbolic aspects of Maya healing has tended to de-emphasize the wealth of medical empirical ethnobiological knowledge possessed by the Tzeltal and Tzotzil, leading generally to the conclusion that highland Maya medicine incorporates a poor understanding of human anatomy, has but a weak relationship to physiological processes, and primarily satisfies psychosocial needs through magical principles (Arias 1991:40–43, Fabrega and Silver 1973:86, 211, 218; Holland 1963:155, 170–171; Vogt 1970:611). Since magical principles have little to do with science, a reading of these works supports a view that Maya ethnomedicine is anything but scientific and that the Maya themselves lack a scientific understanding of health and disease.

Our data on the naturalistic aspects of Maya ethnomedicine lead to quite the opposite conclusion. The extensive medical ethnobiological materials we have collected demonstrate that the highland Maya have a remarkably complex ethnomedical understanding of the physiology and symptomatology of particular health conditions. In addition, they have identified a wide range of plant species that treat the symptoms associated with these conditions. Our findings show that

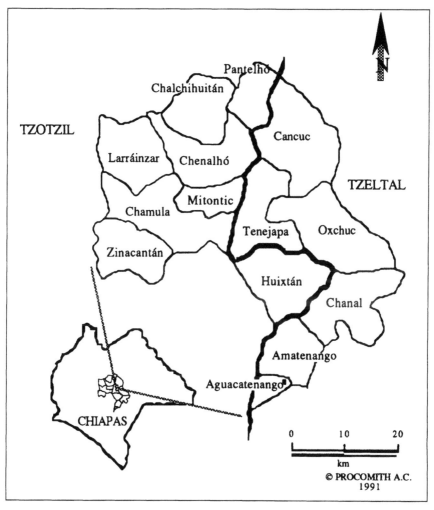

Figure 7 Location of the Tzeltal and Tzotzil Maya populations of the central highlands of Chiapas, Mexico.

almost all of the medicinal plant species in the highland Maya pharmacopoeia specifically target individual health conditions. In this sense, highland Maya traditional medicine is an ethnoscientific system of traditional knowledge based on astute and accurate observation which could only have been developed on the basis of explicit empirical experimentation with the effects of herbal remedies on bodily function.

Gastrointestinal Conditions

The data we present to support these claims are drawn from one portion of the highland Maya ethnomedical system—the gastrointestinal diseases. These materials result from a bi-national, multidisciplinary research program on the medical

Figure 8

Ethnoepidemiological Reports of Health Problems Grouped by Major Ethnomedical Category among the Highland Maya

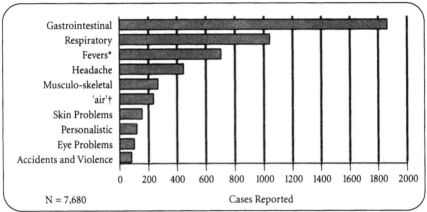

†'Air' is the literal translation of the Maya term *ik'*, which is used in much the same way that 'air' is used in Spanish American traditional medicine and 'wind' is used in Chinese traditional medicine. External air, normally cold in sensation, is perceived to enter the body and cause disease symptoms.

*Fevers is a Maya ethnomedical category indicating a series of conditions accompanied by an abnormally high body temperature.

anthropology, medical ethnobotany, and ethnopharmacology of the Tzotzil and Tzeltal, two Maya-language-speaking populations of approximately 500,000 people in the state of Chiapas, Mexico. Data were collected during the period 1987–1991 and are based on comparative research carried out in eight Tzotzil and six Tzeltal Maya municipalities of the central highlands (see fig. 7).

We have chosen to focus on gastrointestinal problems because of the relative epidemiological and ethnomedical significance of these health conditions for the highland Maya. Figure 8 presents the major patterns of primary health problems drawn from retrospective ethnoepidemiological surveys conducted in the fourteen Maya municipalities shown in figure 7 from 1989 through 1992. Since these are retrospective self-reports of health problems (see below for description), these data are indicators of the cultural saliency as well as the perceived frequency of illness.

Our medical anthropological, ethnobotanical, and ethnopharmacological analyses of data relevant to gastrointestinal conditions allow us to make the following major claims:

1) the medicinal plant species employed by the highland Maya for the treatment of the recognized classes of gastrointestinal diseases are condition-specific in that a limited set of distinct species is selected as treatment for particular gastrointestinal diseases;

2) the Maya explanatory models of gastrointestinal conditions show how the pharmacological properties of plant species relate to the symptoms that these health problems manifest, and the folk diagnoses sometimes allow for inferences about pathogenic agents;

3) a major proportion of the plant species employed show demonstrable bioactivity that is likely to account for their purported pharmacological importance among the highland Maya.

Methods

Medical Ethnobotany

Information on the principal medicinal plants used in the treatment of gastrointestinal conditions is drawn from two primary data sets:

Data Set 1: Informant responses to botanical collections, comprising some 7,000 collection numbers representing approximately 1,650 species grouped into 750 genera and 150 families.

Over a fifteen-month collection period in 1987–1988, medicinal plants were collected in nine highland Maya municipalities with the assistance of adult informants who were recognized as knowledgeable in traditional herbal medicine. A total of 351 informants (305 males and 46 females,[1] including 190 laypersons and 161 healers[2]), were interviewed by Maya field investigators in the local municipal dialects of Tzeltal and Tzotzil. Medical ethnobotanical data on each plant collection were transcribed in standardized printed collection notebooks that included information required of standard botanical vouchers (description of the plant species, locality of collection, altitude, name of collector, collection number, date, and name of the informant aiding the collector). In addition to these data, medical ethnobotanical information was collected on the following variables: 1) native name(s) of the plant, 2) health condition(s) that the plant is said to treat, 3) plant part(s) employed in the preparation of remedies, 4) methods of preparation, 5) methods of administration, 6) quantities of all ingredients used, 7) doses, and 8) duration of treatment. These data comprise a database of 6,959 records.

Data Set 2: Informant responses to a subset of 204 of the most frequently collected plant species of the full inventory of 1,650 species.

In order to obtain comparable data from a large number of individuals throughout the region, we developed a standardized set of stimulus materials representing the most frequently collected plant species in the region (i.e., species collected six or more times in several municipalities). Samples of these species were mounted on 8 1/2-inch-by-11-inch botanical mounting paper, sealed in clear plastic covers, and placed in six large standard three-hole notebook binders, ranging from the most commonly collected species to the least commonly collected. These portable herbarium specimens (referred to as the "traveling herbarium") were taken by our Maya assistants into each of the fourteen municipalities in the study area and shown, one by one, to a total of 126 knowledgeable informants (63 males and 63 females, between eight and ten informants in each of the fourteen municipalities). Interviewees were asked to name each specimen, specify the health conditions it was used to treat, indicate any other plant species mixed with it in the preparation of the herbal remedy, and to note why the particular species was thought to "have the power to cure." The results of this survey comprise two large databases of nearly 30,000 records.

Ethnomedicine

Prioritization and discussion of the gastrointestinal conditions are based on ethnomedical data drawn from two additional sources:

Data Set 1: Structured Maya explanatory models surveys on the following variables: 1) cause, 2) onset, 3) signs, 4) symptoms, 5) complications, 6) prognosis, 7) seasonality of occurrence, 8) special groups affected, 9) treatment, and 10) curing resources.

Two types of explanatory models surveys were carried out with in-depth interviews. *Individual interviews* were conducted in Tzeltal or Tzotzil by Maya field investigators with a total of 15 males and 11 females from the Tzeltal municipalities of Amatenango, Cancuc, Oxchuc, and Tenejapa, and the Tzotzil municipalities of Chamula and Larráinzar. *Group elicitation sessions* were conducted with participants from the Tzeltal communities of Aguacatenango, Amatenango, Chanal, and Tenejapa, and the Tzotzil communities of Chenalhó, Huixtán, Mitontic, and Pantelhó, and included 38 males and 16 females. These materials comprise databases of approximately 2,500 records.

Data Set 2: Retrospective ethnoepidemiology surveys concerning most recent health problems and their treatments.

Ethnoepidemiological surveys were conducted in five households from each of twenty hamlets in each of the fourteen municipalities yielding a population sample of more than 7,000 persons. The interview instrument consisted of a standard three generation genealogy, with ego's generation representing male and/or female heads of household or ego(s) and their siblings. Individuals in the first ascending generation represent parents of ego(s) and individuals in the first descending generation represent children of ego(s). In these surveys, all recent self-defined illness events and their treatments were recorded in Tzeltal or Tzotzil for all household members. This ethnoepidemiology database consists of 7,680 records.

Ethnopharmacology

Data on the potential pharmacological activity of the botanical species in our research are drawn from our own laboratory studies and/or from published literature sources.

Pharmacological analysis: Carried out on those species prioritized as the most important plants used by the Maya in the treatment of gastrointestinal and respiratory disorders.

Plant extracts were tested for activity against *Escherichia coli, Staphylococcus aureus, Candida albicans*, on KB and P388 cancer cell lines, and for spasmolytic effects on guinea pig ileum. More than 300 extracts representing some 150 species have been evaluated experimentally for antimicrobial, antifungal, phototoxic, cytotoxic, and antispasmolytic activity. Tests for antimicrobial activity are indicators of pharmacological effects that potentially kill or inhibit etiologic organisms (Rios et al. 1988). The cytotoxicity tests determine the ability of plant extracts to destroy or prevent reproduction of cancer cells. These effects or other toxic reactions may be initiated or enhanced by exposure to ultraviolet light, which would be determined by the photoxicity studies (Meckes-Lozoya, in press). Spasmolytic activity demonstrates whether the herbal medications might provide symptom relief by slowing intestinal peristalsis thus reducing both diarrhea and the pain associated with this condition (Lozoya et al. 1990).

Literature searches: Carried out using the large databases of NAPRALERT (Beecher and Farnsworth n.d.), as well as other relevant publications on traditional Mexican medicine available at the Laboratory on Pharmacology of Medicinal Plants of the Mexican Institute of Social Security.

Results

Criteria for Selection of Gastrointestinal Conditions in Maya Ethnomedicine

Three criteria were established for the selection of gastrointestinal problems for further analysis—frequency of report, informant consensus of naming, and specificity of signs and symptoms. Frequency of epidemiological report is based on prevalence records from our ethnoepidemiological surveys, on incidence records from regional clinics and hospitals, and on frequency of treatment reported in the ethnobotanical surveys. Informant consensus is established by shared patterns of disease class recognition as can be inferred by the distribution of specific disease names within and among municipalities. Specificity is determined by reference to unambiguous clinical diagnostic features of the particular gastrointestinal condition.

For the highland Maya, illnesses that directly affect the gastrointestinal system are grouped into three major classes encompassing eight conditions that can be demonstrated to meet the criteria outlined above. The major gastrointestinal conditions include the diarrheas (*watery, bloody,* and *mucoid*), the abdominal pains (*abdominal pain, epigastric pain, abdominal distention* [gases], and 'mother of man', a condition associated with the abdominal pains and resulting primarily from acute gallbladder disease), and worms (see table 1).

Frequency of Report

Ethnoepidemiologic prevalence is demonstrated by noting the relative frequency of reports of gastrointestinal diseases (fig. 8) and by the number of reports within the general class of gastrointestinal conditions (fig. 9). As seen in figure 8, gastrointestinal diseases have the highest frequency of report for the highland

Table 1

Major Gastrointestinal Conditions and Their Key Characteristics

Category	Key Characteristic
tza'nel	diarrhea (normally watery)
ch'ich' tza'nel	bloody diarrhea
sim nak'al tza'nel	diarrhea with mucus
k'ux ch'ujtil	abdominal pain
k'ux o'tanil	epigastric pain
me' winik	abdominal pain, probably of biliary origin
pumel	abdominal gas and bloating
lukum	intestinal worms

(category in dialect of Tzeltal of Tenejapa)

Table 2A

Key Signs and Symptoms of the Most Important Diarrheas and Worms

Signs and Symptoms	watery	bloody	mucoid	worms
liquid stool	•			
bloody stool		•	+/-	
mucoid stool		+/-	•	
pain in small intestine	•			•
pain in large intestine		•	•	
colorectal pain		•	•	
dorsolumbar pain		•		

Table 2B

Key Signs and Symptoms of Gastrointestinal Pain

Signs and Symptoms	abdominal distention	abdominal pain	epigastric pain	'mother of man'
intestinal pain	+/-	•		
epigastric pain			•	
subclavicular pain				•
periumbilical pain				•
distention	•			

Figure 9
Reports of Gastrointestinal Conditions in Ethnoepidemiological Surveys

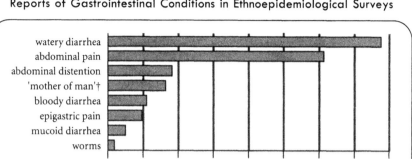

†gallbladder disease

Maya. They represent almost one-quarter of all reported illness, with more than 1,800 of the 7,680 persons interviewed in the ethnoepidemiology surveys reporting recent gastrointestinal afflictions. Figure 9 shows the breakdown of these reports according to the frequency of each kind of gastrointestinal disease. Simple watery diarrhea is the most frequent problem followed by generalized abdominal pain. These are followed by abdominal distention, and 'mother of man' (gallbladder disease), bloody diarrhea, epigastric pain, mucoid diarrhea, and worms.

Consensus of Naming

There is relative consistency of naming for gastrointestinal conditions. Epigastric pain and worms are known by a single term throughout the highlands. Mucoid diarrhea, bloody diarrhea, and abdominal pain are each identified by a single primary term plus one rare variant. Abdominal pain is known by one primary term (reported in eleven municipalities), a secondary term (three municipalities) and a unique variant (one municipality).

Specificity

In many cases, the linguistic constituents of the Maya names for gastrointestinal conditions indicate the key diagnostic criteria by which conditions are recognized. For example, the name for 'watery diarrhea' is *ja' ch'ujt* ≈ *ja' ch'ujt'* (in Tzeltal) ≈ *ja' ch'ut* ≈ *a' ch'ut* (in Tzotzil) or 'water belly,' indicating that the stomach contents are like water. The expression *ch'ich tza'nel* 'bloody diarrhea' refers literally to blood in the feces.

From the explanatory models surveys, we can identify a distinct set of key criteria as well as accompanying signs and symptoms that are consistently described by the Maya. These signs and symptoms, which represent ethnomedical diagnostic criteria, are shown in Table 2A (diarrheas and worms) and Table 2B (abdominal pains).

Table 3

Five Most Frequently Mentioned Species Used in the Treatment of Bloody Diarrhea, Tzeltal Traveling Herbarium

Botanical Species	Number of Reports
Crataegus pubescens	30
Calliandra houstoniana	26
Baccharis vaccinioides	21
Acacia angustissima	20
Calliandra grandiflora	17

Table 4

Three Most Frequently Mentioned Health Conditions for the Five Top-ranked Species Used in the Treatment of Bloody Diarrhea, Tzeltal Traveling Herbarium

Botanical species	Health Conditions		
	1st	2nd	3rd
Crataegus pubescens	bloody diarrhea (30)	mucoid diarrhea (8)	watery diarrhea (7)
Calliandra houstoniana	bloody diarrhea (26)	nose bleed (12)	mucoid diarrhea (8)
Baccharis vaccinioides	bloody diarrhea (21)	watery diarrhea (11)	cough (9)
Acacia angustissima	bloody diarrhea (20)	mucoid diarrhea (5)	eye problem (5)
Calliandra grandiflora	bloody diarrhea (17)	nose bleed (8)	mucoid diarrhea (8)

Criteria for Selection of Species Employed to Treat Gastrointestinal Conditions

Principal plant species used in the treatment of gastrointestinal conditions have been isolated (from data derived from responses to specimens in the traveling herbarium) on the basis of two major criteria:

1) species X is one of the top five species used in the treatment of some particular gastrointestinal condition Y, and

2) gastrointestinal condition Y is the most commonly treated condition associated with species X.

The application of these criteria can be seen by reference to the data presented in tables 3 and 4. Table 3 shows the five most frequently mentioned species

Table 5
Principal Species Employed in the Treatment of Gastrointestinal Disease among the Tzeltal and Tzotzil

Botanical species	Family	Tzeltal	Tzotzil
Acacia angustissima	Fabaceae	•	
Ageratina ligustrina	Asteraceae	•	•
Ageratina pringlei	Asteraceae		•
Allium sativum	Liliaceae	•	•
Arthrostema ciliatum	Melastomataceae	•	
Baccharis serraefolia	Asteraceae	•	•
Baccharis trinervis	Asteraceae	•	•
Baccharis vaccinioides	Asteraceae	•	•
Borreria laevis	Rubiaceae		•
Byrsonima crassifolia	Malphigiaceae	•	•
Calliandra grandiflora	Fabaceae	•	•
Calliandra houstoniana	Fabaceae	•	•
Chenopodium ambrosioides	Chenopodiaceae	•	•
Cissampelos pareira	Menispermaceae	•	
Crataegus pubescens	Rosaceae	•	•
Dahlia imperialis	Asteraceae		•
Foeniculum vulgare	Umbelliferae		•
Fuchsia splendens	Onagraceae	•	
Helianthemum glomeratum	Cistaceae	•	•
Lantana camara	Verbenaceae	•	•
Lantana hispida	Verbenaceae		•
Lepechinia schiedeana	Labiatae		•
Lepidium virginicum	Cruciferae	•	
Lobelia laxiflora	Campanulaceae	•	
Nicotiana tabacum	Solanaceae	•	•
Ocimum selloi	Labiatae	•	
Psidium guineense	Myrtaceae	•	•
Rubus coriifolius	Rosaceae	•	
Smallanthus maculatus	Asteraceae		•
Sonchus oleraceus	Asteraceae		•
Stevia ovata	Asteraceae		•
Tagetes filifolia	Asteraceae	•	
Tagetes lucida	Asteraceae	•	•
Tithonia diversifolia	Asteraceae	•	
Verbena carolina	Verbenaceae	•	•
Verbena litoralis	Verbenaceae	•	•
Vernonia leiocarpa	Asteraceae	•	

employed in the treatment of bloody diarrhea as determined by a number of medicinal use reports in the Tzeltal traveling herbarium survey.

Table 4 presents data on the three most frequently mentioned health conditions associated with the five species in table 3. As can be seen by comparing tables 3 and 4, the five most frequently mentioned species used in the treatment of bloody diarrhea are confirmed as the preferred species for the treatment of this condition vis-à-vis other conditions that might be treated with these species.

On the basis of these two criteria, thirty-seven species have been identified as the principal gastrointestinal remedies used among the Tzeltal and Tzotzil. Seventeen of these species are named by both groups.

The thirty-seven principal gastrointestinal species recognized among the highland Maya are seen in table 5.

The Tzeltal recognize twenty-four principal species in the treatment of gastrointestinal diseases. The proportional distribution of medical use reports for these species and their associated gastrointestinal conditions, as determined from data from the Tzeltal traveling herbarium, are seen in table 6.

The Tzotzil recognize twenty-six principal species in the treatment of gastrointestinal diseases. The proportional distribution of medical use reports for these species and their associated gastrointestinal conditions, as determined from data from the Tzotzil traveling herbarium, are seen in table 7.

As can be seen in tables 6 and 7, there is a strong tendency for particular plant species to be favored remedies for specific gastrointestinal diseases. For the Tzeltal, *Verbena* spp., *Baccharis* spp., *Psidium guineense*, and *Tagetes filifolia* are preferred treatments for simple diarrhea. Remedies for bloody diarrhea include *Crataegus pubescens*, *Calliandra* spp., *Acacia angustissima*, and *Baccharis vaccinioides*. Two species are important in relieving the symptoms of mucoid diarrhea, *Cissampelos pareira* and *Lepidium virginicum*.

Some five species are highly regarded as medicinals for abdominal pain, with tobacco (*Nicotiana tabacum*), *Ocimum selloi*, and garlic (*Allium sativum*) among the most important. (A specimen of garlic was not included in either the Tzeltal or Tzotzil traveling herbaria. However, ancillary data collected subsequently indicate that garlic is of major importance for both groups in the treatment of abdominal pain). No species stand out as unique for the treatment of epigastric pain or abdominal distention.

Lobelia laxiflora and *Fuchsia splendens* are favored species in the treatment of the pains associated with gallbladder disease. The Tzeltal treat intestinal worms with the well-known remedy *Chenopodium ambrosioides*, although the less common *Helianthemum glomeratum* is of major significance as well.

Similar use patterns are noted for the Tzotzil, where *Baccharis* spp., *Ageratina* spp., and *Lantana* spp. are favored plants in the treatment of simple diarrhea, along with *Lepechina schiedeana*, *Borreria laevis*, and *Verbena carolina*. As with the Tzeltal, remedies for bloody diarrhea are drawn from medications prepared with *Crataegus pubescens* and *Calliandra* spp., in addition to *Psidium guineense*, *Sonchus oleraceus*, and *Byrsonima crassifolia*. No unique species is reported for the treatment of mucoid diarrhea.

Table 6

Proportional Distribution of Medical Use Reports for the Principal Species Employed in the Treatment of Gastrointestinal Disease by the Tzeltal Maya

Botanical Species	Percent of All Reports for Specific Gastrointestinal Condition											Tot N
	D	BD	MD	AP	AD	EP	GD	W	V	MISC GI	OTHER	
Verbena carolina	49	12	9	16	4		1				7	67
Baccharis serraefolia	47		17	6				6			25	53
Baccharis trinervis	47	6	8	16	2					2	18	49
Verbena litoralis	41	11	14	24	8				2	2	10	71
Psidium guineense	35	21	12	3	3			4		7	16	75
Tagetes filifolia	34	2		7	5			7			46	59
Crataegus pubescens	12	51	14		2					2	17	59
Calliandra houstoniana	4	48	15	2	2						35	54
Acacia angustissima	7	43	11	2	2					7	28	46
Calliandra grandiflora	9	31	15							2	44	55
Rubus coriifolius	7	30	11			2					48	54
Baccharis vaccinioides	21	29	6	6	6					3	31	72
Cissampelos pareira	18	21	23	14	5					2	16	56
Lepidium virginicum	12	3	23	18	16		6			5	10	60
Nicotiana tabacum				32	4	14	1			1	48	79
Ocimum selloi	6			31	8	25			4	2	23	48
Tagetes lucida	6	2		27	25	2	13				25	52
Ageratina ligustrina	20	2	4	26	22					2	26	50
Lantana camara	18	11	7	22	8					5	28	72
Lobelia laxiflora		5		5	3	17	38			3	31	42
Fuchsia splendens	5			3	3		37				52	38
Chenopodium ambrosioides	15	1		2	1		1	57	11		11	74
Helianthemum glomeratum	15			1	1			51		1	28	47
Arthrostema ciliatum	7			9	7				36		41	56

Key to Table: d = (simple) diarrhea, db = bloody diarrhea, md = mucoid diarrhea, ap = abdominal pain, ad = abdominal distension, ep = epigastric pain, gd = probably gall bladder disease, w = worms

Table 7

Proportional Distribution of Medical Use Reports for the Principal Species Employed in the Treatment of Gastrointestinal Disease by the Tzotzil Maya

Botanical Species	Percent of All Reports for Specific Gastrointestinal Condition										Tot N
	D	BD	MD	AP	AD	EP	GD	W	V	MISC GI	OTHER	
Baccharis trinervis	57	9	8	3	2	4	2	2			13	95
Baccharis serraefolia	53	7	6	9	5	1	1	1		2		88
Ageratina pringlei	51	4	12	14	2	3	2				11	94
Ageratina ligustrina	46	4	6	14	1	4	7		3		13	97
Helianthemum glomeratum	45	3	6	9		5			3	2	28	65
Lepechinia schiedeana	43	7	1	4	4					1	23	69
Borreria laevis	34	11	6	11	3	5	3		3		24	79
Lantana hispida	26	10	10	9		2	3				41	105
Lantana camara	35	20	12	10	1	2	2			1	1	122
Verbena carolina	32	23	5	5		6	3				27	111
Calliandra grandiflora	30	27	2	1	1	7		1		1	29	84
Calliandra houstoniana	25	31		1	1	8				1	34	83
Psidium guineense	23	34	16	1		5	2				28	112
Byrsonima crassifolia	13	34	5	4	2	5	2			1	32	91
Sonchus oleraceus	11	46	1	4		5	1	2		2	28	83
Crataegus pubescens	7	49	1	5		4					33	75
Verbena litoralis	19	20	4	4	1	14	4	1	1		24	114
Stevia ovata	14	2	3	58		2	9		2		12	65
Nicotiana tabacum	5			32	21	6	2	1		2	32	127
Baccharis vaccinioides	22	1	2	30		8	2			1	35	116
Tithonia diversifolia	21			31		7	11			6	24	71
Smallanthus maculatus	12	3		30	1	10	13		1		17	69
Foeniculum vulgare				2	1	82	1		1	1	14	18
Tagetes lucida	15	4	1	10	9	43		1	1		18	80
Dahlia imperialis	7	1			5	36	17			1	32	75
Chenopodium ambrosioides	8	1		1		1	1	58	5		25	171

Key to Table: d = (simple) diarrhea, db = bloody diarrhea, md = mucoid diarrhea, ap = abdominal pain, ad = abdominal distension, ep = epigastric pain, gd = probably gall bladder disease, w = worms

Nicotiana and *Allium* are also favored treatments for abdominal pain, as among the Tzeltal, but *Stevia ovata, Baccharis vaccinioides, Tithonia diversifolia,* and *Smallanthus maculatus* are also important. *Foeniculum vulgare* and *Tagetes lucida* are the principal Tzotzil species used for epigastric pain, along with the inner-nodal fluids found in the stems of *Dahlia imperialis*. As with the Tzeltal, *Chenopodium ambrosioides* is the single most important species in the treatment of worms.

Bioactivity and Potential Pharmacological Effect

There are two primary pharmacological reasons for which the Maya might have selected these plants as important in the treatment of gastrointestinal disease—their possible antimicrobial effects on the causative disease agents and/or their physiological effects on the mechanisms of symptom causation. The bioactivity found for most of the gastrointestinally important species supports both of these hypotheses.

Principal Species Used to Treat Diarrhea

• *Ageratina* spp. Our laboratory observations show that the methanolic extracts from the leaves and stems of *A. ligustrina* have strong spasmolytic properties, as studied on isolated guinea pig ileum, and a clear cytotoxic effect on cell cultures of KB and P388. The antimicrobial activity of these products on *Staphylococcus aureus* is also evident and increases under ultraviolet light, indicating the existence of a phototoxic effect. These results suggest that the medicinal use of this plant for the treatment of diarrhea and abdominal pain might be related to the presence in the species of as yet unknown compounds with spasmolytic and antimicrobial properties. There are no pharmacological reports in the literature on this particular species of *Ageratina*.

• *Baccharis* spp. Medical pharmacological studies of *B. conferta* go back to the nineteenth century, based on the widespread use of the species to treat stomachache (Martínez 1891). However, few pharmacological studies of the various Mexican species of this genus have been reported. In our studies on intestinal peristalsis in guinea pig ileum, we found that a strong spasmolytic effect was produced by methanolic extracts from the leaves of *B. serraefolia* and *B. vaccinioides*. Cytotoxic activity was also evident on KB and P388 cell cultures, while a slight antimicrobial effect was observed on *Staphylococcus aureus* cultures. *Baccharis trinervis* was found to suppress the gastric motility in guinea pig intestine. The strong spasmolytic activity and mild antimicrobial effect of these plant extracts might explain the medicinal use of the species in the treatment of diarrhea and other gastrointestinal diseases. The hypothesized physiological mechanism is inhibition of luminal gonodatrophins. Gonadotrophins are responsible for the secretion of excess water into the gut and therefore the production of watery diarrhea.

• *Verbena* spp. The methanolic extracts from the leaves and stems of *Verbena carolina* and *Verbena litoralis* demonstrate a phototoxic activity on *Staphyolococcus aureus, Escherichia coli,* and *Candida albicans* cultures. How this phototoxic activity may relate to the medicinal use made of these species requires further investigation.

• *Psidium* spp. Our laboratory research on *Psidium guineense* shows this species to be antimicrobial. Work on *P. guajava*, a closely related species also used by the highland Maya, as well as most other tropical countries, shows remarkable antidiarrheal properties.

The pharmacological and chemical investigation of *P. guajava* began several years ago, when Colliere (Colliere 1949) reported an antimicrobial effect of the aqueous extract from leaves of this plant when applied on *Staphylococcus aureus* colonies. Hypothesizing the possible presence of an antibiotic in *P. guajava*, Khadem and Mohamed (Khadem and Mohammed 1958) isolated the flavonoids from the leaves of this species: quercetin, evicularin, and guaijaverine, attributing to these compounds the antimicrobial effect observed on *S. aureus* cultures. Additional studies on the antibacterial activity of the aqueous extracts from the leaves of *P. guajava* were later performed, extending experimentation to other cell cultures. An inhibitory action on cell growth was observed (Nickel 1959; Malcom and Sofowora 1969; Khan 1978; Singh 1984). However, in recent studies performed by our group, it has been shown that in the dosage range currently accepted for this type of test, *P. guajava* extract shows antimicrobial activity solely on *S. aureus* cultures. This effect is attributed, at least partially, to the presence in the extract of the flavanoid quercetin, whose activity on cells and bacteria in vitro is well known.

Studies of the antimotility property of guava leaf infusions as a possible explanation for its widespread use as an antidiarrheal remedy have only recently begun. In experiments with laboratory animals it has been shown that the leaf extracts of *P. guajava* produce a mild sedative effect (Lutterdot and Maleque 1988) and inhibit intestinal peristaltism (Lutterdot 1989; Lozoya et al. 1990). It is currently thought that the antimotility property of these products is also due to the action of a group of quercetin-derived flavonols (Lozoya et al. 1992).

In clinical evaluations, guava leaf infusions have been administered to a considerable number of patients with acute diarrhea. Comparing its effect to that of the antidiarrheal medications normally used in clinics and hospitals, the results of the guava treatment were notably significant, demonstrating the medicinal value of this herbal resource.

• *Lantana* spp. Studies conducted by our group indicate a clear spasmolytic effect on isolated guinea pig ileum and phototoxic activity in *L. camara* and *L. hispida* (see also Hunt and McCosker 1970 for discussion of the phototoxic products in the leaves). The methanolic extracts from *L. hispida* leaves were found to have a cytotoxic effect.

The ethanolic extract from the leaves administered to rats and dogs produces uterine relaxation, intestinal musculature stimulation, and arterial hypotension (Sharaf and Naguib 1959) that may be related to its spasmolytic properties. Other investigators have reported that antimicrobial effects were detected in the essential oil obtained from the leaves (Avadhoot and Varma 1978; Avirutnant and Pongpan 1983; Ross and El-Keltani 1980). These properties have been related to the monoterpenes-rich content of the essential oil (cineol, geraniol, limonene, menthol, alphapinene, and others). The homeostatic action of the leaves has also been validated (Wanjari 1983).

Hepatotoxic effects of *L. camara* have been well-documented, however (Uppal and Paul 1971; Pass and Seawright 1979; Sharma et al. 1980; Sharma et al. 1981; Sharma and Makkar 1982, 1983), and these results would argue against the safety of the species as a plant medicinal.

• *Helianthemum glomeratum.* In our experimental studies with methanolic extracts from the leaves and roots of *H. glomeratum*, we observed a powerful antimicrobial activity inhibiting growth in *Staphylococcus aureus, Escherichia coli,* and *Candida albicans* colonies. Such antimicrobial effect might explain the use of the species as a remedy for gastrointestinal disorders.

• *Borreria laevis.* In our screening experiments with methanolic extracts from the leaves and stems of *Borreria laevis* we observed an antimicrobial effect on *Escherichia coli* and *Staphylococcus aureus* cultures. This effect was notably intensified by exposure to ultraviolet light. The extracts from the stems show stronger bactericidal activity.

• *Acacia angustissima.* Several years ago it was reported that the ethanolic extracts from the leaves of this species inhibited the growth of malignant tumors (sarcoma WM256) in laboratory animals. This effect was attributed to the flavonoids cyanidine, fisetinidine, and robinetinidine present in the leaves and stems of the plant (Hammer and Cole 1965). The hemagglutinant property of aqueous extracts from the seeds was also investigated and attributed to the plant's content in lectins.

In our screening experiments with *A. angustissima*, we observed that the methanolic extracts from the leaves and stems possess a mild antimicrobial effect on *Escherichia coli* and *Staphylococcus aureus*, as well as a cytotoxic effect inhibiting the growth of cell cultures of KB and P388, while the methanolic extract from the stems demonstrates a strong spasmolitic effect (80 percent inhibition of the peristaltic reflex) in tests done on isolated guinea pig ileum. A spasmolytic effect was also detected in extracts obtained from the stems. These findings coincide with observations reported in the literature regarding other species of the same genus (Ayoub 1983, 1984, 1985; De Oliveira et al. 1972; Dhar 1968; Hagos 1987; Mohamed and Ayoub 1983; Trivedi and Modi 1978). The spasmolytic properties and the antimicrobial effects of the different *Acacia* extracts might explain the medicinal use of this plant for the treatment of diarrhea.

• *Byrsonima crassifolia.* In spite of the widespread use of this plant, our pharmacological evaluation of the methanolic extracts of its bark only allowed us to conclude that they possess a mild antimicrobial activity on *Staphylococcus aureus* cultures. Further pharmacological studies are clearly needed given the high degree of consensus in the traditional Maya medical system on the efficacy of the species as a treatment for severe diarrhea.

• *Crataegus pubescens.* In our pharmacological screening studies on the properties of the methanolic extract from *C. pubescens* bark, we only detected a moderate antimicrobial effect on *Staphylococcus aureus* colonies, an activity that intensifies with exposure to ultraviolet light.

• *Calliandra* spp. No studies of *C. grandiflora* and *C. houstoniana* are found in the literature. In our studies we detected the phototoxic action of methanolic extracts from the seeds, leaves, and stems of *Calliandra grandiflora* as well as its cytotoxic effect on P388 cells.

• *Cissampelos pareira*. In the screening tests, using methanolic extracts from the stems of this species, we confirmed the relaxant effect on guinea pig intestinal musculature, as well as an antimicrobial activity on *Staphylococcus aureus*, *Escherichia coli*, and *Candida albicans*. There are early references to an antibiotic effect of *C. pareira* (George and Pandalai 1949).

A relaxant effect on the smooth musculature of the intestine was demonstrated with both the ethanolic extracts from the leaves and stems and the fraction from the alkaloids that abound in the species (Feng and Haynes 1962; Mokkhasmit and Ngarmwathana 1971; Roy and Dutta 1952). Cissampareine, found in this species, possesses cytotoxic activity (Kupchan et al. 1965).

• *Lepidium virginicum*. Our laboratory work on *Lepidium virginicum* is as yet incomplete.

• *Rubus coriifolius*. The methanolic extracts from the leaves and stems of *R. coriifolius* showed an antimicrobial effect on all cell cultures used in our studies, as well as a moderate cytotoxic activity. The medicinal use made of these species might be related to their antibiotic property.

Principal Species Used to Treat Abdominal Pain

• *Nicotiana tabacum*. Our screening studies showed a strong antispasmodic effect produced by methanolic leaf extracts on isolated guinea pig ileum, suggesting a relation to the treatment of abdominal pain and bloating through relaxation of the smooth muscles of the intestine.

• *Allium sativum*. The chemical investigation of *A. sativum* began many years ago with the identification of the plant's numerous sulfurated compounds: ajoene and derivatives, allicin ym13053, m13187, and diallylsulfurated derivatives. Flavonoids of the quercetin and kaempferol group have also been isolated, and recently the presence of arachidonic acid and prostaglandins (A1 and AA1) have been reported. The medicinal effects of this plant in gastrointestinal maladies are related to the action of prostaglandins, whose pharmacological implications are the object of current studies.

• *Foeniculum vulgare*. There are several experimental reports on the activity of the essential oil from the leaves and fruits of this species. The oil possesses antibacterial and antifungal characteristics (Iliev and Papanov 1983, Leifertova 1979), as well as spasmolytic and anti-inflammatory effects (Alkiewicz 1983; Mascolo and Autore 1987; Shipochliev 1968). The plant's concentration in monoterpenes (cineole, citral, estragole, eugenol, fenchol, and others) is very high, and partially explains its bactericidal properties. The species also contains such flavonoids as quercetin, guaijaverine, quercitrin, and other glycoside derivatives which determine the antispasmodic and relaxant effects on the smooth musculature of the intestine (Akunzemann 1977; Ghodsi 1976; Harborne 1984).

• *Tagetes lucida*. In the studies carried out by our group, we observed that the methanolic extract from the flower inhibits growth of *Staphylococcus aureus*, *Escherichia coli*, and *Candida albicans* cultures. This effect, considerably expanded by exposure to ultraviolet light, is also found in extracts from the root, stem, and leaves when irradiated with ultraviolet light. This suggests the presence of phototoxic compounds in the whole plant. No spasmolytic activity was detected in any of the extracts.

• *Ocimum selloi*. We have carried out no laboratory work on extracts of *O. selloi*. However, studies on the phytochemistry of the genus indicate that the essential oil found in the plant that is likely to have pharmacological activity is eugenol, an active antibiotic with low toxicity and irritability (Morton 1981:775).

• *Stevia ovata*. Our data show that *S. ovata* is antimicrobial and antispasmolytic. The leaves of a closely related species, *S. serrata*, are known to contain an important number of sesquiterpenes, the most salient of which are chamazulene, the steviserolides (A) and (B), and cristinine. It has spasmolytic and anti-inflammatory properties (Calderon and Quijano 1989; Salmon et al. 1973, 1977). The leaves also contain flavonoids (quercetin and hyperoxide) (Rajbahandari and Roberts 1985), and the roots contain sesquiterpenes derived from alpha-longipinene (Bohlmann and Suwita 1977).

• *Tithonia longistrata*. This species has been shown to be spasmolytic and antimicrobial.

• *Lepechinia schiedeana*. We have not yet conducted laboratory analyses of *L. schiedeana*.

Principal Species Used in the Treatment of Worms

• *Chenopodium ambrosioides*. Among the numerous *Chenopodium* species that have been studied chemically or pharmacologically, *C. ambrosioides* and *C. ambrosioides* var. *antelminthicum* are the two major species with medicinal value, due to their oil rich in ascaridol (1,4 peroxide-p-menthene-2) and p-cimenol, limonene, alpha-terpinene, metadin, and pinokaureol (Gallego et al. 1965). Ascaridol has been considered for years the antiparasitic active principle present in *Chenopodium*, especially useful in the treatment of intestinal parasitosis due to its action against nematodes (*Ascaris lumbricoides*) and helminths (*Necator, Ancylostoma*, and *Trichuris trichiura*).

The antimicrobial action of the essential oil was demonstrated and is attributed particularly to p-cimenol, another compound abundant in *Chenopodium* oil (Okasaki 1958; Ross and El-Keltani 1980). This antimicrobial effect was observed in our screening studies with the methanolic extract from the leaves of this plant. No cytotoxic effect of this plant extract was detected in KB and P388 cell cultures.

Principal Species Used in the Treatment of Pain Associated with 'Mother of Man' (probable) Gallbladder Disease

• *Lobelia laxiflora*. There is no experimental bibliographic information on the pharmacological properties of *L. laxiflora*. The pharmacological screening studies carried out by our group showed that the methanolic extract from the bark has a cytotoxic activity on cell cultures of P388.

• *Fuchsia microphylla*. Our laboratory work shows that *F. microphylla* is antimicrobial against *S. aureus* and *Campylobacter jejunni*. No studies have yet been conducted on *F. splendens*.

Table 8 presents a summary of these findings from our laboratory and the pharmacological and chemical literature for the principal gastrointestinal species discussed above.

Of the thirty-two species tested for bioactivity in our laboratory experiments, 75 percent show antimicrobial activity. Half of the species tested also exhibit medium to strong antispasmodic properties. Elsewhere in the literature, three species are reported as having sedative, anticonvulsive, or central nervous system effects, and two would affect bleeding through hemagglutination and homeostatic activity.

Specificity of Pharmacological Effects to Maya Use

It is important to note that since the pathogens tested were not, in most cases, those specific to diarrheal etiologies, application of our laboratory results must be inferential. The greatest effect across all species was on *Staphylococcus aureus*, the gram positive bacteria. To establish potential effectiveness on etiologic agents, we must first correlate agents identified by biomedicine with the symptoms specified by Maya ethnomedicine. Furthermore, we are aware that causal agents are extremely difficult to identify in diarrheal disease, primarily because most stools demonstrate mixed infections.

However, it is also the case that specific signs and symptoms, such as the key characteristics identified by the Maya, are produced by a finite, identifiable set of etiologic agents.[3] Furthermore, there is increasing evidence that, in diarrheal disease, clinical evaluation alone is sufficient to establish etiological diagnosis and justify treatment, at least for some agents[4] (see Chatkaeomorakot et al. 1987; Ericsson et al. 1987; Huq et al. 1987; Mohandas et al. 1987; Taylor et al. 1986).

According to Guerrant, "The leading etiologies [of diarrheal disease having serious impact on fluid and electrolyte balance] are enterotoxigenic *Escherichia coli*, rotaviruses, *Shigella*, and *Campylobacter jejuni*" (1983:25) and, in some settings, *Giardia* and amoebic infections. He further points out that it is useful to

> separate acute diarrheal illnesses into two groups according to pathogenesis and site of disease in the intestinal tract. The first, arising from the action of enterotoxins or viral agents that impair absorption ... [result] in a noninflammatory, often *watery, diarrhea*. The second type arises from destruction or invasion of the ... [intestine] by organisms ... or by cytotoxins, that

Table 8

Results of Pharmacological Screening and Literature Search on Principal Species Employed in the Treatment of Gastrointestinal Conditions

Species	S. aureus	E. coli	C. albicans	spasmolytic	UVA phototoxicity	KB	P388
Acacia angustissima	+	+		+	+		+
Ageratina ligustrina	+			+	+	+	+
Allium sativum							
Baccharis serraefolia	+			+	+		+
Baccharis trinervis				+			+
Baccharis vaccinioides	+			+			
Bidens squarrosa	+						+
Borreria laevis	+	+			+		+
Byrsonima crassifolia	+						+
Calliandra grandiflora	+	+			+		+
Chenopodium ambrosioides	+	+	+				
Cissampelos pareira	+	+	+	+			
Crataegus pubescens	+				+		+
Foeniculum vulgare					+		
Helianthemum glomeratum	+	+	+		+		+
Lantana camara	+			+	+	+	+
Lantana hispida	+			+		+	+
Lobelia laxiflora							+
Nicotiana tabacum				+			
Ocimum selloi	+						
Psidium guajava	+	+		+			
Psidium guineense	+	+					+
Rubus coriifolius	+	+	+				+
Smallanthus maculatus	+			+			+
Sonchus oleraceus				+			+
Stevia ovata	?	?		+			
Stevia serrata	+			+			+
Tagetes erecta	?	?		+	+		
Tagetes lucida	+	+	+		+		+
Tithonia longistrata	+			+	+	+	+
Verbena carolina	+				+		+
Verbena litoralis	+						+

Table 9

Common Diarrheal Disease Agents by Class and Stool Characteristics

Class	Agent	Bloody	Mucoid	Watery
Amoebic	*Entamobea histoytica*	yes	yes	
Bacterial	*Campylobacter jejuni*	yes		
	Salmonella spp.	yes		
	Yersinia enterocolitica	yes		
	Salmonella		yes	
	Campylobacter jejuni		yes	
	Clostridium dificile		yes	
	Shigella		yes	
	Escherichia coli		yes (E.P)	yes (E.T.)
	Vibrio colera			yes
	Staphylococcus sp. E.T.			yes
	Clostridium perfringens			yes
	Giardia lambia			yes
Viral	Rotavirus			yes
	Norwalk like-parvaviruses			

produce an inflammatory dysentery in which the stool may contain *blood* or *pus*. (Guerrant 1983:25, emphasis ours)

In a later paper, Guerrant and colleagues (Guerrant et al. 1986) present an inclusive list of etiologic agents grouped according to three primary symptom categories that are quite similar to those recognized by the Maya and which may be characterized by basic stool characteristics, namely, bloody, mucoid, or watery. Keusch (1983:49) notes that the enterotoxic bacteria, which can produce watery stools, include *Escherichia coli, Salmonella, Shigellae, Campylobacter,* and *Yersinia.* Potentially invasive bacteria include the shigellas and salmonellas and probably *Campylobacter jejuni* and *Yersinia enterocolitica.* All of these organisms can produce bloody stools, as can *Entamoeba histolytica* and *Escherichia coli. Shigella* is probably the agent most classically associated with mucoid diarrhea. However, *Campylobacter jejuni* and *Entamoeba histolytica* can also produce mucus-containing stools. The rotaviruses are relatively uniquely associated with severe dehydrating diarrhea with frequent vomiting. On the basis of their classifications, we can construct table 9 as the potential correlation of pathogenic agents with the key criteria of Maya ethnomedicine.

Table 10

Organs and Probable Manifestations in Cases of Abdominal Pain

Maya condition	Organ	Site of Pain	Probable Radiation
epigastric pain	esophagus	retrosternal	precordial, left shoulder, arm and hand
	stomach	epigastrium	right front hemithorax, right iliac fossa, dorsolumbar
	duodenum	epigastrium	right front hemithorax
'mother of man'	biliary duct	epigastrium, right colon	right colon, scapula, right shoulder
	pancreas	epigastrium	dorsolumbar at mid-waist
	liver	right colon, epigastrium	dorsolumbar, right shoulder
abdominal pain	jejunum, ileum	periumbilical	the entire abdomen
	cecum and appendix	right iliac fossa	right thigh, epigastrium
	colon	hypogastrium	the entire abdomen
	spleen	left descending colon	no radiation
	rectum and anus	anus and pelvic floor	hypogastrium

Unfortunately, the pharmacological information is not sufficient to specify activity against all organisms. We can generalize, however, from positive effects on *Staphylococcus aureus* to all gram positive bacteria and on *Escherechia coli* to all gram negative bacteria (*Salmonella, Vibrio cholera, Shigella*) and other members of the family Enterobacteriaceae (*Yersinia enterocolitica*). Most species utilized by the Tzeltal and Tzotzil show activity against agents potentially producing watery diarrhea. We are unable to specify bioactivity against viral, protozoal, and amoebic agents at this time, and these agents need further study.

In many species there is concomitant physiologic activity to alleviate the signs and symptoms associated with the diarrheal event, such as the severe tenesmus of bloody and mucoid diarrhea or the reduction of blood loss from the intestinal lumen. The spasmolytic effects would clearly serve to calm the excess intestinal motility that rushes the stool through the bowel and produces liquid stools.

While it is difficult to specify differential areas of pharmacological activity in the various kinds of gastrointestinal pain, use of the Maya categories of abdominal pain to correlate with biomedical concepts of the organs associated with pain may lend some insight to the mode of action and conditions treated. Column 1 of table 10 presents the Maya classification of the gastrointestinal pains. The remaining

Table 11

Association of Pharmacological Effects with Types of Gastrointestinal Pain

Abdominal Pain	
Plant Species	**Physiological effect**
Ageratina ligustrina, Allium sativum, Baccharis vaccinioides, Lantana camara	spasmolytic
Nicotiana tabacum	spasmolytic CNS depressant, anticonvulsive
Stevia ovata	antispasmolytic, anti-inflammatory
Epigastric Pain	
Plant Species	**Physiological Effect**
Foeniculum vulgare	anti-inflammatory, antispasmolytic
'Mother of Man'	
Plant Species	**Physiological Effect**
Lobelia laxiflora	anticancer

columns present the biomedical perspective on the abdominal organs associated with the condition (column 2), the locations (column 3), and sites of radiation (column 4) of pain associated with those organs from a biomedical perspective.

To the extent that pain locations vary, they would predictably correspond to the organs and sites listed in the table. The most common Maya description is that the condition of abdominal pain affects the entire abdomen. Biomedically, this most likely implies intestinal pain, more specifically of jejunoileal or colonic origin. Gastritis may be due to parasitic infections common in the area. The pain might occur as a direct result of the parasite(s) or indirectly due to damage to the intestinal wall by the parasites. Also, peptic ulcers could locate in the lower esophagus, the stomach, duodenum, or jejunum. Cases of peptic ulcer are more likely to be classed as epigastric pain by the Maya, rather than as generalized abdominal pain. Work done by our interdisciplinary team suggests that 'mother of man' best corresponds with biliary disease, although pancreatitis and possibly liver disease may produce similar symptoms.

Although we cannot distinguish the site of action specific to the condition, of the plants used in the treatment of pains of gastrointestinal origin, as seen in table 11, we can verify potential effectiveness for *Ageratina ligustrina, Allium sativum, Baccharis vaccinioides, Lantana camara, Smallanthus maculatus, Nicotiana tabacum, Stevia ovata,* and *Foeniculum vulgare. Lobelia laxiflora* demonstrates no activity for pain in our pharmacological tests. However, since the condition 'mother of man' is most closely associated with biliary disease, and a small percent of cases proceed to develop carcinoma, it is interesting to note the anticancer activity of this plant. Yet to be tested for potential pain treatment effects are *Ocimum selloi* and *Tagetes* spp.

Of the species used in the treatment of worms, *Chenopodium* is a well-known anthelminthic. While *Helianthemum glomeratum* demonstrates broad antibiotic effects, studies are lacking on its potential effect in the treatment of worms.

Conclusions

Gastrointestinal problems represent a major component of the ethnomedical system of the highland Maya who demonstrate a sophisticated understanding of gastrointestinal disease manifestation and treatment. Our data show that the plant medicinals employed by the Maya are condition-specific for symptom-based, linguistically recognized gastrointestinal illnesses. Most of the plant species selected by the Tzeltal and Tzotzil Maya as treatments for gastrointestinal conditions, especially for the various classes of diarrhea, are potentially effective against the pathogens implicated as some of the pathogenic agents underlying these specific diarrheal diseases and/or affect the physiological processes associated with diarrheal episodes. The potential therapeutic efficacy of Maya herbal medications is suggested by laboratory studies on the pharmacological properties affecting the agents associated with the ethnomedically identified clinical signs and symptoms, which correlate closely with biomolecular medical disease categories. Furthermore, the systematic association of medicinal plant species with specific folk diagnoses supports the stability and specificity of Maya gastrointestinal disease classes.

One can claim that the Maya ethnomedical system is an empirical system of knowledge based on precise and accurate observation. In all likelihood, it is a system of medicine that was and continues to be elaborated on the basis of explicit experimentation with medicinal plants. A number of the plant species selected by the highland Maya as efficacious in the treatment of gastrointestinal disease merit more detailed laboratory and clinical study as natural products with high pharmaceutical potential.

Notes

Funding for this research was provided by the National Science Foundation; the US Agency for International Development (grants BNS 87–03838, BNS 90–03673); the Government of the State of Chiapas, through the strong support of past Governor Patrocinio González Garrido; the University of California Consortium on Mexico and the United States (UC MEXUS); and the University of California at Berkeley's Center for Latin American Studies. This support is gratefully acknowledged.

Our studies have been conducted under collaborative agreements with the Chiapas State Department of Public Health, the State Autonomous University of Chiapas (UNACH), the Chiapas Institute of Culture, the Chiapas Institute of Natural History, and El Colegio de la Frontera Sur (formally the Centro de Investigaciones Ecológicas del Sureste). Herbarium voucher specimens were collected in Tzeltal-speaking municipalities by Juliana López P. and Martín Gómez L. (Amatenango del Valle), Estéban Sántiz C.

(Cancuc), Feliciano Gómez S. (Oxchuc), and Juan Gómez M. (Chanal), and in the Tzotzil-speaking municipalities by Marcos Pérez G. and Eleuterio Pérez L. (Chalchihuitán), Carmelino Sántiz R. (Chamula), Lorenzo González G. (Larraínzar), Bartolo Hidalgo V. (Venustiano Carranza), Manuel López H., Mariano Sánchez P., and Lázaro Díaz M. (Zinacantán), under the supervision of Alonso Méndez G. of Tenejapa. Survey interviews were conducted by Alfonso Luna G. (of Tenejapa), Carmelino Sántiz R., and Feliciano Gómez S. Msrs. Sántiz and Gómez collected the plant specimens for laboratory analysis. Domingo Sánchez Hernández (Zinacantán) is responsible for curatorial assistance and preparation of plant samples for laboratory analysis. Victor Jara A. served as program coordinator 1990–1992, Teresa Velasco C. was charged with administration of the research program 1987–1992.

Luisa Maffi provided substantive and editorial comments.

1. Because of the traditional sex roles relating to the division of labor, as well as levels of education (ability to write), all but one of our botanical collectors were males who, because of their sex, were obliged to work primarily with other males. Maya women are prohibited by social convention from accompanying unrelated men "into the woods." However, male bias is offset in the data collected using a standarized set of 204 most common species (described as data set 2), where males and females are equally represented. Furthermore, preliminary analyses of our data do not reveal significant differences in knowledge of medicinal plants by sex, especially for those health conditions of major epidemiological importance.

2. Previous research in the Chiapas highlands indicate that knowledge and use of medicinal plants is widespread and not the exclusive domain of shamanic healers, and our findings support this conclusion.

3. The division of organisms by diarrheal type is inexact. Most organisms can vary in presentation depending on poorly understood factors. Furthermore, most diarrheal cases represent combined intestinal infections involving multiple agents. Even with laboratory analysis, single organisms can seldom be identified as specifically causal.

4. Northrup (1987) states that microscopic exam "can slightly improve the accuracy of clinical diagnosis [of *Shigella* but] ... the extra work may not be worth the effort." He later concludes more generally that "microbiological studies are not currently of high priority in our efforts to control diarrhea" (Northrup 1988).

CHAPTER 3

Science for the West, Myth for the Rest?
The Case of James Bay Cree Knowledge Construction

Colin Scott

Do Cree hunters practice science? The answer to this question would seem to depend on whether one defines science according to universal features, or culturally specific ones. If one means by science a social activity that draws deductive inferences from first premises, that these inferences are deliberately and systematically verified in relation to experience, and that models of the world are reflexively adjusted to conform to observed regularities in the course of events, then, yes, Cree hunters practice science—as surely all human societies do. At the same time, the paradigms and social contexts of Cree science differ markedly from those of Western science—accustomed as we are in the West to a "root metaphor"[1] of impersonal causal forces that opposes "nature" to "mind," "spirit," and "culture," and conditioned as we also are to view legitimate scientific procedure and production as the prerogative of particular professional and institutionalized elites. While there is no a priori reason to expect that knowledge generated out of non-Western paradigms or social processes should be empirically or predictively less adequate, it has been an effect of Western ethnocentrism to construe non-Western knowledge processes as "pseudoscientific," "protoscientific," or merely "unscientific."[2] Western science, in fostering an ideology of knowledge that supports its own elite status, has assisted the exclusion and disqualification of innumerable "subjugated knowledges" (Foucault 1980a).

Indigenous ecological knowledge finds renewed voice, however, in answer to the environmental anxieties of Western industrial societies, as well as aboriginal people's demands to decolonize and to directly manage environmental resources to which they assert primary rights. The Cree of James Bay, Canada, are such a people. Any account of Cree knowledge occupies a context in which jurisdiction for resources is actively contested, in which "science" is invoked both to attack and defend Cree opposition to invasive development projects sponsored by external governments, and in which indigenous knowledge is both advocated and opposed as a basis for deciding development issues.

I do not directly address these political dimensions in this paper. I focus on the more particular task of exploring how practical, empirical knowledge flows from root metaphors (paradigms) that are not generally associated with "scientific" results in Western thought. The exploration focuses on the way in which root metaphors of pan-species personhood, communication, and reciprocity inform literal models of animal behavior and hunting practice; and how the latter reciprocally transmute the terms of metaphor, as experience is interpreted and actions are formulated.

One conclusion of earlier twentieth-century ethnography, in a line leading from Malinowski through Evans-Pritchard, is that in all societies (including Western civilization) practices dubbed "magical" and "mystical" coexist with rational/empirical processes. Both anthropologists were alert to

> the danger of double selection by which ('primitive' peoples) are described entirely in terms of their mystical beliefs, ignoring much of their empirical behavior in everyday life, and by which Europeans are described entirely in terms of scientific rational-logical thought, when they too do not inhabit this mental universe all the time. (Tambiah 1990:92)

Both anthropologists knew that "a person can in a certain context behave mystically, and then switch in another context to a practical empirical everyday frame of mind" (ibid).[3] This legacy poses two problems of immediate concern for anthropology: how do we get beyond the artificial dichotomy that separates Western and non-Western forms of knowledge, simultaneously discrediting and romanticizing the latter; and how are logical/empirical and mystical/magical aspects of thought related, in all traditions?

Perhaps we have begun to see that the distance separating the scientist and the shaman is not so great as was once imagined. But the evolutionary opposition of science for "the West" to myth and magic for "the rest" is far from dissolved; Western self-conception remains profoundly involved with images of rational "self" versus mystical "other." Several trends in late twentieth-century anthropology, to be sure, have continued to erode this dichotomy. Ethnoscientific fieldwork since the 1960s has brought into view empirically elaborate nomenclatures and classification systems from a wide range of "traditional" societies. In the structuralism of Levi-Strauss (1966), these empirical categories, or "percepts," were signs in the bricolage of mythical thought, which could be seen as the "science of the concrete." The structures of mythical thought can produce scientific results because the mind, perception, and the external world share a common "natural" foundation. For Levi-Strauss (as for Evans-Pritchard before him) the structures of reason in myth and magic are not fundamentally different from those of science (1966, 1973). Yet in Levi-Strauss, the science-myth opposition is salvaged via the obscure notion that mythical "signs" are somehow limited and contained by their empirical signifiers, while the "concepts" of science are more free (1966:18–20).

In other scholarship, ecological adaptation in non-Western societies is seen as systemically reinforced by symbolic structures or cosmologies. Reichel-Dolmatoff (1976) demonstrated the formal compatibility between mythico-ritual structures and ecosystemic principles among Amazonian Tukano. And Rappaport (1968, 1979) showed that supernatural categories—and their ritual entailments—in the "cognized model" of the Maring have homeostatic functions and effects within the analyst's "operational model" of highland New Guinea ecosystems. When the structural or functional connections between abstract cosmology and material ecosystems are the constructs of the analyst, however, it can appear as if the "totalizing" view of Western science has captured what remained unconscious or invisible to native subjects. The intellectual processes involved in framing practical knowledge within cosmological categories, from the actors' point of view, remain largely obscure. The adaptation of native cosmologies to their material-historical environments can then appear to be fortuitously functional, a happy congruence of symbolic and material structures (if functionality and congruence are to be believed); or the outcome of blind selective forces; rather than the outcome of theoretical work and proactive environmental management on the actors' parts.

With the upsurge of multidisciplinary interest in "traditional ecological knowledge," models explicitly held by indigenous people in areas as diverse as forestry, fisheries, and physical geography are being paid increasing attention by western science specialists, who have in some cases established extremely productive long-term dialogues with local experts (see Berkes 1977; Johannes 1981, 1989; Nietschmann 1989). The idea that local experts are often better informed than their scientific peers is at last receiving significant acknowledgment beyond the boundaries of anthropology. Anthropologists may find that we have less knowledge to share with local experts than do our colleagues in biophysical sciences about specific domains of local knowledge, and in this respect we may be at an initial disadvantage in striking up mutually interesting conversations with local experts. On the other hand, anthropology is unique in the degree to which it emphasizes the more inclusive cultural contexts of our local teachers and values ways of translating indigenous knowledge that reflect the symbolic and institutional contexts in which the knowledge is generated. If the sharing of knowledge were to be reduced to a skimming-off by Western specialists of indigenous empirical insights, and their mere insertion into existing Western paradigms, then it would be an impoverished and failed exchange that would ultimately contribute to undermining indigenous societies and cultures.

A number of anthropological studies have addressed the way in which mythico-ritual categories are implicated in actors' modeling of social-environmental practice in situational, strategic discourse about material activity (see Feit 1973, 1978; Nelson 1983; Scott 1989; Brightman 1993). It is this general issue to which the present paper contributes, through consideration of the ethnography of James Bay Cree hunting knowledge. I do not argue that all mythical and ritual symbolism is necessarily directed toward some logic of practical social or environmental knowledge. But I want to highlight that central, recurring propositions within these symbolic discourses are better understood in this way than as mystical precepts; and indeed, our understanding of practical knowledge cannot be adequately formulated without reference to the root metaphors most vividly condensed in myth and ritual.

Here I will proceed in three steps: first, to discuss the significance of ordinary experience within Cree cosmology and epistemology; second, to consider how the "figurative" language of metaphor interacts with the "literal" language of practical/empirical experience (i.e., how paradigms relate to the ordering of empirical experience); and third, to present examples from Cree goose hunting that illustrate the alternation between literal and figurative aspects of knowledge, each providing context and definition to the other.

Signs in Cree Epistemology

As Hesse (1980), following Black (1962), has argued is true for scientific practice, "literal" or "observation" languages are shaped by the use of metaphor in theory —models are expressions of metaphor; and description, the literal reporting of observed regularities, is not independent or invariant of changes in explanatory models. Observation language "like all natural languages is continually being extended by metaphoric uses" (122). As certain root metaphors become conventionalized, as certain paradigms persist, their presence in observation language becomes less noticeable—they become literally implicit in the empirical description of experience. So it is that we may be largely unconscious of the metaphysical paradigms that underlie our own understandings of the world, while those of other knowledge traditions strike us as exotic, improbable, even "superstitious."

It is only in moments of unusual reflexive insight, for example, that modern Westerners are conscious of the extent to which a (meta-)physics of impersonal forces imposes itself on our perception of "nature." So embedded are the Cartesian myths of the dualities of mind-body, culture-nature, that we tend to privilege models of physical causality, rather than relations of consciousness or significance, in our perception even of sentient nature. It is true that we have begun to culturalize animals in animal communications studies, and to naturalize culture in anthropological ecology. But our conventional attitude is to assume fundamental differences between people and animals, while exploring the nature of their connections. The Cree disposition seems rather the converse: to assume common connections among people, animals, and other entities while exploring the nature of their differences. The connectedness assumed by the Cree reminds me of what Gregory Bateson (1979) has termed the "pattern which connects,"[4] patterns of dancing, interacting parts within larger patterns, the stories "shared by all mind or minds, whether ours or those of redwood forests and sea anemones," the "aesthetic unity" of the world.

In Cree, there is no word corresponding to our term "nature." There is a word *pimaatisiiwin* (life), which includes human as well as animal "persons." The word for "person," *iiyiyuu*, can itself be glossed as "he lives." Humans, animals, spirits, and several geophysical agents are perceived to have qualities of personhood. All persons engage in a reciprocally communicative reality. Human persons are not set over and against a material context of inert nature, but rather are one species of person in a network of reciprocating persons. These reciprocative interactions constitute the events of experience.

Again, there is no Cree category for "culture" that would make it the special province of humans. Cree do, however, have terms that resemble notions of "sign vehicle" (*chischinawaachaawapihtawaawan*) and the "meaning/interpretant" (*iishchiishwaamaakan*) of a sign. Cognates of these terms, incorporating the morpheme -*chis*-, evoke the ubiquity of signs in experience. They include *chischaaimaau* (s/he knows [him/her]); *chischaaitamuun* (information, knowledge); *chischinutihaau* (s/he leads, directs, guides [him/her]); *chischinuwaasinaakusuu* (s/he is used for a sign or s/he gives a sign); and *chischiwaahiicheu* (s/he prophecies) (Mackenzie 1982). Animal actions, particular qualities and features in the bodies of animals, weather, dream images and events, visions, and religious symbols all fall within the Cree notion of "sign," with signs constituting knowledge or guidance for actors. Not only humans, but animals and other nonhuman persons send, interpret, and respond to signs pertinent to various domains of human action: hunting success or failure, birth and death, and, implicit to these, the circumstances of reciprocity between persons in the world.

Signs, then, are part and parcel of action, perception, and experience—of life itself. The term *pimaatisiiwin* (life) was translated by one Cree man as "continuous birth." Consciousness (*umituunaaichikanich,* glossed by the same man as "mind and heart, thought and feeling") is at the threshold of unfolding events, of continuous birth. One consequence of this construction of the world is that an attitude of dogmatic certainty about what one knows is not only untruthful but disrespectful. There are many signs of recurrence and regularity in experience, but interpretations cannot be certain or absolute. To expect a definite future outcome on the basis of signs in the past or present, for example, may presume too much about the cooperation of other persons. Someone (human, animal, or spirit) could even retaliate by frustrating hunters' intentions.

Relating and Differentiating as Complementary Aspects of Knowledge

Since events are the actions of various mutually responsive persons, human action is subject simultaneously to moral and technical criteria of evaluation. This is a commonly remarked feature of worldview for many egalitarian societies. Roy Wagner (1977, 1981) attributes it to the prominence of figurative ("differentiating") signification in traditional societies, by contrast to the literal ("relational") bent of signification in Western culture. While I think the opposition is misplaced as one of societal types, the theoretical grounds on which it is presented are illuminating for present purposes.

The basic idea is simple enough. In positing relations between things, we depend on some implicit definition of those things; conversely, to distinguish and define the things themselves, we depend on an implicit context of relations among them. The literal and the figurative are complementary and mutually dependent aspects of any knowledge construct. Depending on where we focus our attention when we shape a knowledge construct, we either relate the perceptibly differentiated, or differentiate the perceptibly relational. To choose a common example, for Western science, "natural" objects are implicit in the cause-effect constructions relating them. Objects are combined into total relational patterns that comprise a context (which, at the most inclusive level, we call "nature"). Natural objects and nature at large are experienced as innate; as "naturally" separate from the scientific culture that represents them. Figurative signification, on the other hand, focuses explicitly on defining objects or entities via metaphors that examine the similarities and differences among them; but implicit in those metaphors is a relational context (relations of reciprocity, for example, among many non-Western peoples).

When non-Western peoples focus on the analogies among themselves and other phenomena in the world, they tend to precipitate their own conventional social context (e.g., communicative reciprocity) as the innate character of phenomena in general. Western science tends no less to precipitate its own conventional social context: surely the technical mastery of an objectified nature is metaphorically connected to centralized social hierarchy and control. The separation of culture from nature depends not on a preponderance of literal-mindedness per se, but on which metaphors are used to frame the literal.

The complementary of the literal and the figurative help us to realize that the distinction between myth and science is not structural, but procedural. Myth, in a narrow and derogatory sense, is the dogmatic application of constituent metaphors as literal truths. There is myth, in this sense, in all science. At the same time, no science can embrace the world except through the creative extension of metaphors to emergent experience. We rework our metaphors as our models address particular contexts of experience. Myths in a broader, paradigmatic sense are condensed expressions of root metaphors that reflect the genius of particular knowledge traditions.

Let us return to the point that the interdependence of the figurative and the literal entails the integration of moral and technical aspects of knowledge. The Pacific cultures discussed by Wagner, much like Cree culture, view the world as an innate realization of a conventional social order of reciprocity. For Cree, as we have noted, communicative exchange is extended so ubiquitously to nonhuman domains that it constitutes a root metaphor or paradigm for knowledge in general. Myth, ritual, dreams, and hunting scenarios all express respectful solicitude as the preferred relation among "persons" in the hunter's world. The mental and physical activity of the hunter is directed at maintaining standing in this network, by being generous and respectful to humans and nonhumans, and by ensuring that what is received is in correct proportion. Where moral standards of positive reciprocity are deviated from, its negative corollary ensues: in all contexts, generosity dwindles in response to disrespect and greed.

But the viability of these common premises depends on the rigorous discernment of differences among persons. "It is," as Wagner (1977:398) has said, "by maintaining a precise awareness of these differences, by differentiating himself and by differentiating the various beings in an appropriate manner, that man precipitates (or from the actor's point of view, invokes) a beneficent relational flow." Here I want to emphasize the creative use of metaphor in interpreting events in specific practical contexts, and to illustrate how permutations of the metaphor of communicative reciprocity vary situationally with the material phenomena that serve as its signifiers.

Metaphors of Eating and Sexuality: Distinguishing Human and Nonhuman

In certain sacred contexts, the identities of hunters and animals are so passionately condensed in the metaphors connecting them, that the aspect of similarity virtually eclipses the aspect of difference. But in most practical contexts, the differences between hunters and animals as reciprocating agents are in the foreground. The definition of these differences flows from quite deliberate relational models that connect hunters and game metonymically—in consumer/consumed complementarity and in cause-effect orders that include other environmental agents such as winds, tides, and topographical features.

Reciprocity among humans is distinguished from reciprocity between humans and animals. In the first instance, biological structures of human reproduction signify a fundamental separation and asymmetry between human community and animal community, in respect of the former consuming the latter. The justification of this asymmetry is no trivial matter. Several Cree myths are concerned with human sexuality as a metaphor for the killing and eating of game, and vice versa. When humans get the terms of their metaphors confused and begin marrying animals or eating other humans (that is, failing to differentiate correctly), the results are impossibly comic or tragic (Preston 1975, 1978). That an animal be available for human consumption is an index of respect and love between hunter and animal; to contemplate the consumption of other humans is horrifying.

The metaphoric juxtaposition/separation of humans and animals is the occasion for much humorous discourse linking the pursuit of sexual partners to the pursuit of game. Hunting and sexuality share a vocabulary: *mitwaaschaau* can mean both "he shoots" and "he ejaculates"; *paaschikan* can refer to both "shotgun" and "penis"; *pukw* to both "gunpowder" and "sperm"; and *spichinaakin* to both "gun sheath" and "condom." But analogy, along with humor, is as much about separation as about similarity. The *atuush*, or "cannibal" figure subverts this separation of human from animal, of sex from food. In one bawdy myth, a cannibal copulates with a woman hunted by his son, before roasting and eating her reproductive organs. In consequence, he consumes his own sperm. He and his son, greatly weakened, are nearly overcome by the superior spiritual power of true human beings.

Another myth illustrates the necessity of killing animals, while respecting certain parameters for doing so. It concerns a supernatural character, Chischihp, who never ate. Chischihp thinks of the food animals whom he loves as his "pets," or "dogs." In this he differs from Cree hunters. Hunters also refer affectionately to certain species as their "pets," but normally it would be the species that an individual hunter is privileged to kill with unusual success. Chischihp would never have begun killing his "pets" had he not met two human sisters on a river journey and desired them for wives. They accept his proposal, but insist that he kill beaver and moose for them. When he objects, they threaten to abandon him. He relents, kills the animals, and eventually, surreptitiously, begins to eat some of the meat himself. However, he goes from excessive abstinence to excessive indulgence, with both the animals and the women, in his imperfect conversion to human status. When he returns to his village, he selfishly hides his wives, preventing their attendance at a public dance. At the dance, he adorns himself with the fatty internal organs and membranes of the moose. These parts are esteemed food delicacies, and their ostentatious display is grossly disrespectful of the animal gift. His wives, for their part, respond with infidelity. Chischihp discovers them sleeping with a lover, whom he promptly murders. He is now classified as a *pwaat*, a subhuman person who lurks at the margins of true human community and who shares some attributes of cannibals. Through treachery, he escapes the wrath of his village, drowns his wives, and is himself transformed into a species of edible waterfowl, the form in which he is known to Cree hunters today.

Human reproduction, then, demands the consumption of animals as positively as it prohibits the consumption of other humans; but there are respectful parameters for both interspecies consumption and intraspecies sexuality that are specific to the form of "reciprocity" in question.

Knowledge Construction in Cree Hunting

I want to go on now to illustrate how the literal interpretation of animal behavior in the environmental context impels the further figurative differentiation of human and nonhuman persons. Cree hunters continually refer to human and animal capacities as interpretants of one another. The family structure, leadership, memory, and communication processes of animals are all explored as analogs of corresponding human qualities, both individual and social.[5] Here I focus on Canada goose hunting and resource management strategies. Goose hunting is a major ritual and economic event during the exceptionally rich fall and spring migrations along the coasts of Hudson and James Bays. Geese as objects of knowledge are extremely important both ritually and economically—they account for as much as one-quarter of all annual subsistence production for coastal Cree.

There are advantages to the Cree paradigm of a sentient, communicative world that transcends but includes humanity. It has oriented Cree to aspects of animal behavior that Western science, inured by Cartesian metaphors of mechanical nature, has admitted rather belatedly. Lorenz (1979) observes that for the "higher" animals, the expression of emotion involves substantially the same neuromuscular system as in humans. Geese possess some quite "human" affective qualities, including loyalty, jealousy, and grief; furthermore, these qualities are manifest in the context of striking similarities in courtship, mating, and the rearing of young. "The family and social life of wild geese exhibits an enormous number of striking parallels with human behavior," Lorenz observes; "Let no one think it is misleading anthropomorphism to way so" (192).

Lorenz finds a greater gulf separating the rational faculties of geese and humans. Yet here, too, ethologists are finding that animals classify elements of environment with some sophistication:

> Animals create a taxonomy appropriate to their species and ecological niche. Thus predators, for instance, distinguish different categories of prey—by size, appearance, odor, and other signifiers—thus forestalling wastefully indiscriminate attacks. Vice-versa, many potential prey distinguish among different kinds of predators as we observe from their use of sundry warning signs, variations in their flight-distances and flight-reactions. (Sebeok 1975:93–94)

Sebeok has argued that for animals, as for humans, aesthetics are intimately linked to the extraction and reconstruction of structures from salient environmental features, "even when the process or the product is disunited from its proper biological context" (1975:61).

The interpretations of Cree hunters suggest that geese are quite apt at learning in what contexts to expect predation, at learning to distinguish predatory from nonpredatory humans, and at communicating appropriate behavioral adaptations to other geese. In other words, there is substantial flexibility for geese to "reinterpret" environmental signs, and this learning is communicated among geese to ramify socially.

It is therefore important for hunters to arrive at precise estimations of goose learning and communication, particularly in relation to themselves as predators. It is heuristically useful but not in itself sufficient to assume, on the basis of the culturally pervasive paradigm, that capacities of intelligence and communication are shared by geese and humans. Hunters need to know more about what is shared and what is different, and in what measure. The more the respective capacities of geese and humans are specifically formulated (i.e., differentiated), the more "literally" they contribute to effective hunting scenarios. The interpretation/modeling of hunting experience is an ongoing refinement of hunters' knowledge of the specific capacities of geese, and the basis for adjustments in hunting practice.

Hunters arrange landscapes that will be attractive and nonthreatening to geese, while exercising caution so that geese will not learn to associate unusual details with the possible presence of hunters. Decoys and goose calls are iconic approximations by hunters of the semiotic landscape of geese. Hunters recognize differences among species of geese. Canada goose decoys must be realistic in profile and must be kept heading into the wind, properly spaced, with decoys appearing in both feeding and alert postures. Generally speaking, greater numbers of decoys are more effective. Snow geese are less sensitive to profile or number of decoys, but respond strongly to color. Two or three white plastic buckets or white rags displayed prominently on a hillside, in conjunction with calling, are sufficient.

The honking of geese is imitated to get the attention of an approaching flock once it is near enough to spot decoys. When the geese have seen the decoys and "made up their minds" to fly over to land, hunters stop calling, or switch to two or three long, low "welcoming" calls at gentle intervals. Calling should be used sparingly—novice hunters must learn both to imitate goose calls accurately, and to know when not to call. If a flock in the air has chosen not to respond to calls but to continue on and away from blinds and decoys, hunters should cease calling, because geese may recognize that such calls are unnatural. When there is less wind, the geese hear calls (and mistakes) more keenly. When a hunter is especially skilled at calling geese, others may prefer to keep silent to reduce the risk of detection.

Neither the semiotic conventions of geese, nor their interpretation and manipulation by hunters, are static. Geese, like hunters, are said to "know the land." Their ability to recontextualize certain perceptual features as signifying the presence of hunters is the potential undoing of the latter, as geese "get wise." For this reason, hunters' precautions to minimize visual and auditory signs of their own presence go well beyond the use of blinds at actual hunting spots. Camps are kept at some distance from concentrations of geese, and are well-hidden in the bush. Snowmobiles and chainsaws are not used near concentrations of geese. Ideally, the only birds on the territory that will be immediately aware of hunters' presence are those from small flocks actually fired on at hunting sites. Shooting on calm days is generally avoided, because the sound of shooting carries over a wide area without a wind to muffle and disperse it. Shooting after dusk is also avoided, because the flame visible at night at the end of a fired shotgun is said to terrify geese. Similarly, the use of lights outdoors at night is restricted.

Hunters' experience is that geese will not return to a hunting spot that has been used too regularly, or where they have been frightened badly. The Wemindji community area along the coast of James Bay is divided into several goose hunting territories, each used by up to a dozen hunters from a number of households linked agnatically, affinally, and by friendship. Hunting activities for each group are under the supervision of a senior "shooting boss." Each territory includes a number of viable hunting spots, and all hunters on a territory are expected to use one and the same spot on any given day, allowing all other spots to "rest." Normally, a new site is chosen each day, so that hunting spots are rotated. In this way, the migrating geese will not learn to expect hunters at any particular location, and will be respectfully permitted to rest and feed undisturbed over the majority of the territory on any given day. At hunting sites, when geese are killed, it should be done accurately and efficiently, to minimize disturbance and to avoid the waste of injured birds that escape.

If these precautions are not taken, geese on the territory will grow increasingly anxious about human presence and adjust their behavior accordingly. Even a fraction of geese too badly frightened communicate their alarm to other geese, which could lead to a reduction in the population staying on the territory, or to incremental avoidance by geese of the spot where the fright occurred. Since the same geese are on the migration route in successive seasons, and since young geese are said to learn their habits from their parents, a hunting spot that has been mishandled can take several seasons to recover.

I will give one example of how changes in goose behavior are effected by hunting activity. In early autumn, when the tide is high and the wind is brisk and onshore from James Bay, the geese fly from the coastal bays to the offshore islands in the bay, early in the morning. They do so to feed on berries there, partly because high tides, onshore wind, and rough waves make it impossible to feed on the eelgrass that grows in shallow water in the coastal bays. The berries, now ripened, are much relished by the geese, besides. This relational model is of key importance in hunters' decisions about where and when to locate themselves to wait for geese. When experience fails to confirm the expected relations, amendments to the model ensue.

It has in fact developed in recent years that geese are less prone than before to fly to hunters waiting on the islands, even when there have been plenty of geese in the coastal bays, plenty of berries on the islands, and favorable circumstances of wind and tide. The interpretation of this decline by experienced hunters is as follows. In recent years, population growth and wage and transfer payment income have led to an increase in the number of and mobility of hunters who have greater access to motorized water craft. This means that any hunting location in the coastal community area is generally accessible within twenty minutes to two hours' travel from the settlement.

Significantly, these settlement-based hunters do not independently enter the more management-sensitive coastal bays, where concentrations of geese rest at night and also feed when the tide is low. Only when such a bay is being hunted as part of a rotational strategy, under the leadership of a "shooting boss," are hunters from the settlement welcome to join. But offshore islands can be used with much less risk of disturbance to the main concentrations of geese in the bays, particularly when there is enough wind to prevent the sound of shooting from carrying far. It has developed as a sort of community compromise between full-time hunters based in camps and wage-earning part-time hunters, who are less flexible as to the times they can hunt, that the islands can be used without direct supervision and coordination by shooting bosses. This allows settlement-based wage-earners to hunt on their days off or after work, even when conditions are not suitable for hunting in the bays.

However, because this hunting at the islands has become more frequent and is no longer coordinated in regular rotation, the geese have come increasingly to expect hunters at the islands. Consequently, there have been more geese flying inland instead of offshore when they leave the bays to feed. Hunting geese inland is less productive and more difficult, so geese have learned not to expect hunters there. They therefore fly lower and less cautiously inland than along the coast and offshore islands.

The kind of interpretation just summarized, and the hunting strategies entailed, involve years, in some cases generations, of practical empirical investigation into how much and what kinds of interactions with hunters the geese will tolerate, without withdrawing from strategic locations. The relational models conventionally signified by this experience help to define goose communication as different in degrees and respects from human communication. Literal interpretations precipitate a figurative complement—the differentiation of human from goose communication, or of goose communication from that of some other animals.

Some of the attributes of goose communication per se are best contemplated anecdotally. Here, metaphors that evoke human leadership, speech, and so on, are evident, but the hunting situations referred to are themselves key interpretants of the appropriate extent and application of the metaphor.

A friend and I were sitting in our blinds early one spring. A few larger flocks were heading north, but they were too high and were not coming into our calls and decoys. There was a lake nearby, perhaps a half-kilometer back into the bush but within earshot. We could hear the gabbling of numerous geese there. They hardly flew all day, until in midafternoon one solitary bird came low over the trees toward us. It wheeled to land among the decoys, an easy shot. We fired half-a-dozen shots, somehow missing, and the goose fled back in the direction of the lake. My companion speculated on the consequences of our poor shooting a few moments later: "Probably that goose told the other ones over there: 'If I don't come back, it's okay to come on over.'" No more geese flew our way from the lake that day. Geese, apparently, could communicate to other geese about phenomena that the latter have not experienced directly. "Scouting" among geese is observed in a variety of contexts, and presumably the behavior of these scouts conveys something about attractive versus dangerous situations. A variety of calls and postures, in flight and on the ground, are distinguished by Cree hunters—from messages of invitation to those of caution and alarm.

On another occasion later the same spring, the same companion and I were waiting for geese on an east-west elevated ridge between two coastal bays. The geese, who were leaving the bay to the south to continue their migration northward, would fly over the ridge. The highest, treeless portion of the ridge was perhaps two hundred meters in length, affording hunters a view of approaching flocks. It happened several times that a flock would cross too far east or too far west of where we sat, so we would be unable to get a shot. This seemed random enough, but then one or more flocks following at intervals of several hundred meters would cross at the same spot as the first flock, again evading us. I wondered if winds might account for this regularity, but rejected that possibility because the flocks had crossed at the same points even when they had approached the ridge from quite different trajectories. We were well-hidden, and the flocks were clearly unable to see us. I was ready to attribute the pattern of evasion to episodes of unfortunate coincidence until my companion remarked: "It always seems to happen like that. I guess they know there's hunters around. They see where the flock ahead of them went over, and they see nothing happens to them, so they think, 'Might as well go over there!'"

This incident was used by my host to instruct me in the attentiveness of individual flocks to the activity of other flocks in selecting a course of action. The safe patterns of a few geese are copied by many, in a context where older geese who are the leaders have learned to expect hunters. It became clear to me how avoidance of certain situations and preference for others could ramify socially among geese, resulting in general behavioral changes for the population as a whole.

At the same time, several aspects of goose awareness and communication remain esoteric. The capacity of animals to anticipate some events is considered superior to, and beyond the ken of many humans. Goose behavior of certain kinds is a predictor of approaching weather; geese begin their preparations before hunters would otherwise be aware of impending changes. Or again, in years when the local berry crop has failed, most geese on the fall migration have been observed to fly, very high up, right on past the James Bay coast. What is outstanding to hunters is that the flocks seem not to have to land to know that the feeding is poor. One interpretation of this phenomenon relates again to scouting behavior, although the precise mechanism of information transfer is ambiguous. In other interpretations, it is supposed that animals experience dream images and corporeal symptoms of the kind that can also alert humans to future or distant events—notable but not particularly unusual premonitions in the Cree world.

This commonalty returns us to the premise of a communicative, reciprocative network that unifies the holistic world. This premise is metaphysically prior to the more particular differentiation of persons in the world—and it is at this level that hunters, animals, geophysical forces, and even God are ultimately of one mind, as it were. Consistent with this premise is the notion that encounters, thoughts, dreams, and rituals involving hunter-animal exchange both index and influence the state of reciprocity that obtains at a given point in time.

Animal and Human Reciprocators

I have mentioned the effect of metaphors of eating and sexuality in establishing a fundamental difference between human-human and animal-human reciprocity. I'll go on now to illustrate how ritual delineations of reciprocity between humans and animals contain, at the same time, abstract but quite literal constructions of social-ecological principles. The life cycle of the hunter, seasonal cycles, the social roles of men and women are all marked by ritual phases that reflect these principles. A few examples will suffice.

Early in the spring season, the geese killed on the first day of the hunt are cooked and a small share is distributed by a respected woman elder to everyone in a camp. This is done no matter how few the geese or how large the camp, and regardless of who killed the geese. Then, the following geese are saved for a few days until there are enough so that every man, woman, and child in the hunting group receives one, two, or more geese, depending on how many have been killed. Again, the distribution is made by a respected elder of the camp, and without regard to who killed the geese. The group then feasts, with each household roasting some of the geese it has received. This process is considered to be an "invitation" to the geese, since animals are said to come more readily to hunters who share them with others. Only after this feast is it possible for the household to accumulate geese for its own consumption. Significantly, the feast occurs while it is still to be determined whether the migration will linger, ensuring bounty, or pass quickly. Failure to contribute generously to the feast can account for poor luck in hunting later on.

A hunter in his blind often "smokes to the game" (*pwaatikswaau*), or sings goose songs. Tobacco was a traditionally valued item of exchange, and smoke is an appropriate vehicle of exchange with creatures of the air. Hunters' songs express spiritual and aesthetic aspects of the exchange, and include vivid images of the ways in which geese fall to the hunter. When a goose has fallen, the gift is respectfully admired by the hunter and later received as a guest into the lodge by the women of the hunter's household. The women take care to use every part of the goose possible, to avoid spoilage, and to dispose respectfully of the few remains. The cartilage tracheae, including the windpipe and voice organ of the goose, are hung from a tree branch where, poetically, the passing wind carries their call, beckoning geese in future seasons to renew the exchange.[6] When the migration has nearly passed and the last of the geese are departing, the hunter bids them farewell, expressing the hope that, granted continued life, he will be able to see them again on their return.

This ritual complex advances two general propositions about human-animal reciprocity that are of key ecological concern. The first is that respectful activity toward the animals enhances the readiness with which they give themselves, or are given by God, to hunters. The second is that sharing of animal gifts among hunters is an important dimension of respect for the animals. Both propositions are implicit in ritual enactments of the special obligations of hunters toward game, if game animals are to fulfill their own special role in supplying hunters. And both propositions convey literally understood truths about ecological relations.[7]

What is involved in the first of these general propositions? The empirical availability of geese, as we have seen, varies with their treatment by hunters. The specification of "respectful" treatment in day-to-day hunting is as complex as the many situations of interaction, but the general and key notion is that technical efficiency in killing animals must be balanced by restraint, and that only the latter can really guarantee the long-term viability of the former.

A hunter must strive for impeccable technique, both in the interest of his own security and to avoid undue suffering or disturbance to the animals. A hunter who, in spite of his effort, cannot kill animals will be disappointed, but he should keep trying and not be too disappointed because it is wrong to expect more than is freely offered. Perhaps the hunter is not receiving more because the partner is not in the position to give. The wise hunter directs his efforts elsewhere if, after trying hard, it is apparent that a particular species doesn't want to be caught. On the other hand, it is wrong to accept more than one needs, even if it can be taken with the means at one's disposal. The generosity of a partner can be overtaxed.

There is a rather concise set of symbols to summarize and express this balance between efficiency and self-restraint. First, when a hunter who normally kills perhaps fifty geese in a season suddenly kills, let us suppose, three times that number, it is taken as a sign that the hunter has not long to live. It is not surprising, then, that when an individual hunter has accumulated a larger than average kill early in the season, he sometimes stops for the remainder of the season, or lets a younger and less experienced hunter in the household bring home the geese. Collective restraint is also exercised after a particularly abundant daily kill has been made, when all hunters on the territory let the geese rest for a day or two.

There is a second symbol for the perils of excessive killing—the albino Canada goose. Such geese must not be killed under any circumstances or the hunter will find it very difficult to kill geese thereafter. Thirdly, there is the sandhill crane, a relatively uncommon bird on the east coast of James Bay, that is most numerous when there is an unusual abundance of migrating geese feeding and resting in the coastal bays. It is permitted to shoot the crane, but it must be perfectly done. To miss a shot is a sign of the hunter's impending death or that of a near relative. Fourthly, there is a rarely reported Canada goose that is said to bear a luminous collar over its neck and across its breast. The hunter fortunate enough to see this goose must kill it with a single clean shot and retrieve it almost the moment it touches the ground. Otherwise, the collar dissipates, passing to another goose. But the impeccable hunter may reach it in time and retain the collar as a charm. The geese will thereafter fly to such a hunter, even when he is not well-hidden.

To summarize:

1) Too many geese killed (excessive killing) = Hunter's death

2) Albino goose killed (excessive killing) = Hunter's poverty (curtailment of gifts)

3) Crane attempted but not killed impeccably ("insufficient" killing) = Hunter's death

4) Collar-bearing goose killed impeccably ("sufficient" killing) = Hunter's wealth (abundance of gifts)

This symbolic set signifies that the hunter must practice both excellence and restraint in the killing of game if he is to live well and avoid poverty or death. There is such a thing as "insufficient" killing, expressed in its negative form in the sanction of death when the crane is attempted but failed, and in its positive form in the reward of food wealth when the collar-bearer is impeccably taken. But there is also the possibility of "excessive" killing, expressed in a strong negative form in the sanction of death for killing too much game, and in the milder negative form of poverty for killing an albino. Symbols of the importance of impeccability have "literal" implications where efficiency and excellence in a demanding environment make the difference between prosperity and poverty, life and death. But symbols of the necessity for restraint are equally intelligible in literal terms because the generosity of animals is empirically exhaustible. Geese hunted too noisily or too often in the same place, too much game killed and too little allowed to escape, could lead objectively to dwindling exchange with hunters and to poverty in animal gifts, synonymous in the not too distant past with death.

Let us now turn to the second of the general propositions cited earlier—that positive reciprocity in human society enhances reciprocity with geese. In what literal sense could interspecies reciprocity depend on human reciprocity? Even more important than sharing food, given households' pride in their autonomy, is sharing the opportunities to kill geese or other game. I have already mentioned that the management of a hunting territory is a delicate matter, requiring a cooperative strategy. Cooperation becomes impossible when generosity is not maintained. For example, if a hunting boss "saves" a rich build-up of geese, then hunts it for the sole benefit of his immediate household without notifying other hunters who would normally be entitled to join in, his reputation as a leader is seriously damaged. Other hunters then feel justified in hunting when and where they choose on such a territory, with the possible result that rotational management becomes impractical. The geese become increasingly wary, or move elsewhere, and hunting productivity on the territory declines. The animal gift, in this literal sense, depends quite literally on human generosity, and it is this social knowledge more than any other factor that accounts for restraint and regimentation in hunting. Social and ecological reciprocity are not just formally interdependent, as inferences from the same root metaphor or paradigm, but interdependent also in human practice.

Both propositions depend on situational elaboration of the reciprocity metaphor; one could say that the empirical contexts of both geese and hunters are assimilated to the terms of the metaphor. It is not as though reciprocity applies "literally" to social relations but only "metaphorically" to relations with animals. Neither set of relations can be said to represent the "primary" meaning—both are part of a reciprocating socioenvironmental continuum; but human-animal differences are elaborated and exploited in empirical detail to produce informative permutations of reciprocity across numerous phenomenal domains.

In certain ritual contexts, the identity of the hunter merges radically with that of the animal, a merging accomplished through body-spirit reciprocity. The death of a hunter in a dream is a common omen of an important food animal about to be given in waking life. In a dream, the goose may be a guardian of the hunter's power and essence and may protect him from sorcery. Throughout his life, the hunter receives the gift of geese, and at a hunter's death, it is often a goose that represents his soul on its journey from this life. At the time of a hunter's death, a solitary goose may fly low overhead, or land near the mourning relatives, acting quite unafraid. The hunter's experience of animals as interpretants of his essential self renders all the more poignant the inevitable separation of hunter from prey, and all the more compelling the morality that joins them in the reciprocity of life-giving and life-taking. The transcendence of this tension at significant moments—the death of an animal, a dream encounter, or the death of a hunter—is the experience of the sacred.

Conclusions

The achievements of indigenous ecological knowledge, as illustrated in the case of Cree hunters, are neither mysterious nor coincidental—they result from intellectual processes not qualitatively different from those of Western science. Western science is distinctive not through any greater logical coherence or empirical fidelity, nor any lesser involvement with metaphysical premises, but through its engagement of particular root metaphors in specific social institutional and socioenvironmental settings. Any number of root metaphors, situationally elaborated in the course of practical engagement with the world, may inform rational explanation and the effective organization of empirical experience. Equally, any number of the same metaphors may obstruct effective knowledge through a dogmatic and misplaced literalism.

Knowledge traditions reflect the morality of the social practices and paradigms in which knowledge is framed. Numerous studies have found that the "anthropomorphic" paradigms of egalitarian hunters and horticulturalists not only generate practical knowledge consistent with the insights of scientific ecology, but simultaneously cultivate an ethic of environmental responsibility that for Western societies has proven elusive.[8] If the inclusion of humans in a figurative world of analogous other-than-human persons promotes environmental responsibility, this depends on the condition of reciprocity in the human society concerned—not on any predominance of figurative versus literal thinking. All societies, whether egalitarian or hierarchical, establish metaphorical connections between the social and the environmental. In all knowledge traditions, literal modeling defines and redefines the relations among objects in the world, relations which in turn are assimilated to the meaning of root metaphors as they are applied in particular situations and contexts. Cree hunters are not less concerned than Western scientists with literal interpretation; nor are Western scientists less involved in figurative invention than Cree hunters. The conventional social context of Western science tends to hierarchy and centralized control, however, and this is the morality that is metaphorically projected onto our own relations with "nature." For this very reason, the historical disqualification and subjugation of indigenous knowledge is intimately linked to Western culture's domination of nature.

Notes:

The ethnography and much of the analysis in this paper was previously published as Scott (1989). For the present publication, I was asked to rewrite it with a more general audience in mind. The introduction and conclusions are new, and some technicalities of anthropological semiotics in the body have been clarified or eliminated.

I wish to acknowledge the support of the Social Sciences and Humanities Research Council of Canada for grants supporting this work.

1. Certain metaphors are so pervasive in a knowledge discourse as to constitute what have been termed "paradigms" or "archetypes" (Black 1962). Others have called them "root metaphors" (Pepper 1942, Ortner 1973) or "metaphoric networks" (Ricoeur 1977). In this paper, I use the terms "root metaphor" and "paradigm" synonymously.
2. Not to mention the polemical deployment of other adjectives: "magical," "irrational," "superstitious," "traditional," "primitive," etc.

3. For a thoroughgoing historical review of the anthropological analysis of knowledge and belief, see Tambiah (1990).
4. For Bateson, the "totemic analogy" between the social system of which people are the parts and the "larger ecological and biological system in which the animals and plants and the people are all parts" (1979:155) was a better analogy than the one that likened people, society, and nature to nineteenth-century machines.
5. One might observe that a consequence of this sort of analogical thinking is to anthropomorphize animals, but that would assume the primacy of the human term in the metaphor. Animal qualities react with perhaps equal force on understandings of humans, so that animal behavior can become a model for human relations. Preston (1978:152) has suggested that the goose as exemplar of Cree ideals of social coordination, grace, and composure may be "better" than human.
6. The sequence here described in abbreviated fashion is one variation on a general ritual structure for "bringing home animals" (see Tanner 1979).
7. Feit (1973, 1978) has offered seminal discussions of the ecological significance of respect for animal gifts, to which my own analysis owes a great deal. Animals are felt to be given at times and places in which, by virtue of numerical availability and characteristic behavioral traits, they present themselves to the hunter's weapons or traps with maximum efficiency and minimum struggle. When hunters notice animals becoming scarce and difficult to catch, they say it is because the animals are "angry," perhaps because hunters have been taking too many. The respectful response is to stop hunting the species in question until it once again is more freely given.
8. See Rappaport (1968); Reichel-Dolmatoff (1976); Wagner (1977); Feit (1978); Nelson (1983); Bennet (1983); Scott (1983).

CHAPTER 4

The Savagery of the Domestic Mind

Jean Lave

Introduction

Sociologists and anthropologists of science have begun to study scientific knowledge production in laboratories and elsewhere as everyday practice. This work raises questions about what we mean by "everyday practice." It encourages questions about what becomes of "everyday" everyday practice when science is thus reconceived. And it provokes opposition from many, in part because views of science as everyday practice contest conventional claims. The prevailing belief among scientists and nonscientists alike is that whatever science may be it is most definitely not everyday. Such a belief defines "science" in opposition to the "not science" otherwise known as "everyday life." Philosophical traditions and scientific discourses ascribe to scientists specialized, value-neutral, and powerful thinking and to their work an exceptional character, inevitably helping to reinforce the hegemonic role of science and underlining distinctions between real scientists and the rest—the "others" who are not scientists.

Anthropologists of science find themselves in a tough position, threading their way through varied conceptions of everyday practice that spring from their research on the practice of science, from scientists' understanding of themselves and their work in opposition to "the everyday," and from the traditions of Western thought that undergird the practice of science.

Several years ago, my students and I undertook a study of everyday mathematical practices in Orange County, California (de la Rocha 1986; Murtaugh 1985a, 1985b; Lave 1988). Our main concern was to explore interconnected American cultural practices that generate and sustain core conceptions about rationality. Rather than investigating laboratory science practices, we observed people engaged in daily activities in supermarkets and kitchens—as they shopped for groceries, cooked meals, dealt with quantitative relations while learning the Weight Watchers dieting program, and managed their household finances. This offered us an unusual perspective—that of "just plain folks," (jpfs)—on what is often labeled "everyday problem solving" by laboratory studies of cognition or research on learning in schools. The differences between these perspectives of lab studies and jpfs raised strategic questions about characterizations of "the everyday" and "the inferior other." Both concepts have contributed to ideologies of "science" and "scientific thinking."

Our project differed from the growing body of research on physical scientists in their laboratories in other respects as well. Many disciplinary practices juxtapose "science" and "everyday life" (Lave 1988:4), including discourses on the mind, cognition, representation, problem solving, logic, mathematics, the expert, scientific thinking, and the effects of schooling. Among the social sciences, cognitive science has been an especially important site for the production of claims about the everyday world and its relations with "scientific" knowledge. Cognitive studies depend on, and produce, contrasting views of what are called "scientific" and "other" forms of thinking (Lave 1988). Indeed, the study of functional modes of thinking in this century has been persistently based on an imagined conception of the thinking of "the other," considered the inferior. The "inferior other" is deeply embedded in Western thought: in the practices of more traditional psychologies and cognitive science, in laboratory experiments and computer simulations, and in schools.

Schooling complicates the story. It is difficult to address claims about thinking, knowing, and learning without analyzing the school practices that are supposed to produce good thinkers, knowers, and learners. For the last century at least, schooling has been a mediating institution in which relations among 1) social science theorizing about everyday life, 2) living everyday lives, and 3) the practice of science are mutually implicated in one another. Schooling is a major institutional form in which the claims of cognitive scientists about "scientific" and "everyday" thinking are confirmed (as well as inculcated). Schooling itself often is made to stand in for "science" as the opposite pole to everyday life.

The everyday practices of cognitive scientists leave little room to notice, much less explore, the practices of those typically designated as "the other." When we began our study, we had all of Orange County to ourselves except the Cognitive Science Laboratory at the university. Why so little company? To witness the practices of jpfs requires venturing outside the university into places where "everyday" knowledge is constituted in activity, in situ. To do this requires assuming that ways of thinking and forms of knowledge are historically and socioculturally situated phenomena. Cognitive theory in particular and Western thought much more generally has assumed quite differently. Learning, thinking, information processing, and knowledge representations are taken to be universal ahistorical processes by which all individual humans operate and have always operated. Two consequences are worth noting. First, with views like this cognitive scientists have no incentive to venture beyond the laboratory, and second, distinctions between scientific thinking and the thinking of "the other" are thus naturalized and universalized.

Claims about the nature of mind, mental processes, and powerful thought have often assumed the mathematical nature of all three (de la Rocha 1986; Lave 1988). (Current rejections of rule-governed models of mind in favor of associationism and intuition—[Dreyfus and Dreyfus 1986], propose changes in the value assigned to mathematical structures of mind but not the terms of the debate.) These models take the mind itself to be mathematically structured, take mathematics as the structure of thought processes, and take the mathematical content of reasoning (or mathematical intuition) to be the most powerful content of the most exemplary—read scientific—thinking. Thus, mathematical practices, especially mathematical practices in nonschool settings, offer a promising venue for exploring theoretical issues concerning science, the everyday, and the thinking of "the other."

I have outlined three ways in which our "anthropology of science" has differed from more typical studies in the field. Rather than studying biomedicine or physics, we chose to study social science approaches to mind and mathematics; rather than studying scientists in their laboratories, we investigated domestic practices (though against a background of school and psychology laboratory practices). And that meant that our research and this paper address intertwined American cultural practices involving two sets of "natives." One set are the people whom social scientists investigate (in studies of cognitive processing, problem solving, learning, and cognitive development). The other set are the social scientists themselves (as they engage in the everyday practices of cognitive research).

The following section introduces the domestic quantitative practices of jpfs. The one after explains why we describe these practices as "assembling and transforming relations of quantity in ongoing activity," rather than interpreting and evaluating them in terms of scholastic math puzzle solving. Next we turn to the other subjects of our research, the scientists of mind, exploring the dialectical process by which scientists' theorizing about the mind depends upon assumptions about the "civilized mind" to generate characteristics of the "primitive mind" and vice versa, so that in the end each is defined by how it is not the other. The following section describes how practices of empirical cognitive research similarly generate superior and inferior others. In the last section, we take up where and how the dialectics of competent and incompetent rationality are absorbed into the practices of jpfs.

The Math Problem in Everyday Life

Here is a fairly typical example of mathematical activity by jpfs in the supermarket. The example is intended to illustrate the difficulty of translating everyday math practice into conventional thinking about math problem solving:

The following was observed during a grocery-shopping expedition. A shopper was standing in front of a produce display. She spoke as she put apples, one at a time, into a bag. She put the bag in her grocery cart as she finished talking.

> There's only about 3 or 4 [apples] at home, and I have 4 kids, so you figure at least two apiece in the next three days. These are the kind of things I have to resupply. I only have a certain amount of storage space in the refrigerator, so I can't load it up totally.... Now that I'm home in the summertime this

is a good snack food. And I like an apple sometimes at lunchtime when I come home (Murtaugh 1985b:188).

In the end the shopper bought nine apples. For analysts of mathematical practices this example poses several problems that a conventional math "word problem" does not. There are several plausible answers—9, 13, 21. It appears that the problem was defined by the answer at the same time an answer was developed during the problem, and that both took form in action in a particular, culturally structured setting—the shopper's local supermarket. The shopper's engagement with quantity did not lead to a pause for formal calculation, yet relations between inventory, family apple consumption, budget, and sack size were all reconciled in action.

Quite a substantial body of research on math in everyday practice now exists. It started with Gay and Cole, *The New Mathematics and an Old Culture* (1967); my research on tailors in Liberia; and Posner's (1978) and Petitto's (1979) work with farmers, tailors, and cloth merchants in Côte d'Ivoire. More recently Sylvia Scribner and her associates carried out a path-breaking study of math practices among blue collar workers in a commercial dairy in Baltimore (1982), and her most recent research has focused on CAD-CAM production practices. Carraher, Carraher, and Schliemann in Brazil have studied market-vendor children at work in open-air markets selling garden produce and in school (1982, 1983; Carraher and Schlieman 1982). They have also compared master carpenters to trade school carpenters' apprentices and have made a study of the math practices of bookies taking bets on the national lottery "numbers" game. Geoffrey Saxe (1990) studied Brazilian children's activities selling candy on the street and learning math in school, and Hutchins (1992) has been following U.S. Navy navigation teams on a helicopter transport ship.

Two robust findings appear in all of this work. The same people deal with relations of quantity in quite different ways in different situations. On the one hand, when engaged in everyday activity, they are remarkably accurate in their calculations—even by the (arguably irrelevant) standards of school math practices. Participants in our grocery shopping research averaged 98 percent accuracy in their calculations in the supermarket. Market vendors in Recife in Brazil who had spent very little time in primary school were 99 percent accurate. Dairy workers in Baltimore, with an average of a sixth grade education, made no errors when observed at their work of assembling wholesale orders for dairy products. On the other hand, the same people did much less well on tests designed to be comparable with problem solving in the supermarket, open-air produce market, or commercial dairy. This finding suggests that things are different enough "out there" to pose some interesting questions for conventional views of cognitive theorizing about everyday mathematical practice.

The discontinuity in performance between work and test settings suggests that even mathematical problem solving is situationally specific activity. Other sources support this claim. For instance, people's educational biographies (i.e., how much schooling they have had and how long ago), predict test scores but do not predict everyday performance differences between individuals, which in any case hardly exist. In the end, however, it is the apple example and many like it that provide the more interesting empirical evidence for the situational specificity of everyday math. They show that the math seems part of the flow of activity; that math activity is quite different in different settings; and with it the assembly and transformation of quantitative relations.

The Transformation of Quantitative Relations in the Resolution of Quandaries

De la Rocha's study of participants in Weight Watchers provides rich, complex, situated examples of the transformation of quantitative relations in everyday practice (1986). She followed the activities of nine women, all new members of the dieting program, as they incorporated new measurement practices into their meal preparation over a period of weeks. The diet program emphasized meticulous control of the portions of food consumed. Thus, it promised to generate many opportunities for calculation in the kitchen, and she hoped to see attempts at new kinds of math activity in a setting far removed from school. De la Rocha carried out repeated, intensive interviews with each dieter, including an exploration of the participants' biographies as dieters. She spent lots of time with them as they prepared meals in their kitchens. At the end of the six-week observation period the nine women took part in a variety of arithmetic-testing activities.

The participants also kept diaries of all food items consumed each day. De la Rocha conducted interviews about the process of dieting, asking how each person learned the Weight Watchers' system of food portion control and about the specific procedures she used in weighing and measuring each item.

There was lots of measuring and calculating activity. All dieters calculated portion sizes for about half the food items they prepared, on average, across the six weeks. This statement requires some qualification, however. First, some dieters calculated considerably more often than others. Second, dieters measured and calculated more at the beginning of the new program than near the end. We were especially interested in the "disappearance" of math over time. Third, having carefully coded the measuring and calculating activities of the dieting cooks in the preparation of hundreds of food items, de la Rocha showed that none of several factors that might plausibly account for differences between the cooks' measuring patterns did so. Factors that did not explain their uses of arithmetic included their age, the number of children living at home, the dieters' years of education, the amount of weight they hoped to lose, the amount of weight they had already lost, and their scores on the arithmetic tests.

De la Rocha's analysis of the dieters' accounts of their lives and diets offers insight into the quandaries that compel dieters to engage in the transformation of quantitative relations. Begin with American culture writ large. It turns out to encompass some serious problems of quantity. The abundance of food products in the United States, the ideology of consumption, and the fascination with the self-mastery reflected in a slim physique have provoked an obsession with body weight and its control. For most plagued by it, "excessive" weight is a profound blight deeply affecting self image. Although weight can be remediated by dieting, it is only with great difficulty and often for brief intervals. In translating the determination to lose weight into practical action, the dieter faces the dilemma posed by a strong desire to alter the "distortions" of the body and a craving for the solace and pleasure of food. Dieting is an arduous process that requires not just one decision, but a continual struggle in which the problem of self-denial arises many times each day over many months and even years. Inconsistent commitment leads to backsliding, or the end of a diet cycle, and resulting feelings of failure and depression. From interviews about their history as dieters, it appeared that participants in the project had relatively long-term, consistent resolutions to these dieting dilemmas. Some espoused the view that meticulous control of food portions was the way to control weight. Others expressed their approach to dieting as "so long as you feel hungry you must be losing weight." Each of them put their resolutions to these dilemmas into practice: long-term dieting styles clearly shaped measurement activity differently. Methodical dieters used arithmetic measurement and calculation techniques on nearly two-thirds of the food items they recorded in their food diaries; the "go hungry" dieters measured only a quarter of the food items.

Gaining control over food intake is the central quandary of dieting. But attempts to control food portions come into conflict with other concerns of dieters. The more elaborate the steps to gain control of quantity, the greater the conflict with putting food efficiently onto the table. While the dieter calculates, the family and the evening meal wait. The conflict between dieting rules and efficient food handling appeared most directly to generate the dieters' arithmetic "problems" and clearly shaped the long-term shift from more to less calculation over time. All the dieters responded to this conflict in two ways: by generating reusable solutions to recurring math problems and by finding ways to enact solutions as part of ongoing activity.

One simple example illustrates both. Initially, to find the correct serving size for a glass of milk the dieter had to look up the correct amount in the Weight Watchers manual; get out a measuring cup, a drinking glass, and the carton of milk; pour the milk into the measuring cup and from the cup into the glass; then wash the measuring cup and later the glass. This procedure was shortly transformed into getting out the glass and milk and pouring the milk into the glass to just below the circle of blue flowers, knowing that this would be one cup of milk. This is just one among a myriad of examples, for the cooks invented hundreds of units of measurement and procedures for generating accurate portions (de la Rocha 1986). In the process they made it possible to do less and less measurement and calculation over time. But they continued losing weight at the same rate, so presumably no relevant accuracy was lost.

Importantly for the dieting cooks, math problem solving is not an end in itself. Procedures involving quantitative relations in the kitchen are given shape and meaning by the quandaries that motivate their activity; school math knowledge does not constrain the structure of their quantitative activity, and it does not specify what shall constitute math problems. It is the specific character of action-compelling conflicts that generally determines what will constitute a "problem-in-need-of-solution."

Other characteristics emerged as typical of "everyday" math practice. People are efficacious in dealing with problems of number and space in everyday settings. Their math activity is structured into and by ongoing activity and its settings—its structure unfolds in a situated way. So people do not stop to perform canonical, school-taught math procedures and then resume activity. In the supermarket and kitchen, shoppers and cooks have more than sufficient resources of mathematical knowledge to meet the mathematical exigencies of their activities. They almost never arrive at wrong answers because they stay clear about the meaning of the quantitative relations they are trying to interrelate and what a ballpark solution should look like. Also, having a strong sense of the meaning of what they are doing, they are able to abandon problems they recognize they cannot solve in the time, and for the purposes at hand. Many relations of quantity have closer relations with other aspects of activity than with each other. For example, people frequently go straight from a relation of quantity ("The price is higher this week") to a decision: "Forget it!" Thus, there are many more relations of quantity than there are well-formed "arithmetic problems." Generating problems for themselves, shoppers and cooks also change those problems, resolve them, transform them, or abandon them as well as solve them. At the same time, quantitative quandaries that do not have solutions but only partial resolutions constitute almost all of what is seen as "problematic" in ongoing activity. People are walking histories of their own past calculations, but not of procedures for solving problems. Old results, "answers" if you will, are carried around, but procedures are invented on the spot, as part of situated ongoing activity. Finally, the kinds of activities we investigated do not provide a curriculum for school mathematics: the "assembly and transformation of quantitative relations" is not school-like in any sense.

What I have not conveyed in telling the story of the Weight Watchers study, and summing up some of the ways in which math is different "out there," is the extreme difficulty of capturing "what is going on" in everyday math activity in something like its own terms. The languages and assumptions of math-cognition studies, formal mathematics, and closed-system puzzle/problem-solving processes are not easy to exorcise. They presume that everyday practice is simple, erroneous, routine, particular, concrete—in short, inferior. These assumptions furnish an important contrastive set of meanings to the (assumed) character of the thinking of experts and scientists.

This observation has led me to inquire into the practices and beliefs of the second set of natives under discussion here, those engaged in the theory and practice of cognitive studies—including anthropologists as well as psychologists.

The Absent Mind of the Civilized Savant

Given the long history of attempts to distinguish "the primitive mind" from the "civilized" one, we might plausibly expect that psychologists and anthropologists would have established two essential resources for their inquiries: a canonical set of categories pinning down just what the superior, scientific mind consists of, and an intimate acquaintance with some relevant "natives." Nothing could be further from the truth. Assumptions about each type enable claims about the other, as "superior" minds are constructed with respect to putative "inferior" ones, and vice versa.

The work of Lucien Levy-Bruhl offers a classic example of how this works. He published a book in 1910 whose title has been translated "How Natives Think." A more literal translation would be "The Mental Functions in Lower Societies." It has been widely influential throughout the century and was most recently reprinted in 1985. Levy-Bruhl claims that a set of mental functions divides the thinking of "civilized" people from the thinking of "primitive" people. Primitive people are, in his terms, nonrational, have no interest in logic, and have no concern with the law of negation, the proposition that if something is A it cannot be -A at the same time. "Natives" think concretely and thus create shallow trees, categorical structures with many terminal taxa, rather than deep structures of classification. They participate directly in the world and hence cannot think logically about it. They cannot abstract from nor generalize about their experience. They lack, in short, the distance from the world that "powerful thinking" requires. How can the enmeshed natives survive to function at all? Levy-Bruhl speculates that they have superb memories for detail, a point that Frederick Bartlett tried to substantiate several decades later in Africa.

We must ask where Levy-Bruhl discovered those mental functions. Clearly it is not from direct acquaintance as he never traveled far abroad. It could be that he derived his arguments from scientific treatises, philosophical or empirical, detailing the characteristics of civilized minds. But he says not:

> As far as the mentality peculiar to our society is concerned, since it is only to serve me as a state for comparison, I shall regard it as sufficiently well defined in the works of philosophers, logicians and psychologists, both ancient and modern, without conjecturing what sociological analysis of the future may modify in the results so obtained. (1910:19)

I conclude that if Levy-Bruhl did not draw his inspiration from either the subjects of his claims, or his colleagues, he must have fallen back on the common wisdom of his time. The civilized mind seems elusively defined here, although vital to such a comparative enterprise. The detailed picture of the primitive mind that emerges in his work seems more a specific resource for characterizing the "superior other" than the reverse.

Levy-Bruhl's work was taken up directly by L. S. Vygotsky in Russia, E. E. Evans-Pritchard (whose work on the Azande shows up with unusual frequency in cognitive psychology texts), Frederick Bartlett, and others. Cognitive psychology texts today still take up the opposites of items in the list of primitive mental characteristics (e.g., Anderson 1990; Rumelhart and McClelland 1986). The meaning of "good thinking" is dependent on its implicit contrast with the thinking of unspecified (primitive) others.

In a century of debate about superior and inferior minds, "science" and "scientists" have been the matter-of-course foil for the "primitive." Levi-Strauss in *The Savage Mind* (1966), invents neolithic "science" in contrast to the (unspecified) modern variety. With it, he contrasts the scientific versus the mythical, abstract thought versus intuition, and the use of concepts versus the use of signs. Jack Goody, in *The Domestication of the Savage Mind*, sums up:

> In the simplest terms, [this] is a contrast between the domination of abstract science ... as against the more concrete forms of knowledge ... of 'primitive' peoples. (1977:148)

Cognitive psychologists tout Einstein, great chess players, and famous mathematicians as heroes for school children whom they hope to teach to think similarly. Scientism pervades all levels of educational research.

The opposed low categories against which the scientist and scientific thinking play include any and all marginal, powerless, or stigmatized categories in Western society: the lower classes, women, children, criminals, the insane, and, of course, the primitive. But the central focus is not the same today as it was in 1910. The contrast between civilized and primitive thought was the major preoccupation of the early anthropologists (among others). The more salient contrast today is between the everyday thinking of "ordinary" folks and scientists. Bartlett, for instance, spoke of everyday thought as standing in contrast to scientific thought.

> By everyday thinking I mean those activities by which most people, *when they are not making any particular attempt to be logical or scientific,* try to fill up gaps in information available to them. (1958:164, italics mine)

Here is C. R. Hallpike on the same subject.

> Rather than contrasting primitive man with the European scientist and logician, it would be more to the point to contrast him with the garage mechanic, the plumber, and the housewife in her kitchen. (1979:33)

He thus calls attention to the interchangeability of women with those engaged in manual labor and with primitive man. Each represents the "inferior other" satisfactorily in opposition to the white, European, bourgeois, male scientist.

Barnes has argued that the historical and artifactual basis of the vague characterization of "civilized" and "scientific" thought lies in an antiquated empiricist philosophy of science. He speaks of its role in anthropology in the seventies, and the analysis is equally apt for cognitive studies today.

> Attempts to understand or explain preliterate systems of belief have frequently led anthropologists to compare them with ideal 'rational' models of thought or belief.... It is clear that the form of many anthropological

theories has been partially determined by the ideal of rationality adopted and in practice this ideal has usually been presented as that which is normative in the modern natural science, that is to say modern anthropological theory has been profoundly influenced by its conception of ideal scientific practice. (1973:182)

That is, imagined ideal forms of scientific practice have furnished normative beliefs about the nature of civilized/scientific thought. These depend in turn on definitions of the inferior other. Contemporary ethnographic studies of the practice of science (whether grounded in phenomenological or social practice theories) clearly call into question such conceptions. They also challenge broad claims about what constitutes high and low forms of thinking, theorizing about learning, and educational practice.

Such challenges are urgently needed: Consider the contrast between our description of Weight Watchers resolving problems in their kitchens and another interpretation of their activities. This exchange occurred in the context of debates about proper relations between math as it is taught in school and as it arises in everyday life. The following is a published description of an encounter that occurred with yet another Weight Watcher dieter during the Adult Math Project (Lave 1988:165):

> We posed a problem of quantity to new members of Weight Watchers in their kitchens. The dieters were asked to prepare their lunch to meet specifications laid out by the observer. In this case they were to fix a serving of cottage cheese, supposing that the amount allotted for the meal was three-quarters of the two-thirds cup the program allowed. The problem solver in this example began the task muttering that he had taken a calculus course in college. Then after a pause he suddenly announced that he had "got it!" From then on he appeared certain he was correct, even before carrying out the procedure. He filled a measuring cup two-thirds full of cottage cheese, dumped it out on a cutting board, patted it into a circle, marked a cross on it, scooped away one quadrant, and served the rest. Thus, "take three-quarters of two-thirds of a cup of cottage cheese" was not just the problem statement but also the solution to the problem and the procedure for solving it. The setting was part of the calculating process and the solution was simply the problem statement, enacted with the setting. At no time did the Weight Watcher check his procedure against a paper and pencil algorithm, which would have produced 3/4 x 2/3 = 1/2 cup. Instead, the coincidence of problem, setting, and enactment was the means by which checking took place.

This example has achieved a certain fame in circles where the nature of cognition is debated. Brown, Collins, and Duguid (1989) quoted accurately and then discussed "the cottage cheese problem," and Palincsar, a cognitive and educational researcher, replied. Her comment illustrates how researchers produce claims that, e.g., dieting cooks are "inferior others."

> the article [Brown et. al. 1989] cites with approval the example of some poor soul at [sic] Weight Watchers confronted with the challenge of measuring three-fourths of two-thirds of a cup of cottage cheese. The dieter mounded a pile of cottage cheese, separating first three-quarters of a cup [sic] and then taking two-thirds of that [sic] to arrive at the required half-cup of cottage cheese. The authors regard the dieter's ineptitude with fractions as giving rise to an inventive solution. . . . Instead, it was an act of desperation, born of ignorance. I question whether it was learning at all. Where does this so-called solution lead? Nothing has been learned that could be generalized. (1989)

The domestic savage is alive and well—an impoverished soul, inept at fractions, and unable to generalize—if only in the mind of the cognitive researcher who is only able to characterize the inferior other as what she is not.

Out to Lunch: The Mind of the Other

I have suggested that the theoretical characterization of the "civilized," "scientific" mind seems to have been constituted in the process of imagining the mind of the other. There is also a huge body of empirical research investigating cognitive processing and problem solving. How can this apparently direct empirical investigation of mental activity avoid uncovering the "real thing"—the concrete content of different kinds of mind-in-practice? And how do cognitive researchers generate inferior others, and, if only by implication, their superiors as well?

First, their research is predominantly experimental and designed deductively from idealized, normative models of good thinking rather than from a knowledge of actual practice. R. Rommetveit, in the late seventies, characterized such investigations as inquiries into "negative rationality." Thus, cognitive researchers begin with ideal models of how people should think. For example, Gentner and Stevens point out in their book on "mental models" that they have studied physics problem solving rather than marriage because no normative models of ideal problem solving exist in the latter case (1983). Experimentation would be impossible without such models, they argue. As Rommetveit points out, any investigation designed to explore evidence of "ideal" problem-solving activity is sure to reveal the "shortcomings" of its subjects. That is, experiments devised to reveal forms of thought that reflect normative assumptions (which do not accurately reflect any actual social practice), can only reveal what are interpreted by researchers as "deficiencies" (see Lave 1988, chapter 5). This process creates and confirms a conception of the inferior other and thus affirms the ideal model. At the same time, producing the nonideal thinking of jpfs (the subjects in experiments) for careful inspection by the scientist is part of the process by which "scientific thinkers" generate themselves and their models.

The content of normative models of the logical, rational, representational, and generalizing mind has its own effect on the production of inferior others. It helps to preserve the indirect characterization of the civilized mind by idealizing the separation of thought from action. Rationales for mathematics offer a case in point by viewing mathematics as abstract structure (as the pejorative reading of the cottage cheese example shows). Good mathematical thinking should have the power to extract and formalize structure from concrete particulars. In this view, mathematics is conceived as a move away from situated (read "particular") forms of experience. When everyday math practice involves quantitative relations as a seamless part of its situated unfolding, investigations based on normative models of formal math and scientific thinking interpret this as evidence of the inferior character of everyday mathematical practice, without reference to the intentions of actors; the activity they are engaged in; the located, situated meaning of what they are doing; or how they are doing it. Seamless (though highly effective) everyday math practice suggests to researchers pursuing an understanding of mathematical "thinking" that everyday practice can and should be colonized in the name of formal mathematics.

The Situated Dualism of Mathematical Practice

There are other reasons why social science investigations of everyday activity "confirm" but do not analyze the meaning of inferior otherhood. If the "inferior other" is a myth, it is a myth so deeply incorporated into cultural practices that all of us know how to become its exemplars. Not all the time: the same persons address relations of quantitative structure in very different ways with very different effects when engaged in activities as competent selves and when engaged in activities as inferior others. This is, of course, one way to sum up the findings of the Adult Math Project about the different performances of jpfs in different settings.

Mathematicians know how to display competent incompetence—they make "modest" jokes about being unable to add, subtract, or reconcile their checkbooks. Incompetence at arithmetic in everyday life by nonmathematicians is something else again. We discovered in the Adult Math Project that even denying that someone's math practice is being subjected to a "scientific" gaze causes it to dissolve into displays of schoolish practices less competent than what they would be otherwise. We wondered how jpfs learned to collude in their identification as incompetent others. Schools are certainly deeply implicated, as they legitimize certain kinds of knowledge and selectively eliminate certain kinds of students. I once carried out a small research project with one of my students in a third-grade math class in a bicultural, bilingual school in Santa Ana, California (Hass 1986; Lave 1989). The children liked math and their teacher. When doing "seat work" in a group around a table, they engaged in much sub-rosa mathematical activity together. They hid their use of reliable, familiar counting procedures for solving problems, knowing that the teacher disapproved, in order to create the appearance of competence at procedures for multiplication that the teacher had just introduced. We could see a painful distinction being made in practice, one the children were quite conscious of, between "real math" and "what I do because I'm no good at math." This distinction was generated and elaborated in the organization of school math instruction, not in the division of children's lives between school and the "everyday" of home.

On the other hand, the idiom of math and displays of utilitarian rationality are very commonly employed—but incompetently—to assert the competent self. There are vivid examples of this in the supermarket research. One example concerned an experienced grocery shopper who picked out a package of noodles of a particular size, a choice she had made many times in the past for multiple substantive reasons given the way she cooked. She noticed while shopping with the anthropologist that her choice was not the "best buy." Clearly she felt herself called into question as a competent shopper. She tried to redeem her sense of competence by criticizing her usual choice and promising to choose the more economical package "next time" (even though it would not suit her purposes if she did) (Lave 1988:160–164). This suggests that for jpfs as well as for scientists, certain kinds of "competence" are equated with the "superior" thinking of utilitarian rationality. It further suggests that this unquestioned belief must often lead to incompetent and inappropriate displays of "rationality" by those who do not feel entitled to the real thing.

Conclusions: The Place Value of Other and Self

Research on science as everyday practice has led us to questions about cultural-political relations among sciences of mind, domestic life, and schooling. I have tried to show (through examples involving domestic relations of quantity, on the one hand, and the cultural practices of research about the mind, on the other), that much hinges on assumptions about the exceptional nature of scientific activity and the unexceptional nature of the daily activity of jpfs.

I have tried to show that there are other ways to characterize relations between the quantitative practices of jpfs and mathematicians. For instance, we observed, in the Adult Math Project, jpfs assembling and transforming relations of quantity in the ongoing activities of domestic life in efficacious ways that served remarkably well the purposes for which they were invented. They were not merely pathetic imitations of scholastic mathematics. Scholastic mathematics, dropped into these activities, would have destroyed their purpose and possibility. It seems reasonable to conclude that scholastic mathematics is but one practice among others—but one blessed with an ideological power not given to complementary practices.

This chapter has also called into question dual categories of "good" and "bad" thinking, scientists and jpfs, along with laboratory life and everyday activities, by exploring the manner in which these categories are constituted. Rather than a static bracketing, which simply separates two things, I have tried to show that dual categories are generated dialectically. That is, they involve several simultaneous kinds of relations at the same time. Polar social categories such as "primitive" and "civilized" minds are mutually dependent constructs, elaborated together in "scientific" discourse. Each is a part of the other (pace Levy-Bruhl's law of negation), and is thus embedded in the identities and practices of jpfs and scientists alike. The cultural resources to perform both competent and incompetent calculative rationality inhabit all concerned, though not with homogeneous political effects.

Polar social categorization is a political relation since high and low categories of thinking and thinkers are not equal or symmetrical (see Stallybrass and White 1986). Science and scientists dominate the definition of superior and inferior thinking, and scientists have a stake in sustaining the view that dual divisions (only) separate people into the two categories. Jpfs do not hold the high ground in this respect and participate, in as unreflecting ways as the scientists, in the dominant characterizations of the "inferior other." Further, "sciences" of the rational suppress the recognition of the political character and consequences of their assumptions and activities.

This inquiry into the production of inferior others, and with it the scientific mind, calls contemporary everyday practices into question in several different respects: "primitive" unconcern for the law of negation seems to be a prerequisite for casting polar categories in dialectical terms; mathematicians are just plain folks; just plain folks are virtuosi at sizing up quantitative relations; and sociologists and anthropologists of science, speaking reasonably, offer arguments for disturbing the tidy divisions of high versus low rational orders. We are all more successful at playing our parts in the cultural politics of the world we inhabit than we are in understanding them.

Acknowledgments

The first draft of this paper grew out of a series of conversations with Steve Shapin. Laura Nader deserves thanks not only for bringing together examples of the anthropology of science in this volume, but for organizing an anthropological presence at the meetings of the American Association for the Advancement of Science in the first place. Thanks especially to Paul Duguid for a thoughtful critique at the right time and to Shawn Parkhurst for his perspicacious suggestions, which led me to revise the chapter yet again.

CHAPTER 5

Scientific Literacy, What It Is, Why It's Important, and Why Scientists Think We Don't Have It

The Case of Immunology and the Immune System

Bjorn Claeson, Emily Martin, Wendy Richardson, Monica Schoch-Spana, and Karen-Sue Taussig

"Science matters," we have been told in a recent spate of publications. In *Science Matters: Achieving Scientific Literacy,* "science literacy" is defined as "the knowledge you need to understand public issues ... to put new [scientific] advances into a context that will allow you to take part in the national debate about them" (1991a:xii). Unproblematic as this definition might seem at first glance, the authors contend that by any measure, "Americans as a whole simply have not been exposed to science sufficiently or in a way that communicates the knowledge they need to have to cope with the life they will have to lead in the twenty-first century" (xv). This dire (and, we would like to argue, unfair) diagnosis becomes more understandable when one confronts the extremely narrow and technocratic content of the knowledge contained in *Science Matters* and most other books about science literacy. For example, in media publications spun off from *Science Matters,* readers are given a "pop quiz" that tests scientific literacy, a quiz that the great majority of Americans at all educational levels would fail (Hazen and Trefil 1991b). One question from the quiz is

The blueprint for every form of life is contained in

a. The National Institutes of Health near Washington, D.C.

b. DNA molecules

c. Proteins and carbohydrates

d. Viruses

For Hazen and Trefil, the correct answer, measuring a person's ability to understand public issues and function responsibly as citizens, is *b.* But in a broader, and more socially and culturally informed, definition of science literacy, one might wish people to debate the postulated role of DNA as the blueprint for every form of life. Following the lead of Ruth Hubbard and Eliza Wald in *Exploding the Gene Myth*, for example, one might hope that people would appreciate how little about every life form is actually determined by the gene. Following the lead of Bruno Latour, one might want people to understand how much the establishment of science "facts" owe to the funds and credibility given particular researchers by institutions like the one mentioned in answer *a*, the National Institutes of Health near Washington, DC (Latour 1987).

Even though some publications on science literacy contain reference to social and cultural issues, these issues are usually posed as afterthoughts, or at least as reflections to ponder after the real science is mastered. For example, in *Benchmarks for Science Literacy*, the summary section on "health technology" for high-school students lists six paragraphs of facts about the genetics, immunology, and epidemiology of health. Not until the seventh paragraph do we read that "biotechnology has contributed to health improvement in many ways, but its cost and application have led to a variety of controversial social and ethical issues" (American Association for the Advancement of Science 1993:207).

In our social anthropological research we are uncovering another picture of what science literacy might consist of by asking nonscientists, at all educational levels and from a variety of ethnic and socioeconomic settings, to tell us in their own terms what they know about health and their bodies, in particular about immunity and the immune system.[1] In this paper we will introduce some examples of this knowledge and argue that the definition of scientific "literacy" needs to be broadened to include the existential, metaphysical, moral, political, and social knowledge that is already embedded in people's talk about the immune system and the actions they take to protect it. Below we introduce four stories we were told about the immune system by people in four quite different contexts. We will argue that the people in these stories are engaged in producing what Clifford Geertz has called "local knowledge"; that is, "the artisan task of seeing broad principles in parochial facts" (1983:168) and "stories about events cast in imagery about principles" (215).

Story 1. Professor Keller, a scientist and a professor of microbiology at a large northeastern university, is teaching an undergraduate class on the biology of cancer and AIDS. He seeks to alleviate his students' sense of helplessness in the face of both social expectations and health-endangering illnesses by encouraging students to look within themselves to guide their lives and to find the resources to maintain their health.[2] The class, which he teaches every semester to about five hundred people, is known by students as "the best class on campus."

Professor Keller brings to the class a critical awareness of the limits of Western medical science. In particular, he stresses its lack of understanding of the powers inherent in the human body. He raises questions about the causes of cancer and other illnesses and suggests the possibility that Western medicine cannot answer these questions. He tells stories about miraculous recoveries from illnesses and invites to the class ordinary men and women who have performed feats of self-healing deemed impossible by the medical establishment, thus suggesting everything we do not know about healing. And he criticizes the medical establishment for being "completely interested in keeping all alternatives off the books."

The purpose of the class, then, is to teach students "other ways of thinking" about the body, self, health, illness, and death and to "empower" students by giving them a sense of their internal capacity for control over their lives, a capacity that is ignored or denied by medical science. To this end he uses "biology as ... a common language"; he "talks the language that [the students] are ready to hear" and "interweaves it with stuff that they are not ready to hear, but that they will accept because it is interwoven."

The description of the immune system is part of this common language. As students scribble furiously in their notebooks, Professor Keller tells them about the amazing world within their own bodies: the "dumb" macrophages who "eat foreign objects, stick them out of their own body," and "present" them to "smarter" cells; the "advanced" T-helper cells, "the quarterbacks of the immune system," who "send signals from one part of the immune system to another"; the B cells "who can recognize any foreign shape"; and the suppressor T cells "who are in charge of keeping the immune system within bounds."

Professor Keller stretches students' imaginations by using this depiction of the world of the immune system as a "metaphor for the real power we have" inside us. He tells students that they are powerful, "so much more powerful than [they] even imagine." According to Professor Keller:

> If you believe you've got this, then ... you start believing you're powerful....
> I tell this group, "you have this stuff. Your B lymphocytes are incredible....
> And it's us, and we're really strong." I think they're saying, "Oh." And you almost stand up a little taller and you walk around and say "I'm powerful."
> ... It's almost like the scientific version of thinking about an angel or a protector or something that they probably did in the middle ages.

The amazing world of the immune system is not separated from us; rather, "it is us." To know about it gives us a sense of a powerful self.

Mike took Professor Keller's class two years ago. Now, in his final year of college, he ponders the possibilities that lie ahead. According to Mike:

> I think a lot of people, at that age [their twenties], they want to do something with social justice and they really want to get involved. I think it bothers a lot of people, just with the environment, education, and health care, everything. And they don't know if they should go into being a social worker.... [There's] kind of a pull between that end of the spectrum and the other end of living a realistic, not a realistic life, but I mean you have to make a lot of sacrifices ... if you go that route.

Professor Keller's class "gave you the feeling," he says, that "you're in control" of your life:

> Even before this class, I promised never to do something I really didn't want to do, and if I didn't feel good about it, if I didn't want to do it, I'd go on to something else. And so in taking this class I think it just reinforced that, to really go with a gut feeling, go with what you're happy with, go with what you feel good about.

The class empowered Mike to "go into" and believe in himself.

But Mike believes Professor Keller's presentation of the immune system had little impact on him. Mike could see no connection between himself and an immune system that seemed to be existing independently inside his body. As he says:

> [They] show how the immune system fights off disease. I think they make it sound as if it's only the immune system. The immune system and the disease . . . this happens here, and then this, this, then this happens, then this step, and this type of cell invades here, the virus attacks the white cells. . . . I couldn't relate to that, I felt that that couldn't help me, so I didn't take anything from it. I didn't care about it.

The microprocesses of the immune system may be "amazing [and] overwhelming," but, to him, they are distant, and not empowering.

In contrast to Mike, Elizabeth came to Professor Keller's class with a conception of the world as a series of separate, but interconnected layers. According to her:

> Everything's all connected. . . . Inside me is like a whole other universe. . . . I have like micro cells within me, and then I'll go up to the bones and the organs, put together, and then just after that just my automatic systems, just like my breathing and stuff like that, and then comes my self. . . . And like this building that I'm living in is like a monster and I'm part of . . . that person, and then this building is part of this town, which is like another person. . . .

Her immune system may be a "separate community in there," like her other organs, but she is nevertheless able to relate to it and interact with it.

This sense of connection with her immune system creates the possibility for her empowerment. For Elizabeth, the immune system is more than a mechanism of use only in emergencies to "fight diseases." She interacts with it on a "day-to-day" basis. During their twenty years together, she says, she and her immune system have gotten used to each other:

> It's gotten used to my way of living. My lack of sleep, and my bad eating habits, it's sort of grown to accept it, because I'm not sick the entire school year, but I hardly sleep the entire school year, and I don't eat right. If I were

to get a new immune system, then I would be sick the entire school year until I had it trained that this is me and this is how I function, so help me out a little bit.

Rather than changing her life to meet the needs of her immune system, she has been able to "train" it to accommodate her life. In the context of her conception of the interconnectedness of the world, and her day-to-day interaction with her immune system, scientific knowledge of the immune system becomes empowering. For her, the immune system is, indeed, the "scientific version of . . . an angel or a protector."

Story 2. Two of our informants, George and Phillip, are a gay white male couple in their early twenties. Coming of age sexually in the first decade of the AIDS epidemic, these two men have known gay circles as places in which conversations about the immune system are commonplace. Interest in a possible lover readily becomes an interest in whether or not he has been tested for HIV and whether or not he will wear a condom. Concern over a friend or acquaintance with AIDS easily becomes worry over his falling T-cell count. While Keller's students may gain a sense of empowerment from his lectures on the immune system, Phillip and George, who have witnessed the loss of a generation of gay men to AIDS and who have heeded the safe-sex campaign among gays, speak more of vulnerability. In their descriptions of the immune system and AIDS, they draw heavily upon an idiom of boundaries. Their talk of boundaries, safety, and risk, used in depicting the body and the threat of HIV, is consonant with their descriptions of danger on a different scale, danger that threatens the neighborhood through crime.

Phillip laments that AIDS and a preoccupation with protecting oneself, that is, wearing a condom, is at odds with love, which is an act of letting down one's barriers. He explains:

> a lot of time, and this happens with straight and gay people . . . the issue of love enters the scenario and sometimes . . . when you love someone [you] don't feel that . . . you should have to protect yourself, and I think that not using protection is a result of that, because [you] feel that you love this person, that this person loves [you] and because you love each other, you shouldn't have to hide or protect anything about yourself from this person.

The biology of sex and of HIV mandates a barrier, the condom, between two bodies, while the sociality of lovemaking mandates openness between two people.

In his description of how a fetus might contract HIV from a infected mother, Phillip focuses again upon boundaries, this time describing the function of skin as a barrier to disease. "While the baby's inside the mother, that baby is part of the mother. . . . All her body fluids are coming into contact with it . . . even if its blood type is different, the blood is still coming into contact with this baby. . . . It does not have its fully developed skin on it to protect it. " Because of a breach in its skin through which the mother's infected blood can pass, the fetus is vulnerable to HIV. Compromised boundaries, openings in the skin, constitute vulnerability to HIV.

His partner, George, also emphasizes that caution toward HIV entails not only vigilance against possible points of entry, but also care toward the fluid medium containing HIV. He draws upon an example from his own life, recounting the time he threw away an exacto knife at his office after a coworker cut himself with it. He defends his act to Phillip, who thinks him somewhat paranoid, by arguing that HIV is highly contagious because of its presence in fluids and because of fluid's special quality of permeating boundaries. It is the "bodily fluids" of others, from outside, he suggests, that one must avoid getting inside oneself. "I mean, you know that this disease is transmitted through fluid, o.k., simple.... If I've got an open wound on me, I'm not going to roll around in bed with someone, and like, have all this semen or fluid or whatever have a chance to enter my body, you know." Protesting that he was not paranoid in throwing out the knife, he argues further that he would act within reason if he encountered a person with HIV:

> If I know someone has AIDS ... I will take precautions. I'm not going to be ... total[ly], well, "you stay ten feet away from me, and if you sneeze, cover your mouth, don't get any fluid on me."... I'm not going to be like that, because that would be kind of ridiculous, but on the same chance, I'm not going to welcome the opportunity by ... getting horny and sleeping with the person, [or] ... shar[ing] a needle."

In order to articulate the value of AZT (an antiviral medicine) for someone who has tested positive for HIV, George draws upon a boundary metaphor, describing the medication as at least some protection, however compromised, against the progression of disease. He explains that while this drug may be of only limited help to the person, merely delaying the onset of symptoms of AIDS, that help should not be ignored or criticized:

> AZT is good.... You can't fuck with it because it's the only thing we got.... You know, it's like you're out in the rain, and you see this awning, and you know, its got a few leaks in it, but you'll be a hell of a lot dryer from standing under those few leaks than you will if you stand out in this pouring down rain. There's nothing else there, so you got to go for what's there.

In thinking through issues of health and illness, Phillip and George employ notions of barriers and their permeation by bodily fluids. Talk of vulnerability shifts easily, however, from the level of the body to that of the neighborhood. Just as skin is seen to circumscribe the body, protecting a vulnerable interior from exterior threats, so are streets seen to delineate the margins of a neighborhood, marking good areas from bad. When describing their area of the central city, George and Phillip give street names to outline the "safe" portion of the neighborhood. George notes that not only have gentrification efforts on streets south and west of the lines of safety failed but also that "bad" areas are encroaching upon "good" ones. "They tried to build that up, and a lot of yuppies moved in, but for some reason, they just didn't make it.... That's really all that's there.... I think bad areas are kind of moving in, all the good areas are kind of moving up toward this way."

Phillip invokes an image of a safe, enclosed area, outside of which one faces possible dangers:

> I mean to me, it's almost as though we're a little box.... If you go over past Lake Avenue, you start to get into, you know, an area where you might put yourself at risk, you get down towards Packard Street going north, you're starting to put yourself at risk.... If you go toward Central Street, that same thing applies, and it just pretty much ... it's a nice little square, your safe, little, cozy square right here. If you leave it, you're increasing your chances of crime, not to overly worry about it often, but I walk around all the time, I mean I'm always aware, I always keep my eyes and my ears open.

Two related themes are apparent. First, boundaries mark off or contain an area of safety. In one's neighborhood, one knows what streets to avoid and at what streets "bad" areas appear to begin. Second, if one knowingly crosses such borders, one must assume the risk of harm. Cross the street and you may be asking for trouble.

The same could be said of "risky" sexual practices and contracting HIV. Indeed, Phillip warns:

> If you're going to have sex, you can play it safe if you want, but just ... be aware of what you're playing with ... and condoms aren't 100 percent certain.... If it's right for you, go ahead and do it, but just be willing to deal with the consequences.... You know, if you're going to go around and act like a slut, and then catch it, don't scream "it's not fair, it's not fair, it's not fair," you know, it's not fair for anyone, but you were well aware of what you were doing at the time.

Others share the concerns of Phillip and George about crime in the neighborhood. They also hold similar cognitive maps of dangerous and safe zones in the area. Residents and business owners, for instance, have organized a citizens' patrol of the neighborhood streets at night to supplement the surveillance provided by the police department. A fatal shotgun blast to the face of a young gay man who was leaving an after-hours nightclub in late spring of 1990 was the event instigating the patrols. In this murder, people saw a number of the dangers that the neighborhood unfortunately hosts. Some gays and lesbians considered it another example of the increasing violence directed toward the homosexual community during a time marked by a fear of AIDS. A few considered it the plight of a neighborhood not cohesive enough to exclude such "outsiders" as criminals.

Crucial to the initial stages of the organization of this citizens' group were debates over what area of the neighborhood the patrol would cover. The limited number of participants to schedule for patrols constrained the size of the beat, as did the decision not to place patrols on the "questionable" streets, those being the most peripheral. Patrols were not to walk east of Lake Avenue or west of Central Street. To go beyond these limits was to put oneself in considerable danger, particularly if on foot. Within the perimeter, however, patrols would monitor the street for crime, calling in the police to investigate and remove, if necessary, suspicious characters from the neighborhood.

Concern over bad areas spilling into good ones and criminal elements circulating within the neighborhood prompted residents to be keenly watchful of activities in their area. Phillip and George are similarly vigilant against the threats that may lie outside their "little, cozy square." Apprehension about the breach of boundaries, however, also marks their appraisal of the threat that HIV poses to the body. Mindful of their safety on the streets and cautious about any contact with "bodily" fluids that may contain HIV, Phillip and George elaborate the social significance of boundaries.

Story 3. Another informant, Bill, also made references to boundaries in his discussion of the immune system, relating "borders" of the body to those of nation-states. Bill is a white man in his thirties. He lives with his wife, originally from Argentina, in a small row house. Theirs was one of the first racially integrated neighborhoods in Baltimore. The residents are both renters and owners and many live openly as gay couples. When asked why he and his wife chose this area of Baltimore, Bill explained,

> I think affordability is part of it, although I think probably mostly sort of progressive community type stuff.... Affordability, community, yeah not only views of people but also the fact that people did stuff here, they were active. We wanted to live somewhere where there was a community association we could get involved in stuff if we wanted to, that sort of thing.

Bill's concern with individual rights and autonomy were evident throughout our interview. Significantly, his interest in decentralization was reflected in the way he described his work and also became apparent when he began to talk about the immune system. Bill has a degree in architecture but found that office-oriented work did not appeal to him. In graduate school he took a course on creating low-income housing in developing countries. He found this approach more satisfying. This is how Bill described his interest. "In a developing country context, housing is, is all user. I mean the best [that] the public realm, the government and stuff, can do is maybe push a few laws aside or give somebody a scrap of land, but other than that it's whatever, whatever the user or the squatter or whatever can do for themselves." His research took him to Sri Lanka and Chile.

Working on a housing project for the city of Boston, Bill felt involving the community in decisions was, as he put it, a "more progressive way to deal with this notion of developing houses." He now works as a development consultant for nonprofit organizations, including the family center in his neighborhood. Bill's fascination with questions of local autonomy and centralized power were also plain when we asked what he knew about the immune system. At first he was reserved and responded, "I don't think I think much about the immune system per se. ... I mean the immune system is all these little white blood cells running around eating up all the bad shit in your body right? That's my understanding of the immune system technically speaking."

As Bill oriented his thoughts around a metaphor of organized systems of authority, however, he became excited and the immune system took on a larger significance. He went on to say,

> I mean there's a natural policing thing that happens in the body right? And, I mean it's an incredible policing thing, when you think about it, or at least it's incredible in my mind, because it's a system-wide authority that works. I don't know of any system-wide authorities that work in our culture, so this is amazing right?

Having compared the immune system to a social system, Bill has given it a culture that can be compared to our own. If the metaphor of the police state makes the immune system comprehensible to Bill, at the same time it also becomes a commentary on political organization. It allows him to relate different states of the body to different political systems. As he exclaimed, "I'm sure there are diseases [in which] that the immune system ends up destroying the body. ... It's more like an Argentina or a Bulgaria kind of immune system."

Through his use of a body-country metaphor Bill reveals his conceptions of both bodies and nations as bounded and independent. When asked how he would describe to a child how the body repairs, or takes care of itself, he answered:

> It would [be] an explanation that would involve having to explain that the body is a self-contained system, that it has its own discipline. ... I suppose one would want to use a metaphor, you know. ... Obviously using Argentina or Bulgaria would be an ineffective metaphor. You would have to think of something in the child's world that was a comfortable notion of a closed, self-nurturing system.

As a closed system, the body must protect its boundaries. The body when sick, like a nation at war, is threatened by what Bill describes as, "a foreign organism ... that is competing for the body's resources in some way or another." Having described the body as a social system, Bill also encountered the moral dilemmas of the social world. What about the individual's rights? What about cultural relativism? In reference to a *Time* magazine cartoon drawing of a virus and a white blood cell fighting, Bill said:

> I was thinking about this notion of good guy versus bad guy. It's really pretty silly because, I mean the virus is just trying to make a living right? It's trying

to bring up its family just like the white blood cells are, so there's not really ... a clear good and bad in it, unless you take somebody's point of view of the larger ... organism, then you have a point of view.... It depends on where you position yourself, where you sit.

By using a social metaphor to describe the body, it becomes a system with logic, an aggregate of players with motivations and intentions. Earlier in the interview Bill explained, "It's easy to talk about policing systems and white blood cells being the good guys and all this kind of good stuff, and then making metaphors around countries, these political things are easy things to talk about and understand for me."

The nation-body metaphors that Bill uses not only enable him to understand the immune system. As Bill suggested earlier, his descriptions of the body also become a commentary on the social world. The words he uses to describe the world of the body also orient his understanding of social interaction and global politics. The role of the politician can move from the global politics of the cold war to the global politics described in the immune system. Looking at a micrograph of bacteria being showered with enzymes, Bill exclaims:

Oh that's neat! This is great. I mean this is like star wars and I don't mean Reagan's version. This is like a battle, an action shot right? This thing blowing the other one away. That's great. See, if the people that are so good at making those kind of stupid political arguments, like Reagan, so good at actually convincing people about things like star wars? If he would concentrate that level of creative fabrication [on these] kind of issues, I think this would be a better world to live in.

Later in the interview, discussing the effectiveness of prevention in the control of AIDS, Bill returned to the issue of star wars. We asked, "How effective do you think prevention would be?" Bill responded,

From what I understand of the disease, again which isn't much, but the way in which it is spread, it sounds like it could probably be very effective.... People don't talk much about prevention, I don't even know what something like that might look like. It's interesting though actually.... Earlier I was ... [speaking about] Reagan with his star wars thing. That was a preventative piece and people really sought that out right? 'Cause it was the notion of things going up and intercepting ... right? And that actually, I mean that's, those are like, that's like a giant condom network, right, in a sense. So that's actually interesting. So maybe there is some hope, maybe there is a political angle on prevention that is every bit as sellable as curing. I don't know.

In order to communicate his understanding of the immune system and orient his thoughts, Bill has chosen the image of the nation-state. To Bill the topic of the immune system is versatile and allows him to move the discussion from health and illness to issues of global politics and the rights of the individual. Discussion of the immune system becomes a forum to think not only about our bodies and scientific "facts," but about political structures and what positions we must take to understand the rights of others. But these are not just arbitrary subjects held together by a central metaphor, they are also interconnected queries of great relevance to Bill. By speaking about the immune system, Bill was able to convey his views on the interrelations of preventative health, national priorities, and individual rights.

Story 4. Unlike Bill, for whom AIDS is one of many questions about individual rights and national structures, for Mara, AIDS and the problems of immunity are of primary concern. Mara is a thirty-one-year-old white woman who works as a technical writer for an engineering firm, volunteers by writing grant applications for a local performing arts group, and lives with her husband in a quiet Baltimore neighborhood. All four of Mara's college roommates have died from AIDS. In 1989 Mara became the primary care giver for another close friend who was sick with AIDS. She was with him when he died in a hospice the day after her thirtieth birthday. He was over six feet tall and weighed seventy pounds when he died. Mara is an AIDS activist. In describing her experience with AIDS she said that "a lot of gay people are very angry, a lot of straight people too. And especially those of us involved in the arts—to watch our world crumble, and nobody even look at us . . . nobody even give a shit . . . and we're fighting for little bones that they throw us." She told us that AIDS, the body, and health are things that she thinks about every single day of her life.

Mara is, by anybody's standards, scientifically literate. During our interview we showed her an unlabeled micrograph of a white blood cell surrounding an asbestos fiber and asked her what she thinks of when she sees such pictures. Her instant and stunning response was, "Oh, that's an asbestos fiber in a white blood cell."

Mara's familiarity with the language of science seems to come from her experience with AIDS. She is extremely well informed about the disease, how it is transmitted, and different possibilities for treatment. She is able to clearly articulate her understanding of what she thinks happens inside the body of someone with AIDS. She told us that what she imagines happens inside the body of someone with AIDS has "changed" with the course of the disease. She said, "I used to feel like it was a total loss of control, and like it was Sisyphus, that you would push it up the hill just two feet and then it will roll back down over you. I used to feel like it was only chaos. Now I feel like it's still chaos but that we have a handle on it. . . . I know that it is war, absolutely, it's war inside the body." She goes on to tell us:

> I've never seen anything like this . . . seen twenty-six-year-old men not able to make it up the stairs. I have never seen diarrhea so bad that, that, you put a mouth full of water in your mouth and swallow it and you, you are having explosive diarrhea. . . . This is the weirdest and the most bizarre thing I can . . . imagine. I mean, every nightmare that I ever would have had of what could happen to the human body has come true, and we're right in it every day.

Mara's concern is about the effects of the disease on society. She believes that AIDS is something society may not be able to withstand. She told us that she feels "like we all have AIDS . . . whether we have the virus or not." When asked to elaborate on her suggestion that everybody has AIDS, Mara said, "I really believe that we all have AIDS. I mean, I don't believe that I carry the virus in my body, and I've proven that to myself by getting tested, but my life has been changed forever by AIDS, and I feel that by the time the crisis is over, if it's ever over, and that I'm not sure about, everyone will be touched by it, directly affected by it."

Mara's answer to a question regarding her view of what the course of the disease will be in the United States and the world also reflects her concern for society. She told us:

> I think we're really in for it. . . . I don't have any hope at all. I used to, but I don't. I think that it's, it's here . . . widespread throughout the general population. . . . We're already seeing it moving heavily into the IV-drug-using populations of color in our urban areas. I think that the next really big hit is going to be teenagers and people in their early twenties. Where you first saw it was in a population that, in general, had access to medical care, had access to knowledge and information about their bodies, and in general works, so the course of their illness is a little different. . . . I think that in the inner cities we're seeing a holocaust, and I don't . . . have any idea how, how it can be slowed down or stopped at this point, unless the government took on a real heavy campaign.

While Mara sees the effects of AIDS moving easily through society, she doesn't think that the virus is very contagious. She describes the disease as "just a virus." She said:

> That's one of the things I say . . . like when people give value judgments about people with AIDS, I mean it's just a virus, the virus . . . happens to be transmitted by people having sex with each other, it's blood borne, so it's . . . specific in its transmission route. But that's all it is . . . just a virus. . . . I think it's very difficult to get it. . . . I mean I think it's impossible to get it unless you engage in unprotected sex, or share needles. . . . I worked with bare hands on bodies of people that were sick. . . . I've been splashed with vomit, I've been splashed with feces, I've been splashed with . . . saliva, tears, sweat, the whole bit, and if anybody was going to get AIDS from having contact, I would have it.

Mara sees her role as an activist as "something that is every day, all day. . . . I feel like I do it every day. But you know, that's what an activist does." She describes her role, telling us:

> I don't talk in a prejudiced way about people with AIDS. Some of it is hands-on, direct care where I'm needed. . . . I also . . . have kind of a special challenge,

being a heterosexual female, voices like mine aren't often heard.... I can go to a Baptist church and say I'm a married, heterosexual female, I'm thirty-one years old, and here's what I've been through. And they will let me in.

She also sees her role on what she calls a "microlevel" and describes her boss who "told AIDS jokes and made comments" when she first started working with him and now she describes him as "real sensitized to [AIDS], and real supportive and ... share[s] with me ... issues of his own, having to do with a brother who died of cancer at twenty." In spite of the horror she finds in the reality of AIDS, she also expresses hope about possibilities for new kinds of collective awareness.

Mara discusses her experiences with death by talking about caring for her friend and being with him when he died. Significantly, although she sees the effects of AIDS as "every nightmare that ... could happen to the human body" and as a "holocaust," she describes her friend's death as "really an amazing and positive experience." She describes his death, telling us that:

He could sort of talk. And then he died, but when he died it was really wild because our eyes were really glued to each other. It was like his eyes died. That was it. His heart kept beating for a while, but it was great. It was like catching the baby.... I felt different the next day.... I have felt different ever since.... He was such a strong, intelligent person.... He died in a really, in sort of a strong way, strange to say but really true. And so there are times, when that strength and that endurance and that ability, it doesn't seem like it's all gone, and some of it, I felt like I got a big hit of that power from him.

For Mara, death is not only about dying. Her concept of death is imbued with a concept of transcendence and involves ideas about strength and birth.

Logically following from her ideas about death, Mara does not see the body as a boundary. Mara's discussion of her relationships with her friends who have died illustrates her ideas about metaphysical relationships. One of Mara's concerns is about keeping the memory of her friends alive. She feels their presence in her life and makes a point of talking about them. In her discussion of her friend's death she told us that:

You know ... I try to keep him, he's the person that I knew the best that died of AIDS, but ... there are other people too that I ... try to remember as they go through all ... this process I always do talk to them ... those people won't be around to be uncles to my kids.... There's a huge crowd of people with whom I hung around in college, and that was real important ... the whole intellectual development.... Everything else was leading up to that time. Those people were very important to me. But they're not there, and I can't pick up the phone and call "Scott" and tell him about the book I just read and I can't pick up the phone and call "Gary" and, you know, ask him a question about my hairstyle. I mean, you know, I can't, I can't do that, and so I feel like we have a responsibility—those of us who have been with these people, in their lives but also through their illness—not only to carry on the stories about them and . . . facts about their [lives]. They don't have children, or people sitting around the fire talking about them. But also the way they died was so unnecessary, and sometimes I really feel like their spirits are really noisy in my ear.

These stories illustrate that science literacy is much more than merely knowing some basic "facts" and simple concepts (Hazen and Trefil 1991a:xix). Individuals use "facts" in very different ways and often make them work with their particular local circumstances as well as express their most overarching views of the world. Bill uses "facts" he has learned about the immune system to construct broader visions about the nature of social and political forms that relate to his political views and actions. Mara weaves "facts" she has learned about the immune system into profoundly moral and metaphysical views of the meaning of life and death. These views enable her to maintain both important relationships in her life and hope of a better society.

Both Bill and Mara relate disease to a state of war, a metaphor that is not uncommon in scientific texts, the popular media, or conversations with our informants. To Mara the horror of war expresses her shock over the effects of AIDS. For Bill, a metaphorical national system of authority in a defended nation-state provides a way to understand immunity as a system.

Using metaphor to conceptualize the body may affect our conceptualization of social situations. Some theorists emphasize the interactive nature of the elements paired in a metaphor, so that when Dante says "Hell is a lake of ice," the hearer extends the association of "hell" to a "lake of ice," thus transforming both elements of the metaphor (Hesse 1961; Black 1962:37). Through the use of a body/war metaphor we may not only be thinking of the body as naturally war-like, but we may also be thinking of the state of war as natural.

While Bill and Mara both make use of a metaphor that associates war with an immutable part of our biological nature (the immune system), at the same time they also suggest that AIDS is not only a biological but a social condition. Bill expressed his belief that a change in national priorities from the arms race to health would make prevention effective in the control of AIDS. To Mara the body/war metaphor not only speaks of a war within the body but of a social war being waged upon bodies. Within the contexts of their larger commentaries, Bill's and Mara's uses of the war/body metaphor signify much more than just ways to understand science.

In these stories people use scientific "facts" to create knowledge about a whole range of topics. For George and Phillip, to talk about the immune system is to talk above all about boundaries. Talk of boundaries appeared in another area of our research, participant-observation in a laboratory pursuing immunological research. In this research setting, the central tenet of contemporary immunology is the ability of the immune system to distinguish self from nonself. As our examples have just shown, when nonscientists talk about the immune system they also talk above all about boundaries. Those between the self and others—spatial, racial, gendered, class, and relational—and how these various boundaries can clash or be superimposed in complex ways. The concept of the boundary between self and nonself is a touchstone for broader social meanings. Since we find such concepts so commonly in interviews, it raises the question of whether the central role of boundaries in current research immunology is not culturally based in its inception.[3]

This position is explicit in Keller's view of his course. He intends much more than "facts" to be conveyed by his scientific account of the immune system. The students respond in kind. Certainly Mike did not take away "facts" about the immune system from his biology course, but rather the idea that in life one should "go into" and believe in one's self.

One might ask if implicit and local knowledge might be integrated into all areas of science, including those covered in overviews such as *Science Matters*. Suppose that the "facts" of science entail a whole worldview that is often left implicit. If it were made explicit, the way would be opened to begin a dialogue with people who might be living with different worldviews. Certainly such a dialogue would fit with Mara's vision of a better society. Quizzes from the experts which we, the public, fail, could give way to conversations about matters in which we are all experts in our own way.

Recognizing that worldviews are inscribed in scientific images would allow scientists to examine the implicit cultural assumptions that are used to explain scientific "facts." What else is being communicated when scientists describe the immune system as a national defense force or a T-cell as a quarterback?

The "facts" of science, important as they are, can never be more than tiny pieces of the maps that people devise to guide them in life. Even if we could all magically be made to "know" the answers to the science pop quiz, the process of our coming to know those "facts" would entail our embedding them in the diverse social, political, moral, and metaphysical meanings with which we construct our daily lives.

We began by asking how scientific literacy might be defined. We have shown that the people we interviewed are highly literate in the enormously complex social, political, moral, and metaphysical aspects of such matters as health or illness. If we were to turn now to a detailed look at Hazen and Trefil's notion of what it would take to be literate about these issues, we might be struck by the narrowly technocratic nature of the knowledge they regard as relevant. We might wonder whether that knowledge would really be sufficient to enable meaningful public discussion of AIDS, for example. This raises the question of whether it is actually scientists, rather than members of the public, that suffer from illiteracy on the range of considerations that need to be brought to bear on these complex human issues.

Notes:

This title is a slightly modified version of the title of the introductory chapter of Robert M. Hazen and James Trefil's *Science Matters: Achieving Scientific Literacy*, (1991a).

1. An extended account of the fieldwork is contained in Martin (1994). All quotations are from observed situations or interviews.
2. We have used pseudonyms for the names of the people we interviewed as well as street names.
3. See G. E. R. Lloyd's *Revolutions of Wisdom* (1987) for how concepts from daily life were taken up into early science.

PART 2

Culture, Power, and Context

CHAPTER 6

The Prism of Heritability and the Sociology of Knowledge

Troy Duster

In the last two decades, the field of molecular genetics has made great strides in decoding human genetic defects. Notable technological improvements have included the ability to designate and mark those defects prenatally, at earlier and earlier points in the pregnancy. Running alongside has been a parallel development—a "drift" toward a greater receptivity to genetic explanations for an increasing variety of human behaviors. These two developments are related only in that scientific advances in molecular genetics have nurtured a climate in which even the weakest "genetic" explanations can take root. The sociology of knowledge is a perspective that can help us better understand the social base and social location of this second tier of genetic claims.

In this paper, the "prism of heritability"[1] refers to what I will describe in greater detail below as the genetic lens through which we peer to try to explain human behavior. In 1990, I noted how in the previous two decades we had seen a substantial increase in the number of scholarly articles, media reports, and lay periodical literature that had used the genetic lens to explain human behaviors, traits, and attributes (Duster 1990:93–95). I reported that the *Reader's Guide to Periodical Literature* from 1976 to 1982 displayed a 231 percent increase in articles that attempted to explain the genetic basis for crime, mental illness, intelligence, and alcoholism during this brief six-year period. Even more remarkably, between 1983 and 1988, articles that attributed a genetic basis to crime more than quadrupled in their frequency of appearance from the previous decade. This development in the popular print media was based in part upon what was occurring in the scientific journals. During this period, hundreds of articles appeared in the scientific literature,[2] making claims about the genetic basis of several forms of social deviance (crime, alcoholism, and mental illness) and intelligence.

At first glance, it would seem that the appearance of both scholarly and popular articles on the most recent genetic claims was an outgrowth of what was happening in the field of molecular genetics. As noted, many important breakthroughs were occurring during this period, including the increasing ability for intrauterine detection of birth defects. However, the resurgence of genetic claims was not coming from those working at the vanguard laboratories in molecular biology or biochemistry. Indeed, given the requirements of up-to-the-minute monitoring to keep abreast of technological and scientific breakthroughs in these fields, it is of special interest to note that the major data source for the resurgent claims was a heavy reliance upon Scandinavian institutional registries dating back to the early part of the century (Kety et al. 1968, 1976; Kringlen 1968; Fischer 1971; Mednick et al. 1984, 1985; Jensen 1969;). The great majority of new claimants were not the researchers responsible for the new developments in molecular genetics.

Scientific Credentials and Recent "Genetic Claims"

The field of human genetics is a combination of very different strategies of research and analysis. The first is population genetics, which was concerned from the start with explaining evolution from a statistical determination of how much observed patterning could be attributed to heritability. The second, Mendelian genetics, is concerned with the search for patterns of observed traits that might be attributable to genetics if and when they conform to certain laws of inheritance. Since many things "run in a family," including wealth, crime, and occupational choice, and since many things also "run in large population aggregates," a plethora of scientific control problems confront either of these first two strategies of research, analysis, and genetic inference.

Of the many human attributes, characteristics, traits, or behaviors that "run in families" or "populations," why do some become prime candidates for genetic analysis and a program of research?

Those making the claims about the genetic component of an array of behaviors and conditions (crime, mental illness, alcoholism, gender relations, intelligence) come from a wide range of disciplines, tenuously united under a banner of an increased role for the explanatory power of genetics. In this paper, I present evidence to document how relatively few of these claims come from molecular genetics.

It would hardly be surprising if social scientists were the primary contributors to articles in social science journals with the renewed resurgence of calls to take a closer look at biological and genetic elements. A review of authors of publications in scientific journals in the field of human genetics would be a fairer assessment of the degree of disciplinary crossover, or interloping. Toward this end, a bibliographic citation list[3] was generated and analyzed for author information, and a search of five major sources was conducted to obtain information about each author's professional background, including academic training and credentials (see table 1).[4] A study of the scientific backgrounds of the authors of these articles would provide one answer to the question of who was making the genetic claims.[5]

It will be noted from the table that the majority of the authors come from fields outside of genetics. Only about one-fourth could be regarded as credentialed in human genetics or cytogenetics, or a genetic field of any kind.

Table 12

Summary of Authors' Credentials*

Area of Specialization	Number of Authors
Academic Degree (Ph.D.)	
Psychology	7
Genetics	6
Biology	5
Biochemistry	5
Chemistry	7
Zoology	10
Unspecified Ph.D. Degree	17
Total Ph.D.	57
Medicine (M.D. and D.O.)	
General	49
Psychiatry	18
Pediatrics	3
Total M.D.'s and D.O.'s	70
Total without doctorate	7
Total Credentials Found	134

*from 107 bibliographic citations in the areas of genetics; subareas deviance and mental health

From one perspective, it should matter less whether one is trained in a particular discipline and more that one is engaged in interesting and important research, using proper methods, controls, and so on, whatever the training. After all, new developments in molecular biology have forced a realignment of traditional disciplinary boundaries. For example, biochemistry (itself a merger of disciplines) is more related to vanguard research in genetics today than twenty years ago. However, even taking such factors into account, well over two-thirds of the authors' research training is far removed from the frontiers of molecular genetics, a very technical field.

It will be noted that the great majority of authors hold the medical degree. A National Academy of Sciences study revealed that as recently as the early 1970s, medical schools rarely offered a single course in genetics to medical students, much less research training (1975:161–179). While the situation has changed a bit in the last period, with more than half the medical schools now requiring one course in genetics, it can hardly be said that this constitutes preparation for research in one of the most technical fields in the natural sciences. Indeed, in the early 1980s, genetic counselors with only the master's degree were getting far more advanced training in human genetics than medical students studying for the M.D.

I am suggesting that the selection of genetic explanations for a range of troubling human behaviors constitutes an appropriation of the imprimatur of molecular genetics to the explanation of these behaviors, where the molecular genetic basis is not well developed, or more frequently, nonexistent. It now is appropriate to raise the question as to why this should be, and to pose a possible answer.

The Sociology of Knowledge

The sociology of knowledge is, in its broadest conception, an approach to the social conditions of thought. It is neither equipped to assess the validity of the truth claim nor to evaluate the pragmatic fit of a worldview. It is specially equipped, however, to explain the generation and promulgation of particular theories. As a field of inquiry, its roots are traceable to Karl Mannheim (1936) and Max Scheler (1926). While neither provided a systematic or single method of getting at the subject, it is possible to distill some key principles from their works. Foremost among them is the perspective that a) individuals are located in groups, b) groups have an interested position toward the existing order of things, and c) the organization of human thought (in individuals) can be often be better understood by discerning the relationship between *a* and *b*.

> People ... living in groups do not merely coexist physically as discrete individuals. They do not confront the objects of the world from the abstract levels of a contemplating mind as such, nor do they do so exclusively as solitary beings. On the contrary, they act with and against one another in diversely organized groups, and while doing so they think with and against one another. These persons, bound together into groups, strive in accordance with the character and position of the groups to which they belong to change the surrounding world of nature and society or attempt to maintain it in a given condition. It is the direction of this will to change or to maintain, of this collective activity, which produces the guiding thread for the emergence of their problems, their concepts, and their forms of thought. (Mannheim 1936:3)

To illustrate the bearing of this perspective upon a "genetic prism," an investigation into the social backgrounds of the respective adversaries of the genetics and I.Q. controversy revealed that those scientists who pursued research claiming that high I.Q. was genetic and was located in privileged social strata were more likely to have come from upper-middle-class backgrounds; in contrast, those pursuing the environmental argument were more likely to have come from lower class origins.[6]

While it is probably rare that social factors so crudely explain the behavior of scientists engaged in research that is socially controversial, this study does suggest that a systematic investigation of the interplay of social factors in the selection of problems for scientific analysis would add to our understanding of the nature and direction of select scientific work. More recently, Harwood (1993) has documented how German scientists from differing social backgrounds had remarkably similar attitudes toward genetic research during the first third of the century.

Framing the Issues and the Sociology of Knowledge

How differing interests shape the questions that in turn become the building blocks of knowledge structures lies at the core of a revived "sociology of knowledge" (Knorr-Cetina and Mulkay 1983:2).[7] In a culture where race and sex are firmly rooted categories of differentiation and sustained stratification, we should expect both common sense and probing inquiry into the intelligence differences between the races (Jensen 1969; Herrnstein and Murray 1994) and into the biological destiny of females (Corea 1985; Tavris 1992).

In late nineteenth-century Italy, celebrated scholars raised questions about the biological differences between law-abiding citizens and criminals, posited the natural superiority of one over the other, and invoked the known procedures of investigation to sustain the distinction and argument (Gault 1932). There is continuity between this approach and the biological study of crime even today, as there is between the study of I.Q. and race and the genetics of mental health issues.

I earlier raised the question, Why should it be the case that those outside the field of genetics are asserting genetic explanations to complex human behaviors, making claims that exceed those of geneticists themselves? A large part of the answer is the driving force that the social concerns of the age provide. The overwhelming concern is with "defects" or problems, such as alcoholism, poor performance in schools, mental illness, and so on. This parallels the early twentieth-century concern with similar matters (Reilly 1991; Kevles 1985). Thus, it is not the genetic evidence that drives the engine of scientific inquiry, but the social concerns that drive the "scientific" to an attempt to portray and explain these social concerns genetically. Again, note that these claims-makers are overwhelmingly nongeneticists, certainly nonmolecular geneticists. Attempts to explain varying rates of illness or disease by racial and ethnic group has produced some remarkably inventive gerrymandering at the borders of science.

Science, Race, and Genetic Explanations

Even with strong epidemiological evidence that heart disease and hypertension among African Americans is strongly associated with such social factors as poverty, there has been a persistent attempt to pursue the scientific study of hypertension through a link to the genetics of race. Dark pigmentation is indeed associated with hypertension in America. Michael Klag et al. (1991) reported the results of a carefully controlled study looking at the relationship between skin color and high blood pressure. He and his colleagues found that darker skin color is a good predictor of hypertension among blacks of low socioeconomic status, but not for blacks of any shade who are "well employed or better educated." The study further suggested that poor blacks with darker skin color experience greater hypertension "not for genetic reasons" but because darker skin color subjects them to greater discrimination, with consequently greater stress and psychological/medical consequences.[8]

The complexity of the interaction between genetic and environmental factors is a given, but it is how they interact and the relative weights assigned to each that is the source of contention in the nature/nurture debates. My concern is with the implicit prism through which one peers and that predetermines the direction of the causal arrow brought by researchers (primarily) outside the field of genetics.

Beyond single gene determinants (autosomal dominant, recessive, or X-linked disorders), human geneticists tend to be wary and cautious of professional assertions of "genetic explanations" of various diseases, illnesses, defects, and social problems precisely because of the broad and often vague designation of the phenomenon, and the not well-understood multifactorial character (Kevles and Hood 1992). Those outside the field who make such claims have always been vitally concerned with the stratification system and explanations of persistent "problems" that attend it (Ludmerer 1972; Haller 1963; Kevles 1985). Skin color does have a direct relationship to the stratifying practices of a social order.

The metaphor of the shifting prism that rotates to refract upon a central issue in the stratification of groups can better help us understand the angle at which the prism is tilted to determine which matters are to be investigated and researched. The metaphor works as well for the subject matter of the newer molecular genetics as for analyses of family trees and population statistics. An understanding of the prism is less available when discussion turns to the new technologies in molecular genetics, in part because of the complete hegemony of the health and medical and science banner, in part because the technical laboratory issues are out of reach and scrutiny of much of even an informed laity. However, it is the halo from the molecular work of the last decades that has helped provide legitimacy to the claimants who have continued to use I.Q. test scores to assert genetic explanations of racial differences.

The "Science" of "Race": Unraveling the Conundrum

A consortium of leading scientists across the disciplines from biology to physical anthropology issued a "Revised UNESCO Statement on Race" in 1995—a definitive declaration that summarizes eleven central issues and concludes that in terms of "scientific" discourse there is no such thing as a "race" that has any scientific utility, at least in the biological sciences (Katz 1995).

> the same scientific groups that developed the biological concept over the
> last century have now concluded that its use for characterizing human
> populations is so flawed that it is no longer a scientifically valid concept.
> In fact, the statement makes clear that the biological concept of race
> as applied to humans has no legitimate place in biological science.
> (Katz 1995:4,5)

But the title of the statement does not restrict its utility to only biological science. Rather, it asks "Is Race a Legitimate Concept for Science?" (Katz 1995) Nowhere is the intermingling of the concepts of science and ordinary social world categories and commonsense thinking so seemingly hopelessly intermingled as in the idea of race and the ongoing debates about the relationship between "science" and "race." This conflation of scientific and commonsense thinking has produced remarkable trafficking back and forth between scientists and the laity, confusing for both laypersons and scientists, about the salience of race as a stratifying practice (itself worthy of scientific investigation) versus a biological taxonomy (primarily unworthy of further pursuit as biology).

By the mid-1970s, it had become abundantly clear that there is more genetic variation within socially used categories of race than between these categories (Polednak 1989; Bittles and Roberts 1992; Chapman 1993; Shipman 1994). The consensus is newly formed. For example, in the early part of this century, scientists in several countries tried to link the major blood groups in the ABO system to racial and ethnic groups.[9] Since researchers knew that Blood Type B was more common in certain ethnic and racial groups—which they believed to be more inclined to criminality and mental illness (Gundel 1926; Schusterov 1927)—this was often a thinly disguised form of "racism." They kept running up against a brick wall, and could find no causal link.

It is not difficult to understand why they persisted. Humans are symbol-bearing creatures that give meaning to their experiences and to their worlds. The UNESCO statement is ultimately about the difference between first-order constructs in science and second-order constructs. Some fifty years ago, Felix Kaufmann (1944) made a crucial distinction that throws some light on the controversy. Kaufmann was not addressing whether or not there can be a science of race. Rather, he noted that different kinds of issues, methodologies, and theories are generated by first-order constructs in the physical and natural sciences versus second-order constructs. For the physical and natural sciences, the naming of objects for investigation and inquiry, for conceptualizing and finding empirical regularities, is in the hands of the scientists and their scientific peers. Thus, for example, the nomenclature for quarks or neurons, genes or chromosomes, nitrogen or sulfides, all reside with the scientist qua scientist in his/her role as the creator of each of these terms, which are thus first-order constructs.

This is quite different from the task of the observer, analyst, or scientist of human social behavior. This is because humans live in a pre-interpreted social world. They grow up, from infancy, in a world that has preassigned categories and names for those categories, categories provided by fellow commonsense actors, not by "scientists" (Schutz 1973). Persons live in the world that is pre-interpreted for them, and their continual task is try to navigate, negotiate, and make sense of that world. The task of the social scientist is therefore quite distinct from that of the natural scientist. While the latter can rely upon first-order constructs, the latter must still construct a set of categories based upon the pre-interpreted world of commonsense actors. The central problem is that "race" is now, and has been since 1735,[10] both a first- and second-order construct. The following joke, making the rounds among African American intellectuals and reported in Roediger (1994:1), makes the point with deft humor:

> "I have noted," the joke laments, "that my research demonstrating that race is merely a social and ideological construction helps little in getting taxis to pick me up late at night."

This throws a different light on the matter of whether race can be studied scientifically. If we ask, is there a consensus among the natural scientists about race as a first-order construct, then the answer since about 1970 is, categorically, no. The UNESCO statement summarizes why this is so, mentioning every level that is significant to the biological functioning of the organism, with two exceptions. We have already noted that scientific research on first-order constructs about race as a biological category in science in the last four decades has revealed over and over again that there is greater genetic heterogeneity within versus between major racial groupings (Polednak 1989; Bittles and Roberts 1992; Chapman 1993; Shipman 1994). One of the two exceptions is that the gene frequencies, as demonstrated in the use of specific polymorphic markers, occur more frequently in certain populations than in others. But this distribution of gene frequencies, though occasionally overlapping with racial groupings, is definitively not only a racially defined issue. For example, Northern Europeans have greater concentrations of cystic fibrosis than Southern Europeans, and both are categorized as "Caucasians." Southern Europeans have higher rates of beta-thalassemia than Northern Europeans. And sickle-cell anemia is found in greater concentration in Orchomenos, Greece than among African Americans (Duster 1990). Such gene frequencies are not a biologically racially defined matter (i.e., racial in the sense of first-order constructs).

Race and Second-Order Constructs

The median family net worth of individuals in the United States who are socially designated as "white" is $43,279. The median family net worth of individuals in the United States who are socially designated as "black" is $4,169. This is a truth that can be determined by the systematic collection of empirical data, and either replicated or refuted—which is to say that it can be investigated scientifically, without reference to blood groups, the relationship between genotype and phenotype, or the likelihood that one group is more likely to be at risk for cystic fibrosis while the other is more likely to be at risk for sickle-cell anemia. Here is why.

In 1939, the Federal Housing Authority's *Underwriting Manual,* which provided the guides for granting housing loans, explicitly used race as one of the most important criteria. The manual stated that loans should not be given to any family that might "disrupt the racial integrity" of a neighborhood. Indeed, the direct quote from Section 937 of the manual went so far as to say that "If a neighborhood is to retain stability, it is necessary that properties shall be continued to be occupied by the same social and racial classes" (Massey and Denton 1993:54). On this basis, for the next thirty years, "whites" were able to get housing loans at 3–5 percent, while "blacks" were routinely denied such loans. For example, of 350,000 new homes built in Northern California between 1946 and 1960 with Federal Housing Authority support, fewer than 100 went to "blacks." That same pattern holds for the whole state and for the nation as well.

Financially, the biggest difference between whites and African Americans today is their median net worth, which is overwhelmingly attributable to the value of equity in housing stock. In 1991, the median net worth of white households ($43,279) was more than ten times that of the median net worth of African American households ($4,169) (Bureau of the Census:1991).

To definitively assert the natural science point that, at the blood group level, or at the level of the modulatory environment for neurotransmission in the brain, there are no real racial differences (read "biological") is to reify a particular version of science and to ignore the social reality of the ten-to-one economic advantage of coming from a white household in America. But to restate the critical point about second-order constructs, that ten-to-one ratio is subject to scientific investigation as well and can by challenged or proven false, or replicated over and over again. Nonetheless, patterned social behavior, by race, leads to a confusion about the role of genetics and the way in which the analyst peers through the prism of heritability at the direction of the causal arrow. Herrnstein and Murray (1994), Wilson and Herrnstein (1985), Jensen (1969), and a wide band of other claimants[11] with no training in genetics make the commonsense mistake of treating race as a biological construct, then reading back through the social patterns (scoring on a test, rate of incarceration) to make inferences about the biological underpinnings of social patterns. For example, Herrnstein and Murray (1994:105) posit the dominance of genetics in explaining I.Q., concluding that "the genetic component of I.Q." is 60 percent. Then, taking I.Q. as the independent variable, the authors extend this analysis to the genetic explanation of social achievement or failure, so that unemployment, crime, and social standing are "explained" by I.Q. Despite a heavy reliance on the genetic explanation of intelligence as measured by test scores, the basic data from the molecular genetics revolution of the last three decades are completely absent.

They posit that genetics plays an important, even dominant role in I.Q., proposing that it is a whopping 60 percent of g, a statistical measure assumed to be related to "general intelligence." Neither Herrnstein (a psychologist) nor Murray (a political scientist) demonstrates sufficient knowledge of contemporary developments in the human biosciences to be aware of a fundamental problem in attributing g to genetics. Even single-gene-determined phenotypical expressions such as Huntington's disease, beta-thalassemia, and sickle-cell anemia exhibit a wide range of clinical manifestations. For multifactorial conditions, of which, incontestably, we must include the evolving thought processes of the brain, the interaction between nutrition, cellular development, and neurological sequencing has been firmly established. Developments of the last decade reveal a remarkable feedback loop between the brain and the "experience" of an environment. We now can demonstrate that a single neuron displays a variety of activity patterns and will switch between them, depending upon the modulatory environment (Harris-Warrick and Marder 1991:41). Anyone aware of current developments in cognitive science knows that a one-way deterministic notion of the firing of the neurotransmitters and subsequent behavior is a deeply reductionist fallacy. Thus, to assign to "genetics" a ballpark figure of any kind, without regard to these well-known interaction effects, is to display a profound ignorance of the last three decades of developments in molecular biology and the neurosciences, most especially since genetics (60 percent, sic) is posited as the most powerful explanatory variable.

The inverse of the attempt to get at a biological construct of race is the assertion, by those committed to a class analysis, that class, not race is ultimately the master stratifying practice in technologically advanced industrial and postindustrial societies. But that is open to empirical investigation. Both the biological construct and the class construct are attempts at first-order conceptions; they appear "more scientific." Yet, there is a fundamental error in the logic inquiry here—that is, not comprehending that when actors in the scene make use of their own sets of symbolic strategies for stratifying practices, that too is the legitimate, even compelling source of a "scientific" investigation.

The debate about whether race or class is "more real" as a stratifying practice can be better understood by reexamining how the issue is framed. What is it about class that makes it more real than race—save for the empirical fact that more people employ it (or do not) as a way of sorting social, political, and economic relations. The power of apartheid in South Africa and Jim Crow in the United States demonstrate that such divisions are as much located in the practices of actors as in the "objective" relations of workers to capital (Massey and Denton 1993). The answer is that, objectively, class relations are governed by an attempt at a "first-order construct, the connections to the workplace. But during apartheid (and even after, there is evidence to believe), whites have had greater access to scarce resources than blacks. This is the "objective" reality of a stratifying practice and no less so because it is a second-order construct employed by those acting in the world. Rather, during apartheid, it was just more of a pattern with notable and obvious replicability (i.e., as a stratifying practice). The problem comes even more clearly into focus in studies that have tried to focus the lens of heritability on explanations of crime and violence.

Studies that have tried to argue the genetic link to crime rely primarily upon incarcerated populations. Yet, incarceration rates are a function of incarceration decisions, a fact that social science research has long shown to be a function of social, economic, and political factors (Coleman 1993; Currie 1985; Skolnick 1966). For example, while the current incarceration rate of African Americans in American prisons is approximately seven times that of white Americans, this is a very recent development (Duster 1992; Dunbaugh 1979). It is a function of several important changes in the last three decades, of which the two most central are the shifting patterns of employment by race (Duster 1995), and the selective artillery of the War on Drugs, also by race.[12]

Since 1933, the incarceration rates of African Americans in relation to whites have gone up in a striking manner. In 1933, "blacks" were incarcerated at a rate of approximately three times the rate of incarceration for "whites." In 1950, the ratio had increased to approximately 4 to 1; in 1960, it was 5 to 1; in 1970, it was 6 to 1; and in 1989, it was 7 to 1 (Duster 1992). In the last decade, there has been an especially sharp increase in the differential incarceration rates by race. In Florida, the annual admissions rate to the state prison system nearly tripled from 1983 to 1989, from 14,301 to nearly 40,000 (Austin and McVey 1989:4). This was a direct consequence of the War on Drugs, since well over two-thirds of these felonies are drug related. The aim of the War on Drugs affected the races quite differently with regard to respective prison incarceration rates. The most astonishing figure is for the State of Virginia. In 1983, approximately 63 percent of the new prison commitments for drugs were white, with the remaining 37 percent minority. Just six years later, in 1989, the situation had reversed, with only 34 percent of the new drug commitments being whites, but 65 percent minority. This is the very period that we find a significant increase in scientific journal articles and scholarly books (Mednick et al. 1984; Wilson and Herrnstein 1985) suggesting a greater role for biological explanations of crime.

As Gould (1981:21–23) points out, one must not fall victim to the polemical trap of concluding that the organization of knowledge structures is entirely or even primarily political. Rather, it is the interplay of science and external forces, much as it is the interplay of genes and environment, that will best account for the outcomes we observe in behavior.

Notes

1. The term "heritability" has a technical meaning in the field of genetics—an accounting for how much variability in the appearance of a trait within a population can be explained by genetic differences among individuals. The more common lay usage from the standard dictionary simply means "capable of being inherited."
2. More than double that of the previous decade.
3. Generated by Medlars II.
4. The five major sources were *American Men and Women of Science*, including the volumes *Physical and Biological Science* and *Social and Behavioral Science* 13th ed., 1986; the *ABMS Compendium of Certified Medical Specialists*, 1986–1987; *Who's Who in America*; and *Who's Who in Science in Europe*. Locating information on the credentials and training of every author proved to be difficult. Despite these problems in tracking areas of specialization for everyone, credentials for 134 of the 200 authors could be found, constituting just over two-thirds of the entire population. In about one-third of the cases, only the degree granted could be determined and not the area of specialization. These cases are not included.

5. The selection of articles was purposefully biased away from anthropology, sociology, psychology, education, and political science journals, since this would have weighted the study more toward social scientists. Instead, the study was aimed pointedly to see who was making claims in the genetics journals. The articles were published in the *American Journal of Medical Genetics, Human Heredity,* and *Clinical Genetics.*
6. It is also true that the most ardent advocates of the racial purity position in the eugenics movement were wealthy white Anglo-Saxons of the "old stock" (Ludmerer 1972:25–30).
7. Barnes (1974, 1977); and H. Collins (1981).
8. There has been extensive commentary in the *Journal of the American Medical Association* editorial pages on this topic. See, for example, the editorial in the April 20, 1994 issue, entitled "Race, Class, and the Quality of Medical Care," and also, from the January 8, 1992 issue, "The Use of Race in Medical Research."
9. For the discussion in this paragraph, and for the references to the German literature that are used here, I am indebted to William H. Schneider, a historian at Indiana University.
10. This was the year that Linnaeus published *System Naturae,* in which he revealed a four-part classification scheme of the human races that has residues in social and scientific theory still today.
11. Those who claim a genetic interpretation of complex human behaviors.
12. For example, in 1954, "black" and "white" youth unemployment in the United States was equal, with "blacks" actually having a slightly higher rate of employment in the age group 16–19. By 1984, the "black" unemployment rate had nearly quadrupled, while the "white" rate had increased only marginally.

CHAPTER 7

Nuclear Weapons Testing
Scientific Experiment as Political Ritual

Hugh Gusterson

> Nuclear weapons are both symbols and pieces of hardware. Their role as symbols is what matters to most people, including scientists, most of the time.
> Michael May, Director of Livermore Laboratory, 1965–1971

Introduction

In the late 1980s, when I was doing fieldwork at a nuclear weapons laboratory in Livermore, California, the testing of nuclear weapons was a focal concern of both the Livermore Laboratory and the American antinuclear movement.[1] The Livermore Laboratory organized itself around the production of nuclear tests and the antinuclear movement organized itself around ending them. Carrying out or ending nuclear tests was, for each community, the mission that facilitated its integration as a community and connected it in contentious antipathy with the opposed community.

There have been periodic attempts to stop nuclear testing since at least the 1950s, and the American antinuclear movement has consistently seen a test ban as its first priority. In the mid-1950s, when the American public was increasingly concerned about the health effects of atmospheric nuclear testing, an antitesting campaign spearheaded by the Nobel prizewinner Linus Pauling gathered the signatures of 10,000 scientists and mobilized hundreds of thousands of other citizens in opposition to nuclear testing (Divine 1978). Public concern abated after the Limited Test Ban Treaty of 1963 pushed testing underground—although the rate of nuclear testing actually accelerated after the treaty went into effect. In the late 1970s, seeking to rein back the arms race, Jimmy Carter attempted to negotiate a complete test ban but failed as détente collapsed after the Soviet invasion of Afghanistan. In the 1980s, however, the United States came under renewed pressure to end nuclear testing. This pressure came, at home, from the Nuclear Freeze movement of the 1980s and, abroad, from the Soviet Union, which under Gorbachev initiated a series of unilateral moratoria on testing in order to put pressure on the Reagan and Bush administrations to end the arms race. Throughout, America's two nuclear weapons laboratories strenuously opposed all initiatives to ban testing (Blum 1987).

At Livermore, nuclear weapons testing has structured much of the organizational and symbolic life of the laboratory community. Diverse everyday tasks, throughout the laboratory and at the Nevada Nuclear Test Site five hundred miles away, have taken on organizational purpose and symbolic meaning because of their contributions to programs converging in the production of nuclear tests. In the 1980s there were about eighteen tests a year. In a test—an "event" in the parlance of weapons designers—a bomb consisting of about 5,000 components must work perfectly in the fraction of an instant before the components are destroyed by the explosion they create. The preparation for such an experiment generates fantastically complex interactions conducted over a period of years by thousands of physicists, engineers, computer scientists, chemists, administrators, technicians, secretaries, and security personnel. A single nuclear test is a kind of busy intersection where individual lives, bureaucratic organizations, scientific ideas, complex machines, national policies, international rivalries, historical narratives, psychological conflicts, and symbolic meanings all come together. This article investigates the controversy over nuclear testing in the late 1980s as it involved the laboratory and makes the case that nuclear tests can be seen as high-tech rituals that are as important for their cultural and psychological meanings as for their technical significance. They have been vital not only in the production of nuclear weapons but also in the production of weapons scientists and in the social reproduction of the ideology of nuclear deterrence.

Deconstructing Testing

Although many Livermore scientists in the 1980s supported arms control treaties such as SALT I, SALT II, and the INF agreement, they were almost unanimous in their opposition to a nuclear test ban. Why was this? Critics of the laboratory say that the opposition is simply a matter of self-interest. In the words of Hugh DeWitt, an internal critic at Livermore, "the laboratories oppose a comprehensive test ban because they want to continue nuclear weapons development—to refine existing designs and do research in exciting new areas such as the X-ray laser" (DeWitt 1986:104).

Livermore scientists themselves have explained their opposition to a test ban differently. In the 1980s, the main reason for continued testing given by both laboratory officials and many warhead designers was that the reliability of the nuclear stockpile could not be assured without continued nuclear testing. One experienced warhead designer, Jack, put it like this:

> I think a lot of people think a bomb is a bomb; build it once and it's there forever. It's not true. If you bought a Cadillac, you wouldn't just stick it in your garage and stake your life that you could start it ten years later if you

didn't do anything other than put air in the tires and charge the battery. I wouldn't bet on it.

Jack added that, as well as assuring the reliability of weapons, tests assure the reliability of the scientists who must ultimately make judgments about weapons reliability.

These arguments about reliability have twice entered the national political debate on nuclear testing. In 1978, when President Carter was attempting to negotiate a test ban treaty, the directors of Livermore and Los Alamos met with him to warn him that the reliability of the U.S. arsenal would degrade without nuclear testing and that teams of weapons scientists would disperse in such a way that it might be impossible to reassemble them if they were needed again (Agnew 1981; York 1987:285–87).

Meanwhile Jimmy Carter received a two-page letter from Norris Bradbury, a former Director of Los Alamos; Carson Mark, former head of the theoretical physics division at Los Alamos; and Richard Garwin, a nationally known consultant on nuclear weapons physics who played a major role in developing the hydrogen bomb.[2] Bradbury, Mark, and Garwin assured Carter that, in their technical judgment, the laboratory directors were mistaken and that the continued reliability of the U.S. nuclear stockpile could be assured without any further nuclear tests. Herb York, Carter's lead negotiator and a former Livermore director, says in his memoirs that "the nuclear establishment's fears were exaggerated. ... We concluded that regular inspections and non-nuclear tests of stockpiled bombs would uncover most such problems and provide solutions to them" (York 1987:286). The issue became moot when the test ban negotiations fell victim to the collapse of détente in the late 1970s.

The reliability issue resurfaced in 1987, however. In response to congressional hearings on a test ban, the Livermore Laboratory provided the U.S. Congress with a report on weapons' reliability prepared by three of its leading weapons designers (Miller et al. 1987). The report gave details of fourteen warhead designs that had needed postdeployment retesting to detect and rectify flaws. In the case of the Polaris missile the problem was serious enough that, some years after deployment, about one-half of the warheads were found to be "lemons" (Wilson 1983:199).

Meanwhile, Representative Les Aspin, the chair of the House of Representatives Armed Services Committee, made a highly unusual move by asking Ray Kidder, a well-regarded Livermore scientist known to sympathize with the test-ban cause, to reanalyze the information in the Livermore report. Kidder was allowed full access to all the relevant classified information. In his own report, Kidder argued that the laboratory's position was unconvincing. He claimed that all the weapons with problems were inadequately tested in the design process, mostly because they were rushed into the stockpile without full testing just before the 1958–1961 testing moratorium, so that the subsequent tests that revealed problems after these weapons had entered the stockpile should more properly be thought of as deferred design tests than as postdeployment reliability tests. He also argued that it is possible to remanufacture warheads in proven ways that render reliability testing redundant, saying that if the laboratories were so concerned about reliability, it would be easy for them to design more robust and reliable warheads (Kidder 1987). We might also add that the number of tests assigned to verification of stockpile reliability has been a small fraction of the total number of tests—probably no more than one test a year (Cochran et al. 1987:44).[3]

What are we to make of this arcane but hardly trivial dispute? We could plausibly argue that the technical judgment of the weapons scientists has been compromised by their vested interests; or we could just as plausibly argue that the weapons scientists are uniquely placed to know about the mechanical reliability of nuclear weapons and that their opponents are technically less informed or have allowed their own politics to color their technical judgments. As one probes the debate, one has a vertiginous sense of standing on shifting ground as the distinction between political and technical judgments—a distinction anchoring the expert case both for and against a test ban—melts into air.

My goal here is neither to judge the honesty of the participants in this debate, nor to provide a definitive judgment of the technical concerns at issue—a task that would clearly be beyond my competence in any case. Instead, addressing this debate more obliquely than directly, I shall follow the lead of a young physicist at Livermore who reminded me that I was an anthropologist and not a physicist. He, having read some of the literature in the sociology of science, advised me to stop thinking of these technical judgments as right or wrong answers to a question and to start thinking of them as interpretations of highly complex and ambiguous information. Instead of seeking a definitive technical judgment, then, we should ask about the processes by which judgments come to be considered definitive and their authors authoritative. After all, part of the argument made by the designers is that Ray Kidder was not in a position, despite his knowledge of thermonuclear physics, to make the technical judgments he made. By means of what processes, then, social as well as technical, does one acquire the authority to make such judgments?

Furthermore, what is reliability?[4] How much weapons reliability is enough, and why is reliability so important? We have been incited to a discourse[5] here in which weapons reliability is taken for granted as an indispensable asset. The proponents of testing argue that one's enemies are less likely to attack if they are sure, and you are sure, that you have reliable nuclear weapons. We might as plausibly argue, on the contrary, that if neither side is particularly sure that its nuclear weapons work, then neither side will have the confidence to attack or pressure the other, so that a test ban might result in low confidence in weapons reliability, which, in turn, would enhance deterrence. We must ask, then, how a social world comes to be constructed such that deterrence depends on the hyperreliability, not on the soft reliability, of nuclear weapons.

In the remainder of this chapter, by asking why and how reliability matters, rather than whether the nuclear arsenal is in fact demonstrably reliable, I offer a cultural explanation as to how and why weapons scientists come to care about weapons reliability; how nuclear scientists become authorized to make expert judgments; and how a ban on nuclear testing strikes at the heart of the collective culture of Livermore's weapons scientists who must therefore, whatever the merits of the respective technical arguments, feel placed in much greater jeopardy by a comprehensive test ban treaty than by any other possible arms control measure.

But first we must understand the testing process itself.

How to Design and Test a Nuclear Weapon

Nuclear tests are done for a variety of purposes: to explore the basic physics of nuclear explosions, to test a new warhead design approach, to recertify the reliability of an old warhead, to test the effects of nuclear explosions on military hardware, to adapt an old warhead design to a new delivery vehicle, or to validate the finished version of a new design. Any individual test may serve more than one of these purposes.

The nuclear weapons design process consists of three phases (Broad 1992:128). In phase 1, *concept study*, designers review the results of earlier tests, looking for ways to push the limits of the basic physics while satisfying the military requirements articulated by the Department of Defense and the Department of Energy.

In phase 2, *scientific feasibility*, weapons designers turn their ideas into specific design proposals that are evaluated by the bureaucracies in the laboratory, in the Department of Defense, and in the Department of Energy. This evaluation process may involve some nuclear testing. It also involves review meetings, often lasting as long as three hours, where reviewing scientists try to poke holes in the proposals. "These are not nice reviews," said the weapons designer Lester.[6]

> They're very critical. I've seen men all in tears. The big reward in our division is to do an experiment, to get your idea tested. It's highly competitive. For every twenty things people propose, maybe one is going to make it onto that shot schedule.

For phase 3, *engineering development*, scientists validate the performance of a finished prototype warhead ready for deployment in the national nuclear stockpile. It is phase 3 that involves the most nuclear testing. William Broad quotes one Livermore designer who described the climax of the engineering development phase by saying, "I have never seen people work so hard. They would sleep in assembly rooms for days, not seeing their families. Many times I saw people with tears running down their face because they had worked so hard" (Broad 1992:196). The designers, who are often working sixty to eighty hours a week at this point, are under constant pressure because other members of the team are waiting on their calculations. They must consult with engineers on the choice of materials for the bomb; with chemists as they choose how to fit the bomb with tracer elements for diagnostic purposes; and with the technicians who, often working with highly toxic materials in glove boxes, must machine the device's components with extraordinary precision. Since parts of the device are machined at Rocky Flats in Colorado, and the device is assembled at the Nevada Test Site rather than at the laboratory, this phase can also involve a lot of travel. Barry recalls travelling to Rocky Flats to watch the technicians machine the plutonium "pit" for his device.

> I looked at the plutonium in its glass case. It was black, with an oxide skin. It's like coal, except that it's hot to the touch because it's constantly emitting radiation. You know plutonium is pyrophoric. The technician scraped the skin off with his gloves, and white sparks flew out. I remember thinking, "Holy shit! This lump of rock is going to go in *my* bomb. *That's* going to be that powerful."

Gradually the test date approaches. Clark remembers his test nearing completion:

> Here was an experiment I had been working on for three years with a cast of hundreds, and watching this thing as it gets all put together laboriously at a couple different places around the country.... So you see this whole thing coming together, gee, it's almost like having a baby or something. It's a comparable length of time and many more people involved in the process.... One of my big fears was that pieces would get misplaced somehow and the wrong end was facing forward. It was a complex thing. It had lots of parts.... By the time you're done this is at least a $5 million deal, and yet I'm thinking this whole thing rests on a few of my late-night computer calculations.

As the device and the diagnostic equipment are being assembled, the test site workers drill the shaft into which the device and its diagnostic canister will be placed. This cylindrical canister, crammed full of sensitive diagnostic instruments and cables, can weigh hundreds of tons. When the canister is ready for final assembly, the nuclear device is transported with an escort of armored vehicles to its rendezvous with the diagnostic equipment. There the device and canister are bolted together ("married" or "mated" in the parlance of weapons scientists) within a portable building, known as a bogey tower, which sits astride the test shaft to shield it from both the weather and the gaze of enemy satellites. Once the device and canister have been lowered into the shaft, they are connected to nearby trailers on the desert surface by thick cable wires that will transmit a chorus of measurements from the exploding bomb underground to the scientists above ground. The canister is then covered with hundreds of feet of sand and gravel as well as coal and tar epoxy plugs designed to keep all the radiation from the blast below ground. Over the last three decades, as they have become more complex, underground nuclear tests have grown more expensive. By the late 1980s they cost $30 million to $60 million each (Broad 1992:91).

Lester remembers finally seeing his device ready to be lowered into its shaft. "You go out to the test site, and there's this huge 200-foot canister filled with all this beautiful equipment and they're about to put it down. That's a real gut-wrencher."

Before dawn the arming party goes out to arm the device. William Broad (1992:87) describes one instance of this process:

> A small group of scientists and security guards ... drove out to a trailer known as the "red shack" to electronically arm the weapon, which had earlier been placed at the bottom of a 1,050-foot-deep hole and covered with dirt. At the red shack, security was tight as usual. Two of the scientists carried a special briefcase and a bag of tiny cubes that had numbers painted on their sides. They alternately took cubes out of the bag and punched the numbers into an "arm enable" device in the briefcase, generating a random code that was sent to the buried weapon on a special electrical cable. The scientists then drove across the desert to the control point in a mountain pass overlooking the test site.... There, in a high-technology complex surrounded by armed guards and barbed wire, they again opened their briefcase and sent the same random code to the weapon. It was now armed.

When the detonation takes place, the scientists see it in two ways: first, on video monitors relaying the picture from a helicopter hovering over the test shaft and, second, as a set of flickering needles on seismographs and oscilloscopes registering in the control room what the diagnostic equipment has picked up from the transitory flare beneath the surface of the earth just miles away.

Here is Clark describing his test:

> We got up at 3:30 or 4:00 in the morning. Drive out. Stars are out. Go out to this remote outpost. Standing around going through the what-if's. "God, what if this happens, what are you going to do then? ..."
>
> And everything went smooth and it went off and they show a picture on the TV screens there of a helicopter hovering above the site and you could actually see dust rising. I mean it's not like you're watching the old atmospheric tests. I mean it's pretty benign really. You can see a shock-wave ripple through the earth.... You're not allowed out to the site until the crater is actually formed, and that can happen in 30 seconds, it can happen in 10 hours. Turned out with mine that it happened in about an hour.... And that was really awesome, standing there with this thing which was at least 100 yards across, and see what I had been looking at on my computer screen for years all of a sudden show up in this gigantic movement of the earth. It was as close as I've had to personal contact with what the force of the nuclear weapon is like. And then eventually some of the data starts to come in and by the end of the day it was clear that—we didn't have all the data, but it was clear it was going to be a

success and it was a very complicated shot. So I knew that would be good for my career.

As the day progresses, the data through which the bomb's performance will be interpreted are retrieved from the diagnostic devices. "We bury these things half a mile underground and you get some electricity out of a wire and some melted glass that's radioactive and go out and analyze those, and you can tell somebody all about what happened," said Jack, marveling at the almost magical nature of the diagnostic technology.

A Ritual Analysis

Nuclear tests, as well as being scientific experiments, are cultural processes that reproduce weapons scientists as persons and enable weapons scientists to play with, maybe even resolve, core issues in their ideological world. In this regard, it is heuristically helpful to think of nuclear tests as sharing some of the characteristics of rituals. Since the comparison between nuclear tests and rituals may seem improbable, even offensive, to some, particularly the scientists who carry them out, it bears a word of explanation. My intention in making the analogy is not to be cute; nor is it to satirize nuclear weapons scientists by comparing them to "tribal savages," nor yet to deny that nuclear tests are rigorously executed scientific experiments. I make the guarded analogy between ritual and nuclear testing because it seems to me to genuinely illuminate the significance of testing for Livermore scientists in a way that affords a new vantage point not only on the vexed debate over nuclear testing but also, more broadly, on the cultural and psychological significance of scientific experimentation in general. After all, it is not for nothing that, when I asked the weapons designer Barry who he considered a real designer, he replied "anyone who's been through the ritual all the way." Obviously there are many ways in which nuclear tests are not at all like, say, the Catholic mass or African adolescent circumcision rituals. Still, as Moore and Myerhoff (1977) argue in their book on "secular rituals," while we should avoid the temptation to mechanically label almost every social process a ritual, ritual analysis can profitably be applied to many events that are not, formally speaking, religious or sacred. In the case of nuclear weapons testing, if we bracket the obvious differences between a scientific experiment and a sacred ceremony, then the comparison with ritual processes brings into focus certain kinds of intense symbolic meaning nuclear tests carry for scientists that might otherwise go unnoticed.[7]

Anthropologists have theorized ritual in a number of different ways. Emile Durkheim and his intellectual descendants have stressed the power of ritual to heal social conflicts. They argue that ritual allows the symbolic expression and transcendence of conflicts, facilitating the intersubjective production of a sense of community (Durkheim 1954; Gluckman 1954; Turner 1969). This sense of community may be experienced most deeply within ritual in moments of mystical transport that Durkheim labeled "collective effervescence" and Victor Turner called "communitas." Another school of thought, influential in Britain, articulating a more psychological function for ritual, has presented ritual as a means of allaying anxiety by simulating human control over that which ultimately cannot be controlled—death, disease, crop failure, and so on (Evans-Pritchard 1937; Homans 1941; Malinowski 1925). Still others have focused on the ability of certain kinds of rituals, "rites of passage," to transform those who participate in them. In rites of passage, the social status of initiates is irrevocably changed as they are indoctrinated with the special, or even secret, knowledge of the initiatory group (Turner 1969; Van Gennep 1909). Finally the U.S. school of cultural anthropology has portrayed ritual as a text—as a means of celebrating, performing, displaying, and transmitting the ethos, symbols, and norms of the particular cultural community that uses ritual to clarify and speak to itself about its values and identity (Benedict 1934; Geertz 1973). In the analysis that follows I draw eclectically on all these traditions in order to illuminate the significance of nuclear testing for Livermore scientists.

I explored the meaning of nuclear testing for nuclear weapons scientists by collecting tape-recorded narratives of nuclear tests. The first thing one notices about these narratives is that, unlike official laboratory statements about the importance of nuclear testing, they say nothing about reliability testing of old weapons. In fact, Clark went so far as to preface his test narrative with the statement, "stockpile maintenance is boring, so I don't do that." The main themes in these narratives have to do with the fulfillment of personal ambition, the struggle to master a challenging new technology, the scientific drama of bringing something fundamentally new into the world, and the experience of community and communitas in the deeply competitive world of nuclear physics.

Nevertheless, although the reliability testing of old weapons is absent from this narrative world, that world is still saturated with a more broadly and diffusely expressed anxiety about the reliability of the weapons. The narratives are largely about reliability, but the scientists frame that concern very differently than the laboratory's official policy statements do.

Overtly, these narratives are about a purely technical process and have nothing to say about the broader political purposes of these weapons or about the system of international relations and international meanings into which the weapons are inserted. It is, however, my contention that these physics experiments as they are narrated do embody a kind of politics, that the techno-political worldview of the weapons scientists is embedded in, experienced through, and simulated by these experiments—that it is in the design and testing of a nuclear device that the abstract cliches comprising the ideology of deterrence become experientially real to the scientists who must live deterrence as a truth.

Initiation

Nuclear tests are not only a means of testing weapons designs. They are also a means of testing and producing weapons designers—the elite within the laboratory. To become a fully fledged member of the weapons design community new scientists must master an arduous, esoteric knowledge, subject themselves to tests of intelligence and endurance, and finally prove themselves in a display of the secret knowledge's power. If a test goes well, and it is a designer's first test, the designer's social status is changed. In the words of the senior designer Seymour Sack, testing is a way of "punching your ticket by having your name associated with a particular test" (quoted in Stober 1990). One designer, Martin, remembers the day after his first test—a particularly challenging one. "It was extraordinary. I was walking around the lab and people were coming up to me, I mean just all over the place, people I didn't even know were coming up to me and shaking my hand, congratulating me on this tremendous achievement." Martin's experience speaks to the fact that nuclear tests do not only test technology; they also test people (see Pinch 1993).

Tests are also the socially legitimated means of producing knowledge about nuclear weapons. This knowledge takes the form of a socially attributed knack for judgment and, more concretely, the view graphs summarizing the results of tests, which scientists display in post-test briefings in the laboratory's Poseidon Room. Thus participation in nuclear tests confers a kind of symbolic capital that can be traded as power or as knowledge (see Bourdieu 1977). The more tests scientists participate in, the more authority they acquire as they move toward the status of senior scientists whose judgment about nuclear weapons is particularly respected and sought after.

It is partly in this context that we should understand the debate about weapons reliability. This debate is as much about the authorization of knowledge and the hierarchical authority of knowers as it is about the reliability of weapons. The laboratory argues that there is only one way to know for sure whether a weapon will work; only one way to train people to know this; and only a very select group who can certify the continuing reliability of old weapons whose parts are decaying, have been replaced, or have been slightly redesigned without another test. Scientists stress the mysterious uniqueness of knowledge about nuclear weapons—knowledge that cannot be learned entirely from textbooks or briefings, knowledge whose uniqueness is marked by its very nontransferability and ultimate nontradeability.

This throws new light on the significance of Ray Kidder's intervention in the weapons reliability debate. He was not only attacking the central mission of the laboratory in saying that the reliability of nuclear weapons could be assured without continued testing and assenting to the prohibition of the ritual by which membership in the laboratory elite is regulated. Kidder was also a physicist intimately acquainted with thermonuclear physics who attacked the whole system of power/knowledge that organizes the status hierarchies and cosmology of the laboratory. This system is based upon nuclear testing as the means of production of both knowledge and power. By suggesting that knowledge could be separated from its local production in nuclear tests, Kidder threatened to tear the social and political fabric of the laboratory world.

We can bring the power/knowledge stakes in the conflict between Kidder and the weapons establishment into still sharper focus if we consider the conflict in the context of Kidder's own scientific biography. Earlier in his career, Kidder had weapons experience and served on a number of weapons design review committees. From there he went on to found and run the Laboratory's Inertial Confinement Fusion (ICF) program, which created microscopic thermonuclear explosions by firing enormously powerful lasers at pellets smaller than a pinhead. In so doing he developed a different technology, one that did not involve the nuclear weapons design process or the use of the Nevada Test Site, to simulate within the laboratory itself the basic product of the weapons designers. More recently it has been suggested that, if there were a nuclear test ban, the laboratory should rely much more heavily on its fusion laser to explore the physics of thermonuclear explosions. Seen in this light, Kidder was not just a critic of nuclear testing; he was also the author of a different social-technological system for producing thermonuclear power/knowledge, and his intervention in the debate on reliability represented a deep challenge to the power/knowledge system of the laboratory's weapons community.

Mastery

As the psychologist Robert Lifton (1982) points out, the very existence of nuclear weapons inevitably raises the question of whether the weapons are under our control or whether we are at their mercy. The issue here is not only whether humans can be relied upon not to use the weapons deliberately, but also whether people are capable of devising failsafe systems to prevent the weapons from exploding accidentally. Lest the latter be thought a farfetched concern, there have been a number of unfortunate accidents involving U.S. nuclear weapons. In one incident, in 1961, a B-52 accidentally dropped two multi-megaton hydrogen bombs on a farm in North Carolina. Nuclear weapons were again lost in accidents involving B-52s in 1966 and 1968, in Spain and Greenland respectively. In both cases the conventional explosive in the weapons detonated and, although there was no nuclear explosion, plutonium was dispersed over wide areas. In 1955, a B-47 accidentally dropped a nuclear weapon as it was landing at Kirtland Air Force Base and, in North Dakota in 1980, a B-52 with nuclear weapons on board caught fire. Recent computer simulations of the latter incident have suggested that, if the wind had blown the fire in a different direction, there might have been at least a conventional explosion dispersing plutonium as widely as at Chernobyl. Other simulations have suggested that the W-79 nuclear artillery shell, formerly deployed in Europe, might, in certain circumstances, have detonated if accidentally struck by a bullet (Drell et al. 1991; Smith 1990).

Despite these mishaps, nuclear weapons scientists have been reasonably confident that nuclear weapons would not explode due to human or mechanical error. I want to argue here that, for scientists at Livermore, the lived experience of nuclear testing is important in fostering this confidence. Just as, according to classical anthropological theory, the performance of rituals can alleviate anxiety and create a sense of power over, say, crop failure and disease, so nuclear tests can in an analogous way create a space where participants are able to play with the issue of human mastery over weapons of mass destruction and symbolically resolve it. Since the stability that nuclear weapons are supposed to ensure—nuclear deterrence—exists so much in the realm of simulations, and since the reliability of deterrence involves the absence of a catastrophe more than the active, direct, positive experience of reliability, nuclear tests play a vital role in making the abstract real in scientists' lives. Nuclear tests give scientists a direct experience of what can only otherwise be known as an absence, bridging the gulf between a regime of simulations and the realm of personal, direct experience.

The issue of human control over nuclear technology is a recurrent theme in nuclear test narratives. Many test narratives involve a sequence of events in which scientists fear that their machines will not behave as predicted but, after a period of painful anxiety, learn that humans can predict and control the behavior of technology. The most exciting narratives are those, as in any story, where the outcome seems in doubt for a while. For example, the scientist Eric told me this story when I asked which tests stood out in his mind as the most memorable:

> The most exciting tests were the ones . . . where we were in great danger . . . of losing the test altogether . . . and through some enormous heroic development of solutions to problems, we were able to save the test.
>
> There was one test in which we finally solved the problem, and it was an electrical problem, by deciding to do what I should call a lobotomy. We had to destroy a component. And so we finally decided that we could, by sending a powerful pulse of electricity down a pair of wires, burn that component out into the condition that it would allow the rest of the system to function. And so then finally we talked this out and we rehearsed it and we practiced it with cables of the right length and components of the right sort and so forth. And, if it had failed, the experiment would have been completely lost, and it was buried. We couldn't have pulled it back to the surface. It would have been just a piece of garbage at the bottom of the hole. . . .
>
> And that particular test, I felt we had to do something to commemorate it, and so within a few weeks I had invited all of the principals here, and we had a party and set up pictures in the backyard of all of the trials and tribulations we had gone through.

Many of the test narratives have complex emotional rhythms where control over the technology and helplessness before it alternate with one another. If the final point of the story is that humans can control nuclear technology, the scientists often learn, on the way, that they must also trust the technology and let go of their concerns. This happens particularly in the period between the placement of the device in the test shaft and the actual test. "This is a hard period for the designer, especially the younger designers," says Lester. "You go through a period where you have a lot of doubts because the computer codes don't cover everything." As Clark put it: "You're kind of helpless after a time. You've got to just take your hands away and hope everything works out alright." This confidence that we can make the weapons do our bidding mixed together with a trusting helplessness before them is, of course, the basic psychology required by deterrence.

Where many of us worry that a nuclear explosion will occur at some point in our lives, Livermore scientists worry that one won't. Over and over again, scientists have the experience of fearing that something will go wrong with the bomb only to learn—in most instances—that it does not. By means of this lived journey from anxiety to confidence, structured by the rhythms of the testing process itself, scientists learn that the weapons behave more or less predictably, and they learn to associate safety and well-being with the performed proof of technical predictability. Then, like Lester, they can say:

> When you're a device physicist, [the bomb] is no more strange than a vacuum cleaner. You don't feel a fear for it at all, and it's not an alien thing. And I understand that to the people that don't do it, it is an alien thing. I felt the same way before I went to the lab.

Lester is saying that, before he went to the laboratory, he too was nervous about the bomb; but participation in the practices of nuclear weapons design and testing has restructured his subjective world so that he now feels in his bones that nuclear weapons are as benign as vacuum cleaners. For many of us, understanding the engineering of a hydrogen bomb will in no way allay our fear that a mad president, general, or admiral will misuse it. If anything, it may magnify that fear. Lester's remark, which implicitly equates safety from nuclear annihilation with technical mastery over the bomb, only makes sense in the context of the practical consciousness embodied in and engendered by nuclear testing.

To put it a little differently: as well as assuring the technical reliability of nuclear weapons, nuclear tests provide in an elusive way a symbolic simulation of the reliability of the system of deterrence itself. Each time a nuclear test is successfully carried off, the scientists' faith in human control over nuclear technology is further reinforced. Seen in this light, the "reliability" the tests demonstrate has an expandable meaning, extending out from the reliability of the particular device being tested to the entire regime of nuclear deterrence.

Life and Death

Rituals in general are often marked by particular kinds of language: an abundance of birth and death metaphors, allusions to mythic or divine entities, and so on. U.S. nuclear weapons culture is full of mythical allusions. Thus we have the Excalibur weapon, the Polaris and Poseidon missiles, and experimental chambers named *kivas*—the name given to sacred ceremonial spaces by the pueblo Indians who live around Los Alamos.

But more striking than the use of explicitly sacred language in U.S. nuclear weapons culture is the absence of metaphors of death and the superabundance of birth metaphors. The pattern is startling, and it goes back to the beginning of the nuclear age, when scientists at Los Alamos, where the nuclear reactor was named

"Lady Godiva," wondered aloud whether the bomb they were about to test would be a "boy" or a "girl" (i.e., a dud). They called the prototype tested in New Mexico "Robert's [Oppenheimer's] baby" and the bomb dropped on Hiroshima "Little Boy." Secretary of State Henry Lewis Stimson informed Winston Churchill of the first successful nuclear test by passing him a sheet of paper saying, simply, "Babies satisfactorily born." Edward Teller cabled Los Alamos the message, "It's a boy," after the first successful hydrogen bomb test (Easlea 1983:103, 130).[8] In subsequent years there were debates as to whether Teller was really the "father" of the hydrogen Bomb or had in fact been "inseminated" with the breakthrough idea by the mathematician Stanislam Ulam and had merely "carried" it.

Now we have bombs that are constructed around fissile "pits." The production of these "pits" may involve the use of "breeding blankets" and "breeder reactors" to produce plutonium—an artificial substance that does not exist in nature. After the bomb has been "married" or "mated" to the diagnostic canister, it explodes and "couples" with the ground, producing "daughter fission products" that go through "generations." Clark referred to the process of bringing this about as "like having a baby," and he talked about the tense decision at the moment of the test as being whether to "push" or not. Another designer told me he has "postpartum depression" after his tests. When the first nuclear weapon was tested, the Manhattan Project scientists referred to the apparatus from which it was suspended as a "cradle." Subsequently the steel shells in which ICBMs sit in their silos became known as "cribs," while missile officers refer to the ICBMs as being connected to control panels by "umbilical cords."

What is going on here? Brian Easlea (1983) has suggested that nuclear scientists are men who are impelled to their work by womb envy—an overpowering jealousy that women can create life and a determination, inflamed by the distance from women and from birth enforced upon men in modern society, to themselves do something as awesome as birth.

There are problems with Easlea's interpretation. The first is that, since birth imagery is applied to so many activities in our society (from writing term papers to growing gardens), we would, following Easlea's logic, have to argue that all these activities are animated by womb envy. The second problem is that a few of the weapons scientists—including at least one who used these birth metaphors quite inventively—are women. Where most cultural feminist theories can fairly easily account for a few women who behave like stereotypical men, Easlea's theory is so closely tied to female reproductivity as an absolute index of difference that this is more problematic for him.

While there can be no doubt that the culture of nuclear testing is a scientific celebration of masculine values, I would argue against using a broadly Freudian strategy of reading these birth metaphors as clues or "slips" that enable the determined investigator to uncover preexisting unconscious motives at the individual level. I prefer to see a shared language as the symptom of a political ideology rather than individual psychology and as a means of shaping individual subjectivities so that people can work and live together (see Foucault 1980a).

The physicists themselves, when I pointed out the abundance of birth metaphors in their discourse, insisted either that I had found a fact without significance or pointed out that all sorts of people apart from weapons scientists make abundant use of such metaphors.

Still, while the use of birth metaphors might seem unremarkable enough in many situations, it is surely more remarkable to find birth metaphors applied to the process of creating weapons that can end the lives of millions of people than to find them used to describe the process of, say, writing a book or a computer program. All metaphors achieve their effect because of the gulf between the literal and the figurative, but in the case of birth metaphors used to describe nuclear weapons development the gulf between the literal and the figurative is great enough that the metaphor is as dissonant as it is evocative. But this is the point. Thus I take the recurrence of images of fertility and birth in weapons scientists' discourse about weapons of destruction as an attempt to cast the meaning of this technology in an affirmative key. In metaphorically assimilating weapons and components of weapons to a world of babies, births, and breeding, weapons scientists use the connotative power of words to produce—and be produced by—a cosmological world where nuclear weapons tests symbolize not despair, destruction, and death but hope, renewal, and life. In this semantic world, the underground transformation of a mass of metals and chemicals into a transient star under the surface of the earth is phrased in images of life and birth. And, after all, in the context of these scientists' practices and beliefs about deterrence, we can see how each nuclear explosion might symbolize for them the fertility of the scientific imagination, the birth of community, and the guarantee of further life. A weapon is destroyed and a community is born.

The scientists' use of such images is also part of a wider laboratory discourse that exchanges and mingles the attributes of humans and machines. This way of speaking in general, and the birth images in particular, create a discursive world where nuclear weapons appear to be "natural." This happens because the discourse fuses, or confuses, the spheres of production and reproduction, depicting machines made by humans as fruits or babies, as if they grew on trees or inside human bodies instead of being assembled in laboratories and factories. Karl Marx argued over a century ago that, in presenting interest and profit as something that naturally accrues to invested capital, as if by breeding, capitalist ideology obscures the way profit is produced in social relationships that extract some of the value of a worker's labor and convert it into the investor's profit (Marx 1867). In the same vein, the scientists' metaphorical cosmology, by assimilating the world of mechanical production to the world of natural reproduction, obscures the social relationships and political choices underlying the design of new nuclear weapons. This semantic system constructs nuclear weapons by metaphorical implication as part of the natural order, and it gives metaphorical vigor to the "realist" assumption that the arms race and the development of new nuclear weapons have a momentum of their own, that "you can't stop technology."

Conclusion

I have argued that, as well as testing the reliability of particular weapons designs, nuclear tests are a means of testing and initiating the designers themselves. Tests also play a crucial role in structuring the hierarchies of knowledge and power at the laboratory and in inducting new scientists into the laboratory social and ideological system. While laboratory officials have argued that nuclear tests are necessary to maintain the technical reliability of the arsenal, and their critics have denied this and accused them of dishonesty, I have suggested that, if one looks at reliability figuratively as well as literally, then testing is vital to reliability as a felt reality.

Many anthropologists have argued that the myths and rituals of nonliterate peoples often contain what we might call, in our own terms, "scientific knowledge" about the world (Evans-Pritchard 1937; Horton 1967; Levi-Strauss 1966). I have here been arguing the complementary converse—that some of our most expensive scientific experiments are saturated with elements of myth and ritual. This is not to say that they are not really scientific experiments. It is to say that there is more to scientific experiments than meets the eye.

These experiments are ritual furnaces in which the abstract cliches that comprise the ideology of nuclear deterrence are forged into subjective truths in the lives of the scientists who design the weapons. Although the cliches legitimating the scientists' work are often transmitted verbally, through discourse, it is in the practical experience of these rituals that these cliches take on the incandescent quality of truth.

Notes

This article is based on fieldwork funded by a Mellon New Directions Fellowship at Stanford and a Social Sciences Research Council MacArthur Fellowship in International Peace and Security. Write-up was facilitated by a Weatherhead Fellowship at the School of American Research and a John D. and Catherine T. MacArthur Foundation Grant in Research and Writing in the Program on Peace and International Security. My thanks to David Dearborn, Steve Flank, Bill Zagotta, and Routledge's anonymous reviewer for detailed comments on drafts of this material, and to colloquium audiences at the Massachusetts Institute of Technology and Arizona State University West for their comments. This article is adapted from my book, *Nuclear Rites: An Anthropologist Among Weapons Scientists* (1996).

1. The Livermore Laboratory, located about forty miles east of San Francisco, is the younger of two nuclear weapons laboratories in the United States, the other being Los Alamos, where the bombs dropped on Japan were designed. Livermore was established in 1952. Its scientists designed the neutron bomb and the warheads for the MX, Polaris, Minuteman II, and Ground-Launched Cruise missiles. For more information on the laboratory, see Broad (1985, 1992); Cochran et al. (1987:44–52); Gusterson (1996).

2. This letter is reproduced in Appendix K of Kidder (1987).
3. For a more detailed exploration of the issues at stake in this debate, see Fetter (1987–1988, 1988) and Immele and Brown (1988).
4. Reliability is officially defined in such a way that, if a weapon certified at 100 kilotons only produced an 80-kiloton yield, it would be deemed unreliable—even though the explosion would still be many times more powerful than that at Hiroshima.
5. On incitement to discourse, see Foucault (1980, chapter 1).
6. The names, and sometimes genders, of all scientists in this article have been changed to protect their privacy.
7. For other attempts to apply ritual theory to science, see Abir-Am (1992), Davis-Floyd (1992), and Reynolds (1991).
8. By contrast, when India's scientists succeeded in exploding a nuclear device, they cabled their government with the message, "The Buddha is smiling" (Markey 1982:xiii).

CHAPTER 8

Political Structuring of the Institutions of Science

Charles Schwartz

Introduction

Two predominant political interests shape the direction of physical science and technology in the United States: the military and large manufacturing industries. The question I have studied over a number of years is this: How does the system work? In particular, what are the institutional arrangements that integrate the expertise of scientists into the reigning political-economic system of this country?

In thinking about such questions, it is important to understand that science is an organized human activity. Scientists work to collect and analyze information about the world around us and then frequently use this knowledge in practical ways. The work of science, when it is not just an exercise in idle curiosity by some individual, is generally integrated into the social and economic structure of society. That is to say, science is not separated from politics. In a world where knowledge is power, one should thus expect that the activities of science are largely under the control and direction of those sectors of society that hold dominant political power.

This external "control" of science is not complete control because scientific research itself has unpredictable outcomes; but it is effective control. It lies in setting the funding priorities—which areas of investigation will be thoroughly and rapidly explored (i.e., well funded) and which areas will be slighted or ignored. This control of science by dominant groups within society reinforces and legitimizes their positions of power and concomitantly places science in a privileged position in American society. Thus, science serves power.

This view of science and society is unremarkable to social scientists. To professionals in the natural sciences, however, these ideas may speak an awful heresy. We scientists are manufactured and delivered into the world with a special protective wrapping—the firm belief that science is neutral. We believe that the search for Nature's truth is pure and that the work of science is free from the contaminations of politics, money making, religion, passion, and prejudice of any kind.[1] Indeed, many young people are drawn into the study of science because it provides a comfortable refuge from the turmoil of human interactions.

This myth of the neutrality of science has been criticized by scholars of various disciplines, including historians, philosophers, and certainly scientists themselves (Bernal 1939; Feyerabend 1978; Hessen 1971). There have also been critical studies of the ideology of science and its integration into the political economy of society (Arditti, Brennan, and Cavrak 1980; Rose and Rose 1977; Yearley 1988). Such critiques include studies of the political malleability of scientists, illustrating the ease with which prominent scientists, as well as the rank and file of the profession, may be bent to serve the interests of those in command of the power of the state (Busch et al. 1991; Haberer 1969; Mukerji 1990). In this paper, I will contribute to such critiques by illustrating some of the concrete ways in which the external political control of science in the United States works through an examination of selected institutions and their leaders in the national scientific establishment.

The science that I talk about most is my own profession, physics, together with the branches of engineering that stem from the main divisions of physics, namely mechanics, electromagnetism, and atomic and nuclear structure. These fields of inquiry and invention have produced the material stuff upon which the industrial and military might of this nation is built.

Physics came to the forefront of science in this country during and after World War II. Over the past fifty years, a national establishment has been in control—not absolute control, but effective control—to stimulate, channel, and exploit scientific work in certain preferred directions. These directions are, by any definition, political, since the bulk of research and development (R&D) funding comes from the federal government and the directions in which they are invested have important consequences for all of us.

The principal method I have used in my work, termed "power structure research," follows the sociological analyses of elite culture by C. Wright Mills (1956) and G. William Domhoff (1967). Their studies elucidate why and how it is that a tiny minority of people (the elite ruling class) control the major decision-making institutions in our society. The time-honored study of "interlocking directorships" (e.g., the president of oil company X also sits on the board of directors of bank Y) shows how various large corporations are tied to one another. Power structure research follows that method to examine other dominant institutions, such as government (appointments in the executive branch), higher education, and the media, along with the major corporations. My study identifies the science establishment integrated into this network.

This paper, summarizing some of what I have learned, focuses on three contemporary institutional settings where scientists come into contact, either explicitly or implicitly, with national politics. The first concerns scientific advisory committees, wherein select scientists from academia work as expert advisors to the federal agencies most involved with the funding of and the applications of science. The second focus is on the two unique laboratories at Los Alamos and at Livermore, which have held a near monopoly of technical expertise in the fashioning of nuclear weapons.

Third, I examine universities (my own turf), where science professors like to think of themselves as engaged in the autonomous pursuit of "pure" knowledge and generally disregard the sociopolitical implications of their chosen work.

As a physicist and a professor, my personal experiences inform this study. While my criticisms of the very people and institutions that have benefited my career might seem ungrateful, my purpose and hope is to encourage more attention and commitment to social responsibility in science. To that end I conclude this paper with some practical principles that address the root problems of science and power.

Federal Science Advisory Committees

Academics in the United States are familiar with the peer review system, whereby a number of established scientists are asked to evaluate and rank the many competing proposals submitted to Washington for research funds within each specialized subfield. There is little politics at this level. However, when one asks how much money should be provided for each field and subfield, then fundamental political choices are being made.

The federal government, in addition to its own politically appointed staff—in the Department of Defense (DoD), the Department of Energy (DoE), the National Science Foundation (NSF), and so on—makes use of a number of advisory committees composed of selected individuals from a variety of relevant institutions. Thus, the army, the navy, and the air force each has its own Science Advisory Committee. Over them sits the Defense Science Board, which shapes the overall priorities for the Pentagon's R&D budget. (Federal R&D has three standard subdivisions: basic research, applied research, and development.) Recent budgets for all federally funded science and engineering work amount to some $75 billion per year; about 60 percent of this is dedicated to military programs.[2]

These science advisory bodies are made up of experts from the most relevant industrial concerns, weapons laboratories and think tanks, some officers from the military services, a few former government officials, and a number of professors from our leading universities. It is widely supposed that this arrangement provides input for shaping government science priorities that is diverse, if not democratic, and balanced, if not unbiased. But this is an illusion. Let me give an example.

The President's Science Advisory Committee (PSAC), established in 1957, was the pinnacle of science advisory bodies in Washington until its abolishment in 1973. Its members enjoyed the highest access to political power and they were correspondingly looked up to by their fellows in the scientific community. A majority of the members on PSAC came from academia, with only a minority from private industry or elsewhere. This committee composition gave the impression that this was a body that could be trusted to act in a "purely scientific" way since they were free from the influence of any "special interest." This impression was an illusion.

The following data were published a number of years ago following my study of the membership of the PSAC (Schwartz 1975). Out of a total of seventy-eight individuals who had served on PSAC over its sixteen-year lifetime, fifty-five members (71 percent of the total) were identified as academics. More than 66 percent of these academic members, however, were also shown to have significant ties to large corporations involved with science and technology. Indeed, over 50 percent of the academic PSAC members were found to have been members of the boards of directors of Exxon, General Motors, IBM, McDonnell-Douglas, DuPont, Xerox, Westinghouse, Dow Chemical, Northrop, TRW, Hewlett-Packard, Raytheon, Squibb, Merck, and other major corporations.

Similar situations were found on a number of other leading science policy bodies. The NSF is thought of as the home of funding for pure academic science. A study of the NSF board members over that same sixteen-year period showed that 79 percent came from academia, the rest from industry, government, private foundations, and other nonprofit organizations. However, it was also found that over 50 percent of these academic NSF board members had important ties to large corporations, as with the PSAC members. This interlocking demonstrates a substantial integration at the top levels of scientific leadership, the federal government, and the major corporations (Schwartz 1975).

How were individual scientists selected to serve in these elevated positions? It was not enough to have earned a strong reputation as a scientist and to be at a respected institution. A well-established "old boys' club" still carefully selects and cultivates initiates into the circles of government science advising. How this system works was described in an interview some years ago by one of these most experienced and respected individuals (Berkeley Scientists and Engineers for Social and Political Action 1972).

He said that there was a good deal of incest in science advising. People with the most experience were reused, and younger people were brought into subsidiary committees where they could learn how to handle things and then gradually move up if their performance was found satisfactory. He listed the criteria as talent, objectivity, and a willingness to work. It was also basic practice for the adviser to accept the idea that he worked privately for the agency or the person whom he was advising—complete secrecy was required even though the scientific recommendations given were often not followed. He stated that the human element—the personal relations between the adviser and the advisee—were very important to the success of the advising process; yet he continually stressed that the advising was strictly objective, nonpolitical, and related only to technical evaluations.

These desiderata (private, secret, personal) are clearly based upon the subservience of the scientist to the authorities presently in power. That this subservience contradicts the nominal ethic of science, which requires independence of thought and action with frequent challenges to existing authority, seems not to bother the chosen advisors.

PSAC was abolished in 1973 by President Richard Nixon. Since that time, the chairs of science advisers to the White House have been filled with individuals selected quite explicitly for their political reliability in supporting the administration's programs. That is, newer White House science advisors see themselves as near cabinet-level political appointees, while the earlier PSAC members thought of themselves as above partisanship.

Table 13

Past and Present Directors

Directors of the Los Alamos Laboratory:

J. Robert Oppenheimer (1942–1945); founder of the laboratory
Norris Bradbury (1945–1970); Ph.D. 1932, first postwar director
Harold Agnew (1970–1979); Ph.D. 1949, with Los Alamos from 1942
Donald Kerr (1979–1985); Ph.D. 1966, with Los Alamos from 1966 (DOE 1977–1979)
Siegfried Hecker (1985-); Ph.D. 1968, with Los Alamos from 1968 (away 1970–1973)

Directors of the Livermore Laboratory:

Herbert York (1952–1958); Ph.D. 1949, first director of Livermore
Edward Teller (1958–1960); founder of the laboratory
Harold Brown (1960–1961); Ph.D. 1949, with Livermore from 1952
John Foster (1961–1965); Ph.D. 1952, with Livermore from 1952
Michael May (1965–1971); Ph.D. 1952, with Livermore from 1952 (away 1957–1960)
Roger Batzel (1971–1988); Ph.D. 1951, with Livermore from 1953
John Nuckolls (1988–1994); M.A. 1955, with Livermore from 1955
C. Bruce Tarter (1994-); Ph.D. 1967, with Livermore from 1967

Table 14

Assistant to the Secretary of Defense for Atomic Energy and Chairman of the Military Liaison Committee (linking DoD and DoE)

Gerald W. Johnson (1961–1963); with Livermore 1953–1961, 1963–1966
William J. Howard (1963–1966); with Los Alamos 1946–1956, Sandia 1956–1963
M. Carl Walske (1966–1973); with Los Alamos 1951–1956, 1965–1966
Donald R. Cotter (1973–1978); with Sandia 1948–1966, 1968–1970, then with DoD, AEC
James P. Wade, Jr. (1978–1981); with Livermore 1961–1965, then with DoD
Richard L. Wagner, Jr. (1981–1985); with Livermore 1963–1981
Robert B. Barker (1985–1994); with Livermore 1971–1983, then with ACDA, now back at Livermore as a high level administrator

Nuclear Weapons Laboratories—Los Alamos and Livermore

In institutions of nuclear weapons research and development, there is an even tighter structure of interlocking, self-selecting leadership that not only works directly for the government, but also has had enormous influence upon it. This particular clique of scientists influenced the whole history of the nuclear arms race, particularly in its political and technical dimensions.

All the nuclear weapons design work for this country is conducted at two laboratories: one located in Los Alamos, New Mexico, and the other in Livermore, California. These may be the largest scientific laboratories in the world. They employ about 8,000 people each and have annual budgets of over $1 billion a year each. Their funds and their mission are dictated by the DoE, acting as a surrogate for the DoD. These are strictly hierarchical organizations, structured like a corporation or the Pentagon, not like a university campus. Promotion results from recognition of leadership and management skills; that is, getting a project done, on time and within specifications. The ability to initiate and sell new programs is highly rewarded. Loyalty to the organization is key to an individual's success.

Table 13 lists all the past and present directors of the Los Alamos Laboratory and of the Livermore Laboratory.

What is significant about this table is that, apart from the founders J. Robert Oppenheimer at Los Alamos and Edward Teller at Livermore, the scientists worked their way to the top from within the laboratory. Indeed, it appears that almost all have been with their laboratory from the beginning of their career. The nuclear weapons lab is not the end of the road by any means, however, as is illustrated by the subsequent career paths of three scientists who were early directors at Livermore.

Herbert York left the lab in 1958 to become the chief scientist in the Pentagon (his official title was Director of Defense Research and Engineering). With the change of administration in 1961, York left Washington and was the founding chancellor of the University of California campus at San Diego. He continued to serve on important advisory committees (e.g., PSAC) and was a key arms control official under President Carter. York's successor as head of the Livermore Laboratory was Harold Brown. He took over the Pentagon post of director of Defense Research and Engineering when York left in 1961. Brown went on to serve as secretary of the air force under President Johnson; during the Republican White House years 1969–1977, he was president of the California Institute of Technology; he returned to Washington as Secretary of Defense in the Carter administration. Brown's successor as director of Livermore was John Foster, who likewise became director of Defense Research and Engineering and eventually left the government for a corporate vice presidency at TRW.

Another key government position is the Assistant to the Secretary of Defense for Atomic Energy. This person also heads the Military Liaison Committee, which links the DoD to the DoE's nuclear weapons complex. I have tabulated all the scientists who have held this position over the past three decades (see table 14). Every one of these officials has come out of Livermore, Los Alamos, or the Sandia Laboratory, which works intimately with these two labs on nuclear weaponry.

Many other scientists at the upper management levels of the two laboratories have moved into a variety of government positions, and frequently these people return to the labs with increased importance owing to their contacts and experience with the government officials who determine the laboratories' budgets. They carry with them not only their technical expertise but also their narrowly constructed sense of values, shaped from a career devoted to the weapons laboratories, and loyalties to the well-being of those institutions. They have played a vital role in shaping the government's options and priorities not only for the budgets of these laboratories but also for the direction of national policy on all aspects of nuclear weapons development and arms control. We have documented their aggressive promotion of the neutron bomb, their relentless lobbying against a comprehensive test ban, and their scandalous campaigning for President Reagan's Strategic Defense Initiative (SDI). The pervasive cloak of military secrecy prevents a thorough study of the extent and character of the interaction and interlocking of the weapons laboratories with the government's policy-making apparatus, but what we do know is very troubling.

A few years ago I had the opportunity to present this data about the self-perpetuating clique of nuclear weapons scientists and their pervasive presence in the government at a Seminar on Controlling Processes at Berkeley. A visitor at that seminar was someone who had spent many years working inside the Pentagon and he confirmed the picture that I had drawn from outside sources of data. He also said that the members of this select group are very conscious of their special sociology. They jokingly refer to themselves, he said, as "the nuclear mafia." I sometimes think that the analogy to "mafia" in describing the sociopolitical organization of science is an insightful one but I will not pursue that topic here.

A final issue concerning the Livermore and Los Alamos nuclear weapons laboratories is the role played by my employer, the University of California. The University of California Board of Regents manages these weapons laboratories under contract with the DoE. This has been a matter of controversy in the university for over twenty years. The university does not really manage the weapons labs in any responsible way; laboratory officials, and their associates in Washington, use the University of California name and reputation in their own self-interest to extend the aura of respectability. Harold Agnew, formerly director of the Los Alamos Laboratory, has acknowledged that this academic affiliation gives them added credibility in Washington.

In 1990, the faculty academic senates on all campuses of the University of California voted—by a 64 percent majority—to phase out this relationship with the weapons labs. The regents, ignoring the faculty vote, decided to renew the contract for another five years. I believe that money is not a major factor in shaping the university's role as patron of the laboratories. It is rather the political and ideological orientation of the Board of Regents. It is not unique to the University of California but is found throughout the country—a historical legacy that the governing boards of our great public universities (not to mention the private ones) are dominated by wealthy and well-connected individuals (members of Mills' and Domhoff's elite ruling class) who rule without restraint and without accountability to anyone but themselves.[3]

Universities

Finally, I focus on university campuses to see where and how the political needs of the military establishment are served by research in and teaching of science. This is more subtle than straight power structure research. The research managers in the Pentagon know about the sensibilities of professors and are careful to keep unpleasant details about warfare concealed behind the purity of scientific challenge and research opportunities.

Let me give one recent example. The Superconducting Super Collider (SSC)—was a planned multibillion dollar particle accelerator that high-energy physicists worked on—until the project was killed by Congress in 1994—as their last best hope for blasting loose the secrets of elementary particles. This was a project in the purest of pure science, to which President Reagan gave his blessing (that is, he approved funding) in 1987. Acting out of curiosity, I wrote to the DoD asking them, under the Freedom of Information Act, to provide me with copies of all Pentagon documents relating to the SSC. This is not a DoD project; it is funded by the civilian science branch of the DoE. Nevertheless, I wondered if the government's internal decision making might have included some broader considerations. After two years of waiting I did receive a document, written from the Office of the Secretary of Defense to the White House just days before President Reagan gave his approval to the SSC. Here is the relevant quotation.

> We have reviewed the SSC proposal and ... support [it].... The SSC project will have many spinoffs for the DoD, especially in technologies required by the Strategic Defense Initiative, including particle beams, information processing, computer control, pulse power sources, and high energy accelerators.
>
> The nuclear weapons community will benefit from the fundamental research on the building blocks of atomic matter. The SSC will provide a valuable resource of scientific personnel. Many of the scientists now in the DoE nuclear weapons laboratory complex received their training while working on particle accelerators. (Office of the Secretary of Defense 1987)

While the Pentagon directly controls the majority of all federal R&D funds and about 80 percent of federal R&D funds in the physical sciences, its presence in basic research support on university campuses is much diluted, amounting to about 25 percent of academics' outside research funding. In certain fields, however, such as computer science, and on certain campuses, it may be the largest player. Just because a research project is funded by an agency other than the DoD does not mean it is free of military implications. The main funding agencies—DoD, DoE, NSF, and NASA—regularly consult and coordinate their research support programs. A 1982 report by the Defense Science Board states, "Research and development in universities is supported by many sponsors, each relying on complementary funding from the other sponsors to leverage its own expenditures."[4]

Existing university policies require that no secret research be allowed on campus; most recipients of this money will assert that they are not designing weapons but merely engaged in basic research. Sometimes the purpose of a particular research program is clearly militaristic, as in the case of SDI or the DoD's computer science funding for automated battlefield systems. Healthy discussion, debate, and even dissent may develop around such projects. Frequently, however, it is not a simple matter to draw connections between research projects conducted at a university and the development of particular end products. Nevertheless such connections can be drawn, especially by people actively working at the frontiers of a given field of research. It appears that most academic researchers find it more comfortable to avoid, rather than face up to, these troublesome questions. This silence is one more mark of the political success enjoyed by the military establishments in directing universities to do what they want.

There are some exceptions. Several years ago a large number of academic researchers protested publicly against the SDI program, pledging to refuse money for work on that dubious military project. This famous protest was, to my mind, the exception that proves the rule of what some observers see as compliant academic science. Again, I reiterate that the nature of the interaction is subtle and indirect. What the military wants from university scientists is not only the results of their research. The training of students in selected areas of science and engineering is also a top priority for the DoD. Most of the money provided in research grants to the campus goes to supporting graduate students and other research staff working under the direction of the professors given the funding. Even though most of this campus research and training is unclassified (i.e., non-secret), basic research with no obvious weapons application, the payoff for the Pentagon comes when these highly skilled and specialized students seek employment outside academia.

As an example of delayed military benefits, consider this statement:

The point at which career decisions, career directions, begin to be set for graduate students is the point at which they decide what direction they are going to go on their dissertation. If they are engaged early in work that is intellectually stimulating to them and that has some promise for the future and is supported by the DoD, it seems to me you are well on the way to having them hooked into that enterprise for a long time. (Rosenzweig 1985)

These are not the words of a military officer or even a civilian on the staff of the Pentagon, but Dr. Robert Rosenzweig, the president of the Association of American Universities, speaking at the 1985 congressional testimony in support of Pentagon funding for universities.[5]

Far too often I have found that scientists in the university don't want to talk about the problem of the militarization of science. They are unwilling to admit that it is a problem that touches them. It seems that many of my colleagues, when confronted with this issue, will automatically draw upon some familiar rationalizations in order to avoid getting more deeply involved. One good way to develop a deeper understanding of the problem of science and the military—and an exercise I have recommended to students in science—is to think about and discuss the following list of rationales that are commonly used. Each has a core of truth, but also a serious shortcoming.

- We need the best in science and technology to maintain our national security.
- Research is essential so that we know what threatening weapons are possible.
- I only work on defensive weapons, not offensive ones.
- If I don't do this work on weapons, someone else will.
- It is better to have weapons work done by an enlightened person like me.
- By being involved in the weapons program I can be an effective influence on the government.
- With nuclear weapons, war is unthinkable and we provide the political leaders with time to resolve the international problems that cause war.
- I am just a scientist doing my job; I stay out of politics.
- I only do pure research; whether it leads to weapons or not is out of my control.
- I take DoD money, but I am just doing basic research, not work on weapons.
- The DoD is the only agency that has money for the work I want to do.
- My research, although paid for by the DoD, is completely unclassified and it is the work I want to do, not what they tell me to do.
- I am fooling the DoD by taking their money for my research, which they would otherwise spend on weapons.
- I don't use DoD money; DoE and NSF fund my research.
- I don't have any government research funds; I am just a physics teacher.

The job market for certain areas of science and technology—most notably physics, electrical engineering, computer science, and mechanical engineering—offers some rude shocks for graduates who do not wish to work on weapons projects. The military aspect of such a career is a life-shaping reality that we educators do not generally bother to tell our students about. Perhaps we prefer to remain unaware of this unpleasantness so as not to spoil the idealization of our academic science. A few years ago I produced a series of booklets containing information on the job market and other military aspects of careers in science and engineering.[6] I had very little success in getting these booklets distributed to students, mostly undergraduates but also graduate students, who were planning to enter these professions. Only a few of my professional colleagues around the country were willing to distribute them to their students, and, not surprisingly, the leaders of the American Physical Society wanted nothing to do with it. Again, silence and compliance are the marks of successful political control.

Conclusion

Here we have seen some details of the mechanisms by which the enterprise of science is enslaved to the dominant political and economic interests of this society. Money and power (status) are the most apparent variables in this process, and select individuals provide further important linkages between the practitioners and institutions of science and their most powerful patrons, the military and large corporations. The prevailing myth of neutrality in science obscures the influence of these external political forces not only from the general public but also from scientists and students of science.

Leading academic scientists serving on high government advisory committees give an illusion of objective and disinterested expertise in the service of national public interest but the revelation of their many interlocking connections with large corporations raises the spectre of significant bias. A study of the country's two nuclear weapons laboratories reveals a tight clique of technical experts, laboratory administrators, and government officials. In the universities, science professors and their students habitually close their eyes to the ways in which their pursuit of "pure" science is channeled to meet the needs of the Pentagon and industry.

Some people respond to this analysis by wishing that scientists might enjoy greater freedom and independence from such political forces. Such a hope is folly. The enterprise of science is a part of society and cannot exist outside the world of people. Science cannot escape the play of politics, economics, morality, and culture. Yet there is still room for scientists to choose which set of human means and ends they would serve with their skills. Furthermore, citizens who usually leave issues of science/politics to the experts or to the officials in power might also take a more active role.

A haunting story surrounds the famous play *Galileo*, written by Bertolt Brecht half a century ago. In the first (1938) version, Brecht focused Galileo's conflict with the church on the principle of freedom for scientists to search for truth. Writing in the growing shadow of Nazism, Brecht fled Germany where truth and freedom were being trampled under the boot of a totalitarian state. He revised the play in 1947, writing now in the shadow of the atomic bombs dropped on Hiroshima and Nagasaki. Deleting the climactic lines calling for more freedom for science, he substituted a plea for more social responsibility in science. "If I had resisted," he has Galileo say, "if scientists had been able to develop something like the Hippocratic Oath of the physicians, a vow to apply their knowledge only for the benefit of mankind! As matters now stand the most one can expect is a race of inventive dwarfs who can be hired to do anything."[7] In the atomic heat of fear, Brecht turned scientists from heroes to villains. How shall we, today's scientists, confront with this double image?

Scientists need to recognize that they are implements of political power and ask the questions of who, why, and how. We should not expect science to be free or autonomous; this would be an ideological delusion. By denying the existence of the powerful beneficiaries of our work, we only sanctify and increase their hidden power. The basic tenet of social responsibility in science is that we must not shirk the difficult tasks of assessing the likely consequences—in human terms—of the work we do, recognizing that we have choices, and taking responsibility for our actions. I offer two proposals for academic science that, while modest in scope, may assist us in these tasks.

First, any external agency seeking to fund a research project on campus must provide a complete disclosure of its interests and purposes in connection with that project. In addition, each professor receiving an external research grant should be required to prepare and make available for general discussion a detailed assessment of the likely applications and consequences of the research work.

Second, science educators have a responsibility to see that their students, especially in areas that have potentially harmful (e.g., military) applications, are well informed about those aspects of the job market. Study of the concepts and practice of social responsibility in science should become a regular part of the science curriculum.

Notes

My thanks to C. Jay Ou for his assistance in revising this paper.

1. Physical scientists in the United States are overwhelmingly white and male, but I will not address that here.
2. See Irwin Goodwin, "Cheers for Bush's 1993 R&D Budget . . ." in *Physics Today*, June 1992, page 55.
3. The DoD has major weapons laboratories managed by six other universities: the Massachusetts Institute of Technology, Johns Hopkins, University of Washington, University of Texas, Pennsylvania State University, and University of California at San Diego.
4. For reference to this and other relevant quotations, see Selvin and Schwartz (1988:6).
5. For reference to this and other quotes and data see the Schwartz booklet series, "Social Responsibility—Information for Students on the military aspects of careers in Physics" (1989), available from the author.
6. See note 5.
7. These lines, absent from some English versions of the play, were noted and translated from the German edition by Rather (1969:132).

CHAPTER 9

Constructing Knowledge across Social Worlds

The Case of DNA Sequence Databases in Molecular Biology

Joan H. Fujimura and Michael Fortun

The Human Genome Project (HGP), which in the United States is largely institutionalized in the National Institutes of Health (NIH) and the Department of Energy (DoE), has as one of its stated goals to map and sequence "the" entire human genome. The human genome refers to the twenty-three pairs of chromosomes (an estimated three billion base pairs of DNA) that are a part of every human cell. The HGP has become a convenient focal point for discussions about human genetics and its medical applications as they have developed over the last decade and will continue to develop in the years to come. Prominent figures in molecular biology have argued for the importance of the HGP as a foundation for progress in biology and medicine. The news media have headlined the almost weekly reports of "discoveries" of genes for conditions ranging from cystic fibrosis, cancer, and heart disease, to schizophrenia, homosexuality, and "cuddling."[1] While recognizing the potential value of knowing which gene or genes are involved in the initiation and development of conditions such as cystic fibrosis, many social scientists, medical ethicists, and bioethicists have apprehensions about much of the HGP. These concerns include the privacy of information, the efficacy of genetic screening and its social impact, the "geneticization" of social and behavioral disorders, and the social construction of definitions of disease and health.[2]

This paper takes another approach to molecular genetic research, looking primarily at the computerized databases that store much of the information produced by this research. This paper is one effort to educate ourselves and other members of non-molecular biology audiences about some of the work involved in this project, the relationships among scientists doing different kinds of work, and the implications for how scientists conceptualize and intervene in nature. Our own work is situated at the intersection of the symbolic interactionist tradition in sociology and the various disciplinary approaches (sociology, history, philosophy, and anthropology) that are part of science studies. The suggestions, proposals, and questions presented here are based on a combination of field observation, interviews with scientists, and published material. We close with some important questions on which our current research project is based.

The HGP and the "human genome project"

We begin with a quick description of the HGP and draw some fuzzy boundaries between it and the "human genome project." We realize that the similarity between these homologous terms can be confusing and awkward. But the distinction is both necessary and problematic, and the simple difference of capital letters keeps the boundaries both visible and fluid.

The HGP is a coordinated effort aimed at mapping and sequencing "the" entire human genome, as well as the genomes of "model organisms" such as mice and the nematode *C. elegans*. In the United States, in the mid- to late-1980s it was the subject of extensive and at times contentious debates within the molecular biology community; among disciplines such as molecular biology, biochemistry, and human genetics; and to some extent between scientists and a broader public, including Congress. Through numerous discussions, government panels, special reports from the Office of Technology Assessment and the National Research Council, and congressional hearings, strategies for the scientific work were agreed upon and political differences (centered largely on whether the NIH or the DoE should lead the project) were ironed out. In October 1989 the HGP became official with the establishment of the National Center for Human Genome Research (NCHGR) within the NIH, under the initial direction of James D. Watson.[3] The human genome project, by contrast, has been an ongoing effort since the late 1970s involving many scientists, public institutions, private corporations, and national governments to fund and carry out the research of mapping and sequencing human genes, and to develop the basic tools and materials of recombinant DNA research. The first genetic linkage maps covering the entire human genome, for example, were produced by scientists working at the biotechnology company Collaborative Research, Inc., and were published in 1986. Another laboratory at the University of Utah, funded by the Howard Hughes Medical Institute, had already published several maps for specific chromosomes, as had other labs in the United States and the United Kingdom. Many other technologies, techniques, and materials such as the automated DNA sequencer, the polymerase chain reaction (PCR), and new cloning vehicles like yeast artificial chromosomes (YACs) were being developed in a wide array of public and private settings years before the highly managed and generously funded HGP was organized. Indeed, it was in large part the increasingly rapid production in the early and mid-1980s of such maps, the genetic sequences and other DNA probes they were based upon, and the availability of increasingly productive technologies and techniques that became the rationale for subsequent calls for the federally sponsored HGP.

At this time, the HGP is focused more on constructing maps (genetic and physical) of the genome than on sequencing the genome. Further improvements in technology and methods have also been given a high priority. This is due in large part to the concerns of some scientists about the enormous expense of resources needed to sequence the entire genome. They want to wait until sequencing techniques have been made more efficient, and therefore faster and less expensive, before embarking on the dedicated sequencing part of the project. In addition, some scientists have argued that large amounts of sequence information are relatively less valuable and interesting than mapping information or the sequences of known genes or gene regions.

What is the difference between mapping and sequencing genes? First, there are several kinds of maps. Using recombinant DNA technologies, researchers have built genetic maps for human chromosomes over the last decade as a tool for locating specific genes, particularly genes whose biochemical product is unknown. Genetic maps are the basis for prenatal screening technologies for diagnosing conditions that have a genetic component. Genetic maps are linear arrangements of markers. These markers are combinations of known genes and "anonymous" DNA sequences corresponding to specific DNA probes. These DNA sequences are called "anonymous" because they are not necessarily part of functioning genes, and knowledge of their precise sequence is relatively unimportant. The distances between these markers do not necessarily represent physical distances but are instead translations of statistical frequencies of recombinations between parental chromosome pairs. The lower the frequency of recombination, the smaller are the distances between the markers, that is, the more tightly linked or the closer together on the chromosome they are (Suzuki et al. 1989:100 ff.). For example, screening kits for cystic fibrosis are constructed to indicate markers that are closely linked to the cystic fibrosis gene. Linkage maps (as genetic maps are also called) are used to locate a gene of interest (e.g., a gene coding for a particular disease) in a particular part of the chromosome by providing a starting point for "walking down" the chromosome to then physically locate the gene of interest.

Physical maps are supposed to represent the "actual" location of a gene of interest on a chromosome. Physical maps vary in their degree of resolution, depending on the technology used in their production. The patterns of bands visible on a stained chromosome under a light microscope are the lowest resolution physical maps. With the use of DNA probes, they can give very general indications of the location of genes on a chromosome. Collections of cloned fragments of chromosomes constitute another kind of physical map. These are pieces of human DNA, cut into sizes of varying length, cloned into the genomes of bacteriophage or other "vectors" that can be maintained in culture, and "ordered" or arranged in linear sequence by means of experiment and statistical algorithms run on computer software packages. At the highest degree of resolution are the DNA sequences themselves, often called the "ultimate" physical map.

Sequencing genes, on the other hand, is the reading out of the nucleotide bases that constitute the genes along the entire genome or parts of it. Sequencing is regarded by many scientists as not very useful and actually wasteful because it means taking the resources to read out repetitive sequences and intervening sequences ("introns"). Introns are often called "junk DNA" because molecular biologists do not currently understand their functions and because they complicate current experimental recombinant DNA work.[4] Instead, many scientists argue that the work should begin with a problem like cystic fibrosis and then focus on mapping the gene (locating its general vicinity on the genome) and then sequencing it once it is "found." Many argue that sequencing is boring, technical work that their graduate students, and even undergraduate students, do not want to do. Others are concerned about current abilities to handle the amount of information that would be presented by sequencing the three billion nucleotide bases in the human genome. Still others, including members of the various groups of the interested lay public, are concerned about the possible impacts of having the entire genome sequenced. We discuss some of these concerns at the end of this essay.

Molecular Genetic Sequence Information Databases

The sequence information we are talking about is coded in terms of deoxyribonucleic acids (DNA) and ribonucleic acids (RNA) and, for proteins, in terms of amino acids.[5] When one logs onto a sequence database, one sees a linear readout of A, C, T, Gs for DNA, A, C, U, Gs for RNA, and A, R, N, etc. for the single letter codes for amino acids.[6] Databases are the computerized version of publications of sequence information, often accompanied by some other biological information. Before the construction of more efficient retrieval software programs for accessing the computerized databases, scientific journals and books published sequence information related to particular topics.

Databases are a major focus of interest and concern to proponents of and participants in the HGP because of the volume of information that has and will continue to be produced by the work of mapping and sequencing human genes. A human genome alone has an estimated three billion base pairs of DNA. However, while the volume of information being generated by the HGP highlights the crisis in the informatics of molecular biology, molecular genetic sequence databases are not a new product of the HGP. Databases of DNA, RNA, and protein sequence information about many different organisms have existed for some time, beginning with considerable support from the Department of Energy (DoE). For example, in July 1988, before the HGP was lobbied into existence, GenBank (at Los Alamos National Laboratory) contained approximately eighteen million base pairs of total DNA sequence information and approximately two million base pairs of human DNA sequence information (Colwell 1989:124). In addition to human genetic sequences, DNA, RNA, and protein sequence information is also available for other organisms, including the much-studied yeast, mice, *Drosophila* (fruit fly), *E. coli*, and nematode, especially the worm *C. elegans*. Indeed, *C. elegans* has been promoted as a model for the sequencing and database building of information on the human genome (Roberts 1990; Schatz 1991).

Besides the sheer volume of information to be stored, sequence databases are also valued because they allow scientists a faster and more efficient method for accessing information used in their experiments and for interpreting experiments. Some of the kinds of analyses scientists can perform using the database system include translation and location of potential protein-coding regions; inter- and intrasequence homology searches; inter- and intrasequence dyad symmetry searches; analysis of codon frequency, base composition, and dinucleotide frequency; location of AT- or GC-rich regions; and mapping of restriction enzyme sites.

Homologies

Homology searches are the most common operations for which sequence databases are currently used. The theoretical definition of DNA sequence homology is agreed upon by practicing scientists to mean that two sequences are similar and that they are evolutionarily related in the sense of having a common ancestor in an evolutionary tree structure.

> When two sequences are homologous, they share common ancestry. In this sense, there are no degrees of homology. Sequences are either homologous or they are not. Many investigators use the word when they mean "similar." Two sequences may be similar by chance, for example. They may resemble each other to a high degree, but they ought not to be very homologous or slightly homologous. Also, two sequences may be 60% identical, but they are not 60% homologous. (Doolittle 1987:35–36)

> Similarity simply means that sequences are in some sense similar and has no evolutionary connotations, whereas homology refers to evolutionarily related sequences stemming from a common ancestor. (Von Heijne 1987:123)

Homology is a theoretical or hypothetical notion. While there is no disagreement about the theoretical definition of homology, in practice scientists cannot usually determine common evolutionary ancestry without other kinds of evidence (for example, paleontological evidence). In practice, researchers usually use only similarity to decide when two sequences might be homologous because paleontological and other evidences are often unavailable. They then go to the "wet lab" to carry out molecular biological and biochemical experiments to examine whether the two sequences produce similar proteins, that is, whether they have similar biochemical functions. They use these "wet lab" tests to confirm or disconfirm their hypotheses about homologous relationships.[7]

The following short example of a homology search indicates the speed and efficiency provided by computer search and match procedures and the subsequent "wet lab" tests that resulted in interesting findings. The example also shows the speed and efficiency valued by researchers.

Several years ago, Joseph Brown ... purified a melanocyte tumor cell antigen. There was only enough material available for a single microsequencer run, and that endeavor only managed 13 cycles. In fact, only 10 of those first 13 amino acids were identified with any confidence. Still a search of our data bank turned up only a single candidate: transferrin.... Brown promptly tested to see if the tumor antigen bound 59Fe. It did, and with the same avidity as transferrin.... Today that protein, which appears to be an important factor in rendering the tumor melanocyte immortal, is known to be about 40% identical with the better known transferrins and is called melanotransferrin.... The take-home message is: A search of even a very short sequence may put you on the right track and save years of work. (Doolittle 1987:17)

However, biochemical experiments take much more time to carry out than do computerized sequence search and match procedures. The difference in time will increase as the amount of information and higher quality software become available. A difficulty arises then with the division of labor between constructing versus confirming theoretical relationships. Suppose a homology search produces a successful match. Who will gain the credit for the finding? Should it be the researchers who locate the match, despite the fact that they provide only half of the match? A current problem is the reluctance of researchers to submit their sequences (or a delay in their submission) to public on-line data bases because of the threat of being scooped by other researchers who might later find matches to their sequences and publish articles based on these matches. A solution to this second problem has been incorporated at GenBank: they provide to the researcher who "hits" on a potentially significant match only the name and telephone number of the scientists who had deposited the first sequence (Anderson 1989). The assumption is that the two laboratories can then share the details of their work and collaborate rather than "scooping" each other's work.

A related problem is not so easily solved. Computational theoretical biologists use computers in their "dry labs" to construct theories of nature. Since they use the sequences submitted by molecular biology and biochemistry laboratories, they are often viewed as "feeding off" the detailed labors of "wet lab" researchers. In one case, theoretical computational biologists submitted a paper to a journal arguing for a functional relationship between two proteins on the basis of some computer work and "thinking." The biochemical researchers who refereed the paper rejected it on the grounds that there was no experimental "wet lab" work to support the "speculation." Given the growing division of labor between computational theoretical biology and laboratory molecular biology/biochemistry, assigning priority and ensuring equity in scientific work is becoming more problematic. However, a more important concern is the longer term consequences of this growing division of labor in biology for producing representations of nature.

A New Paradigm in Biology?

The problem of priority, based on these changes in the division of labor, is just one indicator of larger changes in the organization of biological work. Molecular biology is fast becoming more dependent on tools from mathematics/statistics and information and computer science. It is also simultaneously developing and perhaps changing those other disciplines. For example, these three disciplines form a major part of the infrastructure of what the National Science Foundation (NSF) calls "a broadly based research effort in mathematical and computational aspects of modern biological problems." Our argument is that this infrastructural change translates into complex changes in both the way work is done in biology and in the way nature is represented. We argue that databases currently represent the best place to study the intersection of biology, computer science, and mathematics/statistics and its consequential changes in knowledge construction and representation.

Walter Gilbert, a molecular biologist at Harvard University, argues that sequence databases and software are creating a paradigm shift in biology. Biology is moving from an experimentally based discipline to a theoretically based discipline.

In 1991, Gilbert published a discussion piece entitled "Molecular Biology Is Dead—Long Live Molecular Biology: A Paradigm Shift in Biology" in *Science* (Gilbert 1991; see also Gilbert 1990) in which he argued that molecular biology had become a technological science and was therefore in danger of losing its position as a science, both institutionally and intellectually. "Conventional molecular biology, now used to attack all the problems of biology, dies if it is a list of techniques" (Gilbert 1990:2). Molecular biology is in danger of becoming a "pocket molecular biology," he claimed. As in chemistry, statistics, and computer science, the balance between technology and science is tending toward the technology side. Molecular biology today can easily be framed as a technological expertise or set of skills for servicing other sciences. This was a commonly accepted view as early as 1985, when Terry Stokes (1985) described a fierce debate at the Burnett Institute in Australia about whether to incorporate a molecular biologist into each department to aid but not define the department's problem-solving efforts, or whether to establish a separate molecular biology department that might choose to construct its own problems and not "service" the other departments. This view of molecular biology as a technology, not a science, is even more exaggerated in the 1990s. Gilbert's answer to the malaise, technologization, and institutional battles is to transform molecular biology from an experimental science into a theoretical one. He argued that the sequence information in the databases will become the new "reagents" that molecular biologists manipulate in the computer to construct theoretical models about genes and biological function. These theories may then be experimentally tested in the laboratory. The important part of the work, however, lies in the hypotheses, theories, and models constructed by molecular biologists using bits (bytes) of information in the computerized databases. Other researchers concurred with Gilbert's view. Theoretical sequence analyses of various kinds are important in this context, since many of the relevant patterns are not immediately obvious. The sheer amount of sequence data now available also makes the appearance of a new breed of "theoretical molecular biologists" unavoidable, and if current proposals to sequence "the" human genome (as if this were a well-defined entity) are put into practice, automated data handling and, above all, data analysis must be given a very high priority (Von Heijne 1987:151).

Whether molecular biologists are more technicians than theoreticians/thinkers is an empirical question. If so, then Gilbert's call for a "paradigm shift" is both attractive and threatening. It requires a change in the skills, or a retooling of skills, on the part of many molecular biologists, yet it provides a way for molecular biology to save itself from becoming a service industry. The question about the implications of a shift perhaps should be specified, then: Attractive to whom? Threatening to whom?

Whether Gilbert is right or not, this interdisciplinary effort, or intersection of disciplines and technologies, has already begun. An organizational outcome of the informatizing of molecular biology is a current and future reorganization of the division of labor to include computer scientists, statisticians/mathematicians, and engineers. The inaugural annual report of the NCHGR stated that

> meeting the challenges presented by genome informatics is complicated by the small number of investigators who are available to work in the area. To reach its goals, the NCHGR is attempting to stimulate new collaborations among geneticists, molecular biologists, computer scientists, statisticians and other mathematicians, and experts in specialized integrated circuit design, robotics, and other fields of hardware and software engineering. (1990:30–31)

It does this through short courses, workshops, and interdisciplinary training programs. This retraining of biologists and reorganization of the division of labor in biological work has already begun and will likely speed up in the near future.

> The development of the matrix and the extraction of biological generalizations from it are going to require a new kind of scientist, a person familiar enough with the subject being studied to read the literature critically, yet expert enough in information science to be innovative in developing methods of classification and search. This implies the development of a new kind of theory geared explicitly to biology with its particular theory structure. It will be tied to the use of computers, which will be required to deal with the vast amount and complexity of information. (Morowitz and Smith 1987:3)

Our point is that this reorganization of work also defines a new biology and new biological phenomena. "To use this flood of knowledge, which will pour across the computer networks of the world, biologists not only must become computer-literate, but also change their approach to the problem of understanding life" (Gilbert 1991:99). The interdisciplinary effort is and will continue to be both institutionally and intellectually formative. That is, both institutions and intellectual or theoretical tools and frameworks are being built simultaneously. The HGP, databases, computer programs, new classification systems, new laboratory techniques, new training centers, new interdisciplinary centers, at least three disciplines, and representations of nature are all being interactively constructed.

Standardization Questions, or Where Is the Organism?

The potential paradigm shift in biology leads to some interesting standardization problems. First, in order for the databases to be constructed and to be useful, information is standardized. The sequence databases contain information in terms of the biochemical sequences of DNA, RNA, and amino acids. The sequences are used to represent genes and proteins in terms of a linear description of deoxyribonucleic acids (DNA), ribonucleic acid molecules (RNA), and amino acid molecules of proteins. The sequence information for different types of phenomena is expressed in the same chemical language. This language standardizes the form of the representations of the phenomena. This standardization raises some interesting questions about the consequent representations: What does it mean to discuss atherosclerosis in terms of DNA sequences when physicians and the World Health Organization have their own classifications of (and classification debates about) atherosclerosis? What does standardization of information mean when much of the information used in some of the human genetics databases is from specific groups with good record-keeping, such as certain families of Utah Mormons, certain Amish groups, and Scandinavians? What does standardization of information mean when "the" human genome is constructed of information from many different sources?

Molecular biologists standardize in order to cooperate and collaborate in understanding the functions of DNA sequences. However, at the same time, they come to practice the same routines and use the same tools to represent nature. What then does this say about the representations they construct?

An important contribution of the recent work in social studies of science is the emphasis on the local contingencies, uncertainties, differences, and work processes that are deleted from the representations sent out the door in the form of facts and artifacts.[8] If there are any differences in the processes of lab work, for example, they are eliminated in the informatizing of biological phenomena. It is sequence data in the form of C, T, A, G or A, R, N, (for the single letter amino acid codes) that we see on the computer screen, not the complexities, uncertainties, and guesses of the work that produces the "phenomenon" under study.[9] It is this "information," in its extracted form, that is used to construct downstream representations and artifacts.

In a very real sense, molecular biology is all about sequences. First, it tries to reduce complex biochemical phenomena to interactions between defined sequences—either protein or polynucleotide, sometimes carbohydrates or lipids—then, it tries to provide physical pictures of how these sequences interact in space and time. (Von Heijne 1987:151)

While sequence data is currently combined with three things—genetic map data, results of specific experiments, and tacit knowledge (for example, about whole organismal biological systems)—this situation will certainly change as more work is done using information bits/bytes as reagents. Molecular biologists and evolutionary biologists are increasingly becoming quasi informaticians, quasi computer scientists, and quasi mathematicians/statisticians in order to be able to use the information databases to do biology. They will not only have to be able to manipulate and manage information in computers, they will also have to construct meaning using statistical methods (not experimental methods) as the amount of available data increases as the HGP progresses.

Once the [human genome] map and sequence data have been obtained, extracting useful information from them efficiently and economically will also require automated means. Meaningful patterns in the data can often be revealed only by statistical analysis, at the cost of many repetitive calculations. As the project continues and the quantity of data available for comparison increases, the size of even the simplest comparative analysis will exceed the time available to the human analyst. (National Center for Human Genome Research 1990:26)[10]

We who have studied the sciences have witnessed both the intellectual and institutional status differences accorded to theoretical versus experimental and descriptive work. However, molecular biology has generally escaped this division, as it has primarily been experimental. The longer term consequences of a division of labor in molecular biology between theoretical and experimental research, then, might be a new dominance of representations constructed of bytes of information in computer databases over representations of nature constructed of other kinds of information. Information in bytes are not simply transparent heuristics that will then allow us to bump into reality. They are instead frames for organizing and carrying out the biological work of constructing reality. They are the forms that will be used to construct reality.

Discussion

We are living through a movement from an organic, industrial society to a polymorphous, information system—from all work to all play, a deadly game. Simultaneously material and ideological, the dichotomies may be expressed [as a set] of transitions from the comfortable old hierarchical dominations to the scary new networks I have called the informatics of domination. (Haraway 1990:203)

The sequence databases are a crucial component in the remaking of much of biology in the past decade, and in the years to come. They are one of the knowledge bases of the human genome project, built through the commitment of large resources, and by the committed work of many individuals acting collectively. As we have argued here, they are changing the way biology is practiced, changing the relationships between biologists, computer scientists, and clinicians, and changing what can count as durable claims to knowledge about the world.

The consequences of these actions and the effects of these knowledge claims are not limited to the world of molecular biology. These changes carry important meanings for our conceptions of disease, for the organization of our health research and delivery institutions, for our commitments of public and private resources. We then have to question not simply how these sequence-based technologies will be used, but also what kind of nature we are constructing with these forms of knowledge and, as a result, what kind of society? What kinds of institutional arrangements are necessary for the maintenance and growth of sequence-based technologies? What do they do in a world characterized by profound economic inequalities, further divided along lines of gender and race? Is genetic knowledge capable of bridging these divisions? Are certain types of social relations—less equitable, more rigidly technocratic—more likely to result from the kinds of knowledge and practices that are constructed primarily with sequence information?

Many people in the field of science studies are asking these types of questions, and one set of responses has been extremely pessimistic. From this perspective, because the sequence information coming out of the human genome project is subject to the types of standardization we discussed earlier, it becomes tightly linked to notions of genetic determinism that reinforce various forms of social discrimination and inequality. Abby Lippman has put it quite cogently and compellingly:

> For society, genetic approaches to health problems are fundamentally expensive, individualized, and private. Giving them priority diminishes incentives to challenge the existing system that creates illness no less than do genes. With prenatal screening and testing in particular, the genetic approach seems to provide a "quick fix" to what is posed as a biological problem, directing attention away from society's construction of a biological reality as a problem and leaving the "conditions that create social disadvantage or handicap ... largely unchallenged." (1991:47)

Evelyn Fox Keller has also noted how "genetic determinism has just taken over, ... just swept through our culture" in the last few years (Casalino 1991:118). The sequence-based biology discussed here has undoubtedly played a significant role in such a spread of determinism, making it more likely to think of certain individual and social conditions as being "genetically determined," or at least as having a "genetic component." It becomes easier and "more reasonable" to act, given the technological and social resources that prevail, as if things were genetic.

But to say that genetic determinism as a social and cultural phenomenon is the result of, or is founded upon, sequence-based biomedical research, would itself be a determinist argument, albeit on a different register. Many other cultural, social, and technical forces are at work, and their meanings and trajectories are not always clear or univocal. Consider the much-cited case of cystic fibrosis, the most common "single gene" disorder among Caucasian populations. The accumulation of the kinds of sequence-based scientific practices discussed here has shown the incredible complexity that exists at the molecular level, and nearly 300 variants of the "cystic fibrosis gene" have been characterized and mapped against the further complications of population genetics (Tsui 1992). Thus, simplification and complication have progressed simultaneously, undermining from within plans for widespread population screening based on standardized genetic resources (Kolata 1993). Nevertheless, there remain many circumstances when genetic tests for cystic fibrosis are useful and valued by certain individuals.

Attempts to articulate the genetics of homosexuality provide an example of a different set of complexities. The publication of a paper from a laboratory headed by Dean Hamer at the National Cancer Institute, which linked a genetic marker on the X chromosome to "male sexual orientation," resulted in a great deal of public commentary highlighting a wide range of tensions, difficulties, and ambiguities (Hamer et al. 1993). Statistical methods, polarized definitions of gender and sexuality, and other framing assumptions were debated, as were the political effects of such claims to genetic explanation. Among gay rights activists and organizations, there are markedly different perspectives about the implications and consequences of a "gay gene." Some think that claims to a biological basis for homosexuality would lead to wider public acceptance and greater self-esteem for those who are homosexually oriented, while others argue that further stigmatization and a reinforced sense of biological "Otherness" would be a more likely result (see, for example, Carrithers 1993). No simple biological, moral, or political tale can be read from the genetic sequence.

Both Keller (Casalino 1991) and Donna Haraway (Darnovsky 1991; Penley and Ross 1990) have discussed the ambiguities of the human genome project and some of the responses to it. They have argued (although from somewhat different viewpoints) that like most scientific undertakings in the late twentieth century, the human genome project is characterized by inconsistencies, complexities, unexpected and shifting alliances, inextricable fusions of excitement and fear, desirable results and specious claims. The dichotomies that have been either explicit or implicit in much of the early social studies of science—good science/bad science, theory/practice, pure knowledge/impure application, liberation/repression, reductionist method/holist method, human/nonhuman—are blurring and becoming unworkable and unhelpful. As a result, the categorical responses of "for" or "against" sociotechnical projects have also become increasingly open to question and difficult to maintain. As Haraway has argued, "we can't afford the versions of the 'one-dimensional-man' critique of technological rationality, which is to say, we can't turn scientific discourses into the Other, and make them into the enemy, while still contesting what nature will be for us" (Penley and Ross 1990:11).

As ethnographers who want to contest and participate in what nature will become, we need to observe closely and write critically of the crafting, uses, and effects of sequence databases. As the coding of genetic sequences is increasingly linked to other "codings"—homosexuality, criminality, intelligence, healthy and unhealthy bodies, and social bodies—we will not be afforded the simple positions of "for" or "against." This does not leave us without critical resources, however. Close and detailed ethnographic readings of the collective, heterogeneous work conducted within and across the many social worlds of molecular genetics—laboratories, the media, clinics, families, funding institutions, corporations, computer hardware and software design—can engage the complexities produced by, and producing of, the seeming simplicities of a string of As, Ts, Cs, and Gs.

Notes

An early version of this paper was presented by Joan Fujimura in the session on the Anthropology of Science and Scientists, organized by Laura Nader, at the 1991 American Association for the Advancement of Science Annual Meeting in Washington, D.C. on February 14–19,1991.

1. See, for example, Bishop and Waldholz (1990); Knox (1991).
2. Some current studies of such issues are Duster (1990), Holtzman (1989), Hubbard (1990), and Nelkin and Tancredi (1989).
3. See Cantor (1990), National Research Council (1988), Cook-Degan (1991), Fortun (1991), and Watson (1990) for histories of the HGP.
4. "Introns" apparently do not code for messenger RNA and therefore do not code for proteins. "Exons" are the coding regions of the gene.
5. There are also genetic map databases under construction.
6. What do these letters mean? In order to explain, I summarize and simplify the dominant molecular theory of heredity here for the reader's convenience. Sequence information for genes is in terms of deoxyribonucleic acids (DNA) and ribonucleic acids (RNA) and, for proteins, in terms of amino acids. DNA is the biologists' candidate for the hereditary material. DNA is envisioned as a long, double-helixed molecule composed of chains of smaller molecules of nucleic acids. According to current molecular biology, a gene sequence is the reading out of the nucleic acid bases that constitute the genes on an organism's genome. The human genome, for example, consists of genes located on twenty-three paired chromosomes. A chromosome consists of a single, long DNA molecule plus many proteins. A gene is an ordered sequence of pairs of nucleic acids (called nucleotides). The genetically significant part of each nucleotide is another smaller molecule. In DNA, this smaller molecule is one of four

types—adenine (A), guanine (G), thymine (T), and cytosine (C). Nucleotides containing A pair join specifically (by way of hydrogen bonds) with those containing T, while nucleotides containing C pair join with those containing G. These four base molecules are organized in different permutations to form the DNA. The particular sequence of these base-pairs is currently considered to be the organism's genetic code; that is, the particular sequence determines which proteins are manufactured in the cell. More specifically, a set of three consecutive base pairs of DNA, via messenger RNA, code for one of the twenty amino acids (plus a terminator) which constitute a protein. A protein is a chain of linked amino acids. A protein's specific "nature," or activities, is determined by the specific type and sequence of amino acids in the chain. An average-sized protein consists of 400 amino acids, which means that the gene coding for the protein is at least three times 400, that is, 1200 base-pairs long, plus a few repetitive and (at this time thought to be) non-coding sequences. Finally, a cell's properties are determined by the activities of various proteins, each protein exhibiting one or several particular function(s). This is a very simplistic description of the current dominant view in current molecular biology.

For more extensive discussions of DNA and molecular genetic theory for lay readers, see any basic molecular biological textbook, for example, Suzuki et al. (1989) and Watson (1987). For more critical renderings and alternatives to this unidirectional, "blueprint" view of DNA, see Yoxen (1983), Hubbard (1990), Fausto-Sterling (1985), Keller (1991), and Oyama (1985). They argue that words such as "determine" and "code for" obscure the many other elements within the cell, the organism, and the larger environment which play interactive roles with the genome in "determining" events in the organism's development and activities.

7. However, the last two decades' work in the social studies of science have discussed the problems in assuming that experimental work can confirm or disconfirm theories and hypotheses.

8. See, for example, Cambrosio and Keating (1988), Collins (1985), Fujimura (1987), Knorr-Cetina (1981), Latour and Woolgar (1979), Lynch and Woolgar (1990), Pickering (1984, 1992), and Star (1989).

9. Further, the assumption that this work and processual information can be recaptured in replication is refuted by work like Collins (1985) and Jordan and Lynch (1992) who argue that there is no exact replication of detailed work practices.

10. See also Waterman (1990).

CHAPTER 10

Kokusaika, Gaiatsu, and *Bachigai*

Japanese Physicists' Strategies for Moving into the International Political Economy of Science

Sharon Traweek

Through the anthropological research method called participant-observation I study the settings and events that the multinational high energy physics community physicists construct for themselves and observe the activities, formal and informal, they consider appropriate in those settings.[1] I learn what they believe they need to know in order to act effectively and strategically, whether locally or globally; I then find the patterns in their actions and cosmologies and how all this shifts over time as the ecology of their community changes. The central theoretical questions for my research have been: 1) how knowledge, especially so-called craft or tacit knowledge, is transmitted from one generation to the next in a multinational community that is committed to discovery and in which crucial features of their knowledge are never written; 2) how different styles of research practices emerge and survive; 3) how disputes and factions are formed and maintained; 4) how these practices differ along lines of class, gender, regional and national culture, and national and international political economy; and 5) how national and international political economies are shaped by these physicists and the work they do.

In this chapter I will introduce some of the Japanese high energy physicists' practical concerns with their own international scientific relationships, including why they are concerned with the Japanese government's international political and economic relations and why the Japanese government pays attention to the high energy physicists. I begin by introducing the significance of the Japanese terms for discussing international relations, *kokusaika* and *nihonjinron,* and explaining how Japanese physicists' career strategies, gender and generational differences, scientific language, and funding are embedded in international relations. I then discuss how two different factions in the Japanese physics community try to use foreigners, the Japanese government, and international relations strategically in their debate about the research program at a national basic research facility in Japan. I conclude by exploring the role of funding and military power in Japanese and American high energy physics.

Kokusaika and *Nihonjinron*

Kokusaika is the Japanese word for the very widely discussed and volatile issue of how Japan ought to participate in global society. It is understood that Japan's recently acquired status as an economically wealthy and powerful country brings with it both the right and the responsibility to act in global, political, intellectual, and cultural arenas. The media images of then Prime Minister Nakasone standing alongside Reagan, Thatcher, and Mitterand at the economic summit meetings of the 1980s were an arresting display of these new rights and responsibilities. The Japanese people were fascinated and disturbed by Nakasone's capacity to comfortably shake hands, laugh, and deal with Westerners (our ethnocentric label—the Japanese formal, polite word is *gaikokujin*, people from other countries) in their style, right down to the first-name banter. To begin to understand the Japanese public's consternation, imagine the consequences in the United States if Reagan had behaved in a "Japanese" manner during Nakasone's subsequent visit to Washington to discuss trade issues.

Most Japanese people believe that Japan ought to participate in global society in ways that are consistent with being Japanese. One government response is to fund research that tries to define Japaneseness; anthropologists and historians of traditional Japanese society are very well funded compared to their colleagues in other countries.[2] There is a vast literature on this subject of Japaneseness. *Nihonbunkaron* (essays on Japanese culture) is the more academic and respectable; *nihonjinron* (essays on the Japanese people) is the widely read public literature, for which academics in the social sciences and humanities have considerable disdain (Harootunian 1988). Physicists read *nihonjinron*. Its popularity is in part a measure of how controversial the subject of international relations has become in Japan. Elsewhere I have emphasized that there are, of course, different versions in Japan of Japaneseness, underneath the 125-year-old image of homogeneity propagated by the central government, just as there are very significant regional differences in other countries, such as the United States, France, and India, to pick a random set. In the current Japanese debate it is not surprising then that different, sometimes regional, versions of Japaneseness correspond to different Japanese strategies for acting globally (Traweek 1992a, 1992b, 1989).

Kokusaika and Career Strategies in Physics

Kokusaika and *nihonjinron* are hot topics in the scientific community, both for individuals and for groups. Since the end of World War II, the best young Japanese scientists were expected to go abroad for two or three years at the end of graduate school to work on projects at well-known laboratories and universities in North America and Europe, and then return to Japan and spend the rest of their careers working in Japan with other Japanese. It has been thought important not to acquire foreign mannerisms and ways of thinking during these sojourns. It also has been considered bad form to speak foreign languages too fluently or to read foreign languages with pleasure: I have heard physicists criticized for reading so much "sideways" print. Men discuss whether to marry before their trip abroad: will a wife keep the physicist "Japanese" while abroad, or will a wife get contaminated with "foreign" ways, making the physicist even more "strange" when he returns to Japan. (I never heard of a woman student being selected for trips abroad by her teachers, although many have chosen this route independently, which I have never heard of men doing.) To be strange in Japan (*bachigai* in the dialect of Ibaraki, the area where Tsukuba is located) is to be out of place, an insulting label most would try hard to avoid.

While doing research in Japan during the mid-1980s I was invited to participate in a three-day conference on international relations at Hachioji University Center in the mountains not too far from Tokyo. The seminar leaders, including myself, were participants in or researchers and commentators on international relations: diplomats, journalists, economists, government policy specialists, academics, and so on. The seminar participants were Japanese university students in international relations who had gained the opportunity to attend the rather prestigious conference through academic competitions. I spoke about the day-to-day difficulties and opportunities of international collaborations based in Japan. The students asked me many questions and spoke at length about how uncomfortable they were with individual Japanese who were at ease with foreigners (informally, but still politely, *gaijin*) and with individual *gaijin* who were at ease in Japan; they thought such people were definitely *bachigai* and untrustworthy. These students of international relations were fully aware of the paradoxes implicit in their position; they wanted to acknowledge this as a problem in their society and point out that these feelings were much stronger in older generations. (I have been acquainted with two Japanese whose fathers had distinguished careers in the Japanese diplomatic corps and they each said that their fathers had reported that such attitudes were rife among career diplomats.)

A second conventional way Japanese physicists have engaged with their foreign colleagues has been to invite scientists from countries poorer than Japan to collaborate on research projects based in Japan. This is more true of the universities in the southwest of Japan, from Osaka and Kyoto in Kansai to Hiroshima and Kyushu. Cosmic ray physics and nuclear physics are very strong in this part of Japan; the former has, until the last fifteen years or so, been very inexpensive, and the latter has long had easily identifiable industrial and commercial applications and hence has had support from private industry. Japanese cosmic ray and nuclear physicists—almost all men—have long maintained close collaborative relations with foreign scientists, particularly in poorer countries; they did not need governments to endorse and finance these hitherto rather inexpensive "little science" projects. By now cosmic ray and nuclear physicists have both come to use very expensive equipment, and they have forged even closer relations with private industry, particularly those in southwestern Japan. Those industries in turn encourage the government to fund basic research in the cosmic ray and nuclear fields as a way of funding what in the United States would be called research and development.

For several decades these same universities and physics groups have been identified with, in the Japanese spectrum, left-of-center national and international politics. They saw such politics as consistent with the practice of "little science" and non-collaboration with physicists from rich, conservative countries. During the 1930s physicists in these areas were said to have been active, at great personal and professional cost, in antimilitaristic political groups, and many of the older Kansai physicists I first interviewed in the mid-1970s saw themselves as following their teachers' values as they promoted "democracy" in the organization of their physics research (Tetu 1974; Kaneseki 1974). Given the great emphasis placed on strong intergenerational bonds in the university research groups, it is all the more striking that the nuclear and cosmic ray physics of western Japan have recently become "big sciences" with ties to big industry and to researchers in the United States. Nevertheless, there remains a distinctively western Japanese approach to international scientific relations.

Yet another position on international scientific relations has been that Japanese research groups, especially in high energy physics, should collaborate with foreign groups on experiments conducted abroad and do a portion of the experiment's data analysis in Japan; this path has been followed at those universities, including Tokyo, Nagoya, and Tohoku, which have long enjoyed the higher levels of government funding needed to pay for their capital-intensive data analysis equipment. Physicists in the Tokyo area and the northeast—again, almost all men—have been seen by their colleagues in the southwest as enjoying the benefits of being in tune with the dominant, conservative political forces in the Japanese government.

The scientists who have occasionally collaborated with foreigners abroad seem to me to have worn their foreign experiences both more comfortably and more lightly than their colleagues who have arranged to be abroad for shorter periods of time. They have not minded speaking foreign languages in front of other Japanese and they have openly displayed their fondness for certain foreign foods or certain foreign arts, especially Western classical music. They have enacted a sort of connoisseurship of foreign, especially European, high culture, a posture of power not unknown in many other countries, including the United States; it is certainly not the same as acting or thinking in foreign ways. These scientists have still wielded their power in very Japanese ways and, with notable exceptions, often have seemed far more concerned with their status in the Japanese scientific community than in the multinational arena.

A distinct minority of Japanese physicists—with notable exceptions almost all men—have spent most of their careers abroad, having arranged for themselves work in the countries where they were sent as postdoctoral research associates by their teachers; the majority consider this career path selfish, since one never brings the knowledge home. The Japanese scientists who have worked mostly abroad are disturbing at home for another reason: they know how to act and think in ways their colleagues label foreign (even if foreigners might find their actions and thoughts rather more Japanese). If they return to Japan to work, Japanese who have spent many years as expatriates have to develop strategies for coping with the deep cultural suspicion in Japan with which they and their families are almost always treated (White 1988). Such people are said to no longer have Japanese souls (*ki*) or be capable of exercising leadership in a Japanese way (such as *hara-gei*). (The very fact that these terms are being used by Japanese physicists suggest that they are reading *nihonjinron;* my Japanese anthropologist colleagues were shocked that other academics would be using such literature to make sense of themselves.) The Japanese generally agree that an adult can live abroad up to three or four years and still have a Japanese soul, but not as long as seven or eight years; the intervening years are problematic. Everyone agrees that spending the high-school years abroad leaves an indelible mark; for this reason some Japanese expatriates chose to return to Japan as their children reach adolescence rather than have them forever thought to be outsiders. With the rapid expansion of Japanese international economic influence, many families have been obliged to live abroad for long periods of time, and the government has developed policies to alleviate this problem; nonetheless, the "crisis of returning" to such a culture can still be a very painful experience, and I have heard this discussed many times since the mid-1970s.

Some scientists work very hard at erasing all traces of their time abroad. Other former expatriates choose not to erase or conceal their foreign habits; some even want to insist on their new ways being acknowledged. But most returnees want their hard-won foreign skills to be recognized as useful and important in Japan; they want to be a resource, not a problem. Ten or fifteen years ago this would have been exceedingly fanciful; in the midst of the current debate on *kokusaika* and *nihonjinron* it is more plausible. The Emperor of Japan's second son, who had himself studied at Oxford, married the daughter of an economist who taught at the University of Pennsylvania, and the economist invited some of his former U.S. colleagues to the wedding festivities. Nonetheless, to return after a long stay abroad is a path requiring courage, because one will still face powerful prejudices daily. I must mention that such people can be effective too: they can know both the usual Japanese ways of doing things and other ways as well; if astute and lucky, they can make use of their expanded strategic repertoire.

Finally, there are Japanese women physicists who have constructed yet another intellectual and career path in the borderlands of both the Japanese national scientific community and the international scientific community.[3] I met several midcareer women scientists and engineers who had gone to graduate school in North America and Western Europe during the 1950s and 1960s; many stayed for one or two postdoctoral appointments before returning to Japan ten to twenty years ago. These women are denied access to local power: funding, high status jobs, chairs, and leadership in professional organizations and journals. Because they lack this sort of power, few of the best students want to work with them. Nonetheless, they have developed strategies for surviving in science. What are their resources? First of all, they have multinational networks. They are invited to publish in the international journals and attend the international conferences; they communicate freely and informally with their colleagues abroad, if not at home. They know well how the large multinational collaborations abroad are organized, so that when they propose analyzing a certain set of data tapes from an experiment, they are likely to know what tapes would be thought interesting, but not enough so to have already attracted the attention of the most influential and powerful groups. They know that their meticulous searches through those tapes may only yield mundane results, but that an inventive approach might produce much more.

Second, they have local networks outside the universities. Most women scientists and engineers in Japan, like their counterparts in the United States, have distinctly higher social-class origins than their male counterparts; I think that the sense of entitlement that comes with upper-class socialization enables them to ignore the class-based gender exclusionary practices of their male colleagues. Remember that these Japanese women often chose to pursue their educations abroad without the patronage of their male professors. They do have access, through friends and relatives, to decision makers in industry to whom they turn to propose that research equipment of their design be provided free. They offer in return the opportunity for the companies' engineers to learn to work with unusual technical demands; they also provide to the companies a few years later a corps of students trained to make use of their state-of-the-art equipment. Where did these students come from? These physicists offer undergraduates the opportunity to do research and in turn provide them with the highly desired introductions to some of the most prestigious companies in Japan. Excluded from the conventional resources for doing research, these women have tinkered with the Japanese system in innovative ways.

I have tried to at least sketch for you some of the intellectual and political allegiances and ethos each of these five career paths would entail. The crucial and highly interrelated obstacles and opportunities are institution, physics research speciality, region, time spent abroad, relationship to foreign scientists, gender, and national and international politics. The Japanese physics community now includes significant numbers of people who have followed each of these career paths; those who have followed a certain career path tend to regard all the others with some suspicion and coalitions between them are not common. The Japan Physical Society (JPS) is what all five groups have in common and where they meet. Many high energy physicists told me that a few decades ago "leftist" cosmic ray and nuclear physicists controlled this organization because they always voted as a bloc; most agree that the JPS has changed since those days; many of the scientists active then are now retired.

Language and Generations in the Japanese Physics Community

I want to emphasize that crosscutting each of these issues in the Japanese physics community I have described are significant generational differences. Although everyone raised in Japan learns to want strong intergenerational ties, the major historical upheavals of the twentieth century in Japan—economic depression and colonialism, war, occupation, poverty, and prosperity—have marked each generation quite differently and everyone knows this. The eldest group, the ones who tend to hold positions of national authority, leadership, and power, are in their late sixties and seventies and were educated primarily before 1945. The midcareer group are in their late forties and fifties; they grew up during Japan's postwar poverty and were in school during the U.S. occupation. The youngest ones, in their twenties and thirties, have been educated during Japan's increasing prosperity and world influence. My preliminary investigations suggest that there are also significant differences in social class origins between these generations of physicists, just as there are in the United States.

In addition, these age groups correspond to major changes in science education in Japan. After the Meiji Restoration of 1868, which established the modern Japanese state commissions for each science, faculty visited universities around the world, deliberated, and then decided on which model they thought ought to be followed or adapted for Japan; each discipline was free to decide independently. The textbooks, pedagogy, and even language for the various fields were English in one case, French in another, German in a third, and so on. These diverse models defined education in the sciences in Japan until the occupation of the late 1940s, thereby shaping the eldest generation of physicists active today.

MacArthur at one point decided that the Japanese language was itself "undemocratic" because he thought the number of ideographic Chinese-origin characters (*kanji*) used in Japanese writing prevented ordinary Japanese people from having easy access to their own written language. Although two phonetic scripts are also used in Japanese writing, MacArthur actually proposed that Japanese be written with the English alphabet, a position supported by some Japanese physicists. He eventually changed his mind on that, but he did require that all newspapers and public documents, including all government records, be written using only about 2,000 *kanji*, which were to be taught to all students by the time they had graduated from secondary school. His policy was implemented: in addition to the benefits he envisioned, the complex written vocabulary associated with the educated classes has gradually declined, and fewer and fewer Japanese have direct access to their own country's rich intellectual history. European intellectual life was perhaps similarly transformed with the transition from Latin to the vernacular languages during the Renaissance; my point is that in Japan the transition was sudden and rather recent, and currently differentiates the intellectual life of the elder, governing generation from their juniors. For example, most people in Japan under sixty cannot read the poetry written in the scrolls that customarily hang in the *tokonoma* (special alcoves) of almost all interiors where Japanese gather; young people often install televisions in their *tokonoma* instead.

Another consequence for scientists of the linguistic policies of the U.S. occupation of Japan after World War II was that the diverse models for science pedagogy and research launched in 1868 were conflated to a U.S. model and many scientific terms were changed to Japanese phonetic transliterations from American English. Furthermore, since the occupation, all Japanese children have been required to study written English for several years, making English the only foreign language most students ever learn. This means that most scientists under sixty would find it difficult to read the scientific work of the previous generation or to follow their informal conversations about the scientific work they did during those years.

The third change is very recent. During the mid-1980s, as part of the new Japanese confidence in their prosperity, some of the eldest generation of national politicians, the national leaders, argued that Japan should reconsider some of the policies inaugurated by the occupation. One such change is already in place: funding proposals to the Ministry of Education (which funds all university-based research in Japan) now should be written without phonetic transliterations from other languages. The practical consequence of this change was that all scientists suddenly had to construct *kanji* for current scientific terminology, based on their sense of the meanings of the terms. Japanese words for energy (*e-ne-ru-gi*) and radiation (*ra-ji-a-shu-n*) had to be reinvented. It is almost as if the U.S. National Science Foundation (NSF) suddenly required all proposals to be written in prewar scientific German.

What I learned was that every subfield has developed their own *kanji* for the scientific terms they commonly use, according to their own notions of *kanji* etymology, and that there is little correspondence between subfields' *kanji*, even if in European languages those terms, such as radiation, might be the same across several subfields. The youngest generation of physicists is less daunted by this change than their elders; but in time they too will find the written work and conversations of the previous generations less and less accessible. This particular linguistic change also means that the differences between subfields, such as those in physics, have recently become very large, increasing the distances that already existed. These three phases in the history of Japanese science pedagogy and research strongly differentiate the scientific thinking of the three generations of researchers currently active in Japan today.

Koza, Funding, and Extramural Strategies

Another distinctive feature of the Japanese physics community concerns the Japanese chair (*koza*) system, adapted from the German university system in the years following the Meiji Restoration of 1868. Funding for research from the Ministry of Education is allocated to the *koza*, but capital for new research equipment is traditionally allocated only at the founding of *koza*. During periods of scarce funding, very few new *koza* were founded. As high energy physics was becoming powerful in Europe and North America during the 1940s, 1950s, and 1960s, Japanese basic science research funding was very low; whatever high energy physics research was done was conducted usually within some of the *koza* established for nuclear physics. Those *koza* leaders eventually were willing to use their own resources to encourage those of the next generation to do work unlike their own; most high energy physicists in Japan feel deeply personally indebted to those nuclear physicists.[4]

When a few high energy physicists happened to gain access to the highest levels of government decision making in the mid- to late-1960s, they wanted state-of-the-art research equipment to be built in Japan and funding to staff and operate such a lab. The history of this decision is complex and only small parts of it have been written.[5] One very interesting point is how high energy physicists in Japan gained access to the most powerful politicians. In the United States high energy physicists formed a partnership with the government (especially the war department) during World War II; the senior members of the community have strengthened that partnership in the decades since then in many ways, which have in turn been described by many historians of science. The Japanese high energy physicists had no such special relationship to build upon; instead they worked through friends and family. Some of the high energy physicists became acquainted with men who are now powerful politicians because they attended the same secondary schools. Each prefecture has exceptional schools with competitive admissions; ambitious, smart children, especially those from rural areas, work very hard to gain entry to those schools. Being part of the alumni of such schools remains important throughout one's life. In addition many Japanese feel that the most significant and long-lasting friendships are formed while one is in secondary school.

In Japan the distribution of seats in the national legislative body did not change significantly in the forty-five years following 1945, although in that same time period there was a massive population shift to a very few cities from rural areas all over Japan. A redistribution plan was discussed in the 1991 Diet; public opinion polls indicated that citizens would like their politicians to represent an equal number of voters, but they expected the politicians to accept only a few changes (Landers 1991). (In the period since then the Liberal Democratic Party [LDP] lost its majority—but not its powerful influence—and modest changes have been introduced in the ways politicians are elected.) The political party that had been in power those past several decades, the LDP, drew its strength from those rural seats, which in turn were controlled by the senior, most powerful members of the LDP. As it happens, many of the senior high energy physicists are also from rural areas; some of them have links to LDP politicians either through their home towns or their prefectural schools. By contrast, most academics and most government bureaucrats are from cities. Policies in Japan are usually forged by the LDP, government agency bureaucrats, and business leaders; academics, including scientists, usually work through the bureaucrats. The high energy physicists have been able to gain access not only to the LDP politicians, but also to the business leaders. Since the National High Energy Physics Laboratory (*Ko-Energie butsurigaku Kenkyusho* [KEK]) was first funded, the construction of the new capital-intensive facilities has brought the high energy physicists into close contact with those businesses capable of supplying the laboratory with the required high-technology equipment.

The high energy physicists who wanted new facilities got only some of what they had originally asked for in the 1960s, in part because of the opposition of the Ministry of Education and many physicists outside high energy physics, including nuclear physicists. The Ministry of Education seemed to follow the advice of the JPS, dominated by nuclear physicists, in allocating resources for physics (much as the U.S. Department of Energy [DoE] receives advice on allocating resources from the High Energy Physics Advisory Panel [HEPAP]—which chooses its own members). I have been told that the Japanese nuclear physicists' opposition to a new high energy physics laboratory was triggered by the proposal to situate such a facility outside the traditional university system, which until then had been controlled in physics by the nuclear physicists through their influence in the Ministry of Education, the JPS, and their own *koza*.

At about this time the national government was considering a set of regional economic development plans. It was the brilliant move of some high energy physicists to suggest that many new state-of-the-art laboratories be built in the same economically depressed region, and that the government should not allocate these labs to the existing universities, but should build a new science city, not administered by the Ministry of Education but by another agency as part of a new national and international science and technology economic policy. This proposal gained the support of many politicians and the continuing enmity of the Ministry of Education and almost all university-based physicists. With this gesture it appeared to others that the high energy physicists had repaid the kindnesses of their elders and colleagues in nuclear physics with rejection. The high energy physicists got their lab, KEK, and eventually it was upgraded with the TRISTAN colliding beam facility to an international state-of-the-art laboratory. The cost of this extramural victory was that nuclear physicists consolidated their control of university-based physics. It is as if American high energy physicists, as a consequence of gaining national laboratories like Los Alamos, Brookhaven, the Stanford Linear Accelerator Center (SLAC), and Fermilab, and of gaining access to the DoE, had diminished their influence in university physics departments, and that nuclear physicists had gained control of all crucial resources in universities and the NSF.

The high energy physicists had gained KEK and Tsukuba Science City at the cost of being *bachigai*, a condition that most Japanese take great pains to avoid. They gained the company of new friends: the ruling political party; the businesses supplying the lab; and ambitious, powerful civil servants at some new government agencies who themselves were engaged in territorial conflicts with the staff of the older agencies, including the Ministry of Education. The new long-term plans for economic development in Japan call for two new science cities: one near Osaka is already partly completed and the other will be built near Sendai. The Ministry of Education has even allowed a new kind of Ph.D.: it is not based on reading and reenacting canonical great articles and great experiments, but on apprenticeship in research projects conducted at the new national laboratories, such as high energy physics, space physics, and anthropology. (If anthropology being on this list surprises you, remember that the government wanted Japaneseness defined in order to develop its *kokusaika* policies, and that they had turned to anthropologists for their answer.)

I now offer a terse summary of my far too brief account of all the differences that make a difference in the Japanese physics community: Japanese physicists are divided by region, gender, generation, subfield, institution, national and international politics, ties to government agencies, and conceptual and linguistic ways of representing their subject to each other. Furthermore, these differences are the foundation of long-lived factions in the community. Certainly one could say the same about U.S. physicists, but the distinctions would be trivial by comparison. It is in the context of these highly significant ongoing differences and conflicts about them that Japanese physicists debate about how to participate in international science, now that they are rich enough to have more choices.

TRISTAN as the Prize

The specific question the Japanese physicists are confronting is the future of big science in Japan, especially high energy physics. The TRISTAN facility at KEK, the first state-of-the-art high energy physics research facility in Japan, must be upgraded to remain innovative. The high energy physicists originally gained the funding for TRISTAN by making a compromise with the nuclear physicists: TRISTAN would be built at KEK, but only after a nuclear physics research facility, the Photon Factory, was built, also at the KEK site. I remember hearing about this decision at SLAC from Japanese physicists: some thought the implications were ominous, even then.

The Photon Factory benefited from much of the construction activity for TRISTAN; once TRISTAN became operational in April 1987 the nuclear physicists managed daily to limit any drain of resources, including beam time, from their facility to TRISTAN. During this period I heard many stories from university-based physicists who were not in high energy that nuclear physicists were taking over at KEK and turning the TRISTAN colliding beam facility into a mega-synchrotron radiation facility, a master Photon Factory, like the one then being planned at Argonne National Laboratory in the United States. Not one of the many physicists who talked to me about this project spoke as if the news were secret, surprising, or disturbing.

Still, I was startled. The U.S. and other foreign physicists at KEK thought it was a joke that nuclear physicists could contrive to take over a high energy physics laboratory; in their stories, nuclear physicists only get high energy physics laboratories when high energy physicists vacate what they regard as outdated labs for new ones, as a sort of hand-me-down. I tried to explain to the foreigners all the reasons why nuclear physicists have power in the Japanese science establishment and why the Japanese high energy physicists' power, while quite impressive, has been relatively short-lived. The foreign physicists thought I was confused and naive for believing such stories were even possible, anywhere.

The Japanese high energy physicists, of course, knew the stories I was hearing and they said that, given the power of the nuclear physicists in Japan, they might well get TRISTAN. They also said that if that were to happen, there were some interesting possibilities. The first question was what the high energy physicists could get in return for leaving quietly. I wondered where they might go. They said abroad, to the European Centre for Nuclear Research (*Conseil européen pour la recherche nucléaire*—CERN) and the SLC (the Stanford Linear Collider at SLAC). I finally realized that these Japanese high energy physicists meant that the alimony they could exact for leaving TRISTAN to the rich and powerful owners of the Photon Factory might become their dowry for CERN, the SLC, or yet another suitor.

The nuclear physicists who were the source of this dowry seemed to want the transition in control of KEK to occur without any sign of the laboratory itself being sullied; a quiet, dignified departure would be the negotiating chip of the high energy physicists. They said that the former long-term director, still very influential, might accept the maneuvers of the nuclear physicists because it would be much better for his reputation if TRISTAN, so closely tied to his own name, remained powerful instead of being gradually weakened by funding attrition, even if it ceased being a high energy physics lab and became a nuclear physics lab.

I have wondered if the *bachigai* high energy physicists might find their own funding to update TRISTAN, perhaps by forging another liaison, surprising the nuclear physicists. The increasing wealth of Japan and the increasing worldwide pressure for Japan to assume international responsibilities commensurate with that wealth might be creating new options for the Japanese high energy physics community. Many now expect to get funding for four major projects: a new facility at KEK, as well as significant participation at CERN, Fermilab, and the SLC. Others are skeptical that high energy physicists could gain that level of support from the government, particularly given the budget difficulties in today's troubled economy. My current assessment is that all four projects will be funded eventually.

Gaiatsu and Funding for Physics

Gaiatsu (outside pressure) is a very important strategy in Japan; the strategist raises the spectre of ridicule or even danger if word of some specific situation were somehow to become known to outsiders. For example, I was told by some Japanese women scientists that Japan's 1986 equal opportunity laws for women were passed because of *gaiatsu*: the United Nations' International Decade for Women had nearly passed without Japan conforming to the United Nations resolution on women's rights; Japanese women reminded the government of this *gaiatsu* situation, and the laws were passed with great fanfare although they are, as yet, barely enforced. High energy physicists said they had used *gaiatsu* effectively when President Carter and Prime Minister Fukuda were to meet during the energy crisis of the early 1970s: they, along with the U.S. high energy physicists, proposed to the government that the two politicians sign an accord to support basic research on energy and suggested that a noncontroversial research area would be high energy physics. They succeeded because each side wanted to appear cooperative and resourceful to the other and the physicists had given them an opportunity to do so. The Japanese said they had often heard U.S. high energy physicists talk about using international summit diplomacy in this way, but I gathered that this had been their first foray into such negotiations.

Several high energy physicists pointed out to me that Tsukuba and KEK existed in part because of *gaiatsu*. As Japan was becoming prosperous and under some international pressure to assume the responsibilities of being a world economic power, the high energy physicists had been able to suggest that the support of expensive basic research was common among the governments of North America and Europe, and that Japan would be seen as contributing to the good of humanity by financing a project like Tsukuba Science City where scientists from all over the world might come to do research. At that time basic scientific research was not well supported in Japan, but with the judicious use of *gaiatsu* Tsukuba and KEK were funded, in part because the solution proposed to the problem of outside pressure caused very few political difficulties for the ruling party, the LDP. The only disaffected groups were university professors and the relatively low-status Ministry of Education. The party also gained support from two of its traditional constituents: beleaguered farmers in Ibaraki Prefecture, many of whom had been alienated by the botched negotiations concerning the building of Narita Airport, and high-technology industries, which would receive the contracts for all the new research equipment, according to the constraints built into the funding legislation.

In the nearly twenty years since the decision was made to fund Tsukuba Science City, Japan has become a much richer country with even greater outside pressure to assume the responsibilities of a world economic power. An immediate international problem is to relieve the "trade imbalance" with the United States. If the Japanese high energy physicists had participated in some very expensive research in the United States, perhaps at the Superconducting Super Collider (SSC) that was once under construction near Dallas, this outside pressure might have diminished. High energy physicists and politicians in the United States once argued that the $8 to $11 billion SSC project was needed to maintain U.S. leadership in basic science; until congressional funding was first secured, it had been rather impolitic to suggest bringing in the Japanese. Once the SSC was granted its initial funding there was great pressure to include the Japanese, and the SSC management faced criticism from Congress for not having been more successful in gaining Japanese funding. During the Bush administration, SSC funding became an issue at the highest levels of diplomatic relations between the two countries before the SSC was cancelled by Clinton's administration.

A second challenge for Japan in international relations has been to increase their economic assistance to poor countries. Such assistance now exceeds that of the U.S. government, but the Japanese are still criticized for offering much of their assistance in the form of loans for the purchase of Japanese goods. Recently they have begun to offer stipends for foreign students (especially from east, southeast, and south Asia) to study in Japan. As yet the proportion of foreign students in Japan is quite small, there are almost no support services for such students, and they are usually not enthusiastically received. However, some Japanese scientists are very eager to see this program expanded; for example, there are now about five hundred Asian science students, including graduate students, at Tsukuba University where there is more tolerance of foreigners.

A third case concerns the Soviet Union. In the mid-1980s the Japanese government began to hope that Gorbachev might be willing to settle the old territorial dispute between the two countries. In the final days of World War II the Soviet

Union occupied Sakhalin Island and the Kuriles which the Japanese had considered theirs; no resolution has ever been reached and as a consequence Japan and the Soviet Union have never signed a treaty officially ending their World War II conflict. One day a few years ago as I was driving onto the laboratory site my car was suddenly surrounded by several people dressed like police; I was quite nervous until the man who usually staffed the entry gate identified me. Soon I saw a large black limousine with little Soviet flags on the fenders arrive at the laboratory. The Japanese physicists were hoping to provide the two governments with something to agree upon: a policy to support basic research in science. That effort at détente came to naught. In 1991 Gorbachev actually did visit Japan; the territorial issues were not resolved but several Soviet researchers visited KEK and other labs at Tsukuba as a result. Some Japanese physicists said they hoped that the conversion of the Soviet military-industrial complex to peaceful applications would provide a new opportunity for them to use international relations to bring resources to the laboratory: they want the Japanese government to support this conversion by purchasing high technology for KEK from the Soviets.

Gaiatsu, Military Funding, and Physics

The sharpest difference between the Japanese and U.S. physics communities is with respect to the role of military funding for scientific research, even though both communities appear to eschew any contact with the military and both appear to hold the military in disdain. I will elaborate here on the very different ways the Japanese and U.S. physicists engage in stylized relations with the military and their very different discourses about that relationship. I want to emphasize that the Japanese are acutely aware of the U.S. physicists' military concerns while the Americans appear to know little and care less about their Japanese colleagues' concerns about the military; consequently, I explore those features of the Americans' actions and attitudes that attract the attention of the Japanese.

Japan is under great outside pressure to increase its military expenditures. The constitution requires that the amount be kept under 1 percent of the gross national product (GNP); special accounting practices are used to meet that requirement technically, but if the procedures of other countries were used, the current expenditure variously is estimated to be 3 to 5 percent of the Japanese GNP. Nonetheless, *gaiatsu* continues and various Japanese industries and some military groups exploit this; over the last few years certain Japanese companies were allowed by the government to accept some contracts for Strategic Defense Initiative (Star Wars) research. I wondered if the high energy physicists would be willing and able to exploit this form of *gaiatsu* for their own ends, but I saw or heard no signs of it. At a major conference in the mid-1980s I did overhear a prominent Japanese high energy physicist inviting a well-known American to visit his university and his research project; at a certain point in the negotiations the Japanese physicist said to the American that if he had any association with Star Wars research his colleagues would probably discover it and the situation could become very unpleasant for everyone. The American said that he was not involved in that work and the negotiations continued. Neither had offered his own opinions about Star Wars.

Their silence was more consistent with the actions of U.S. high energy physicists than the Japanese. In my more than fifteen years studying U.S. high energy physicists I have found their posture on military funding generally to be one of intellectual condescension, not one of morality or political values. That is, military funding is funding for applied research: to get it one must at least say that applications of one's work are eventually possible. The U.S. scientists I have studied were very proud to say that they had never done any applied research.

There are two obvious contradictions in their position. One is that about three-quarters of all people who have completed postdoctoral research associateships in high energy physics (i.e., three to six years of research beyond the doctorate) leave the field. (By contrast almost all those receiving Ph.D.s in high energy physics in Japan have remained in the field.) Several senior U.S. high energy physicists have told me that they assume that many of these former junior colleagues are now working on defense-related research. One such former high energy physics postdoc whose research is funded by the U.S. Navy told me, apparently in mock horror, that one of his students had actually wanted to work for the CIA; when I mentioned this to a senior high energy physicist, he merely said with disdain that such behavior among one's students had to be expected if one had any involvement with military funding.

Japanese high energy physicists queried me often on U.S. scientists' participation in Star Wars and other forms of military funding. Just as I had left the Massachusetts Institute of Technology (MIT) in 1986 for Japan to do fieldwork, many committees had been constituted and many fora had been scheduled to discuss the implications for MIT, pro and con, of this greatly expanded funding. Since various notices and minutes of the MIT faculty senate were being sent to me, I could show all that information to the Japanese scientists. They too were scheduling many meetings and constituting committees to discuss the implications of such funding for Japanese science. They were surprised that these topics elicited so little interest among the U.S. high energy physicists at KEK. One U.S. scientist at KEK was quite interested, but he was a nuclear physicist visiting the Photon Factory and he had been doing research in Germany for over a decade.

This difference in opinion was also quite noticeable during the 1991 Gulf War, which I spent in Japan. The U.S. physicists at the laboratory seemed very supportive of the war effort, occasionally expressing surprise that any thoughtful person could be opposed to it. Conversely, the Japanese high energy physicists seemed quite surprised that any thoughtful person could support that war. I noticed that these two groups did not mention their views to each other, although there was considerable discussion within each group. Several Japanese expressed real surprise and disappointment when I showed them U.S. media reports that a significant proportion (perhaps 20 percent) of the high technology in U.S. weapons was made in Japan; they said Japanese media were not carrying such reports. Many mentioned that if world opinion forced Japan to increase the role of its military, the right-wing forces that had ruled Japan from the mid-1930s to the mid-1940s might regain control; these physicists thought that those forces had never been completely repressed, partly because of the U.S. occupation policies, and were still strong.

Another contradiction in the Americans' disdain for military funding is the involvement of many prominent high energy physicists, including Nobel Prize winners and laboratory directors, in Jason, an organization of about fifty scientists, founded in the late 1950s, which evaluates scientific and technological projects for the U.S. government, most of which are military.[6] Membership, which is quite stable, is limited, by invitation only, and, until two women joined not long ago, all male; the group meets in Washington, D.C. every spring and hears a number of presentations about proposed U.S. government scientific and technological activities. The Jason membership then decides which proposals they will evaluate; such selection bestows a certain prestige on those proposals and is actively sought by government agencies. Jason then meets for several weeks each summer (for the last several years in La Jolla, the most affluent part of San Diego, California) to evaluate the proposals in detail; there is a certain competition among the members about who has the highest level of government security clearance, because only those with the highest clearances can work on whatever project they choose. At the end of the summer they provide their evaluations; a few Jason members meet in Washington, D.C. in the fall to conclude that year's activities and plan the next.

High energy physicists who participated in Jason in the late 1970s insisted to me that because their *primary* research was not funded by the military they were able to independently evaluate the proposals they studied for Jason, unlike scientists whose primary work was funded by the military. They argued that this was a service they provided their country as responsible citizens and gave examples of projects negatively evaluated by Jason which would have cost many, many millions of dollars and might have endangered U.S. foreign relations had they been built. They acknowledged that there were personal benefits: intellectual challenges, bright colleagues, prestige, very generous *per diem* stipends and salaries, and the opportunity for their families to spend the summer in the very expensive beachfront resort of La Jolla, which many would not ordinarily be able to afford on even the most elevated academic salaries.[7] Some Jason members have changed their research specialization as a result of projects they studied with Jason, and their work is now primarily funded by the military. In 1990 the current chair of Jason, Curtis Callan, a theoretical physicist on the faculty at Princeton, addressed the ethics of participation in Jason in a newspaper interview:

> We feel we have a duty, that it is important to work on national defense problems. We'd like our effort to have positive effects.... Given the realities of the world and that there is conflict and there are armies and the world is not completely benign, I don't see any ethical problem in advising our government to make sensible and effective choices. (Graham 1990)

I have elaborated here on Jason because I found that senior and midcareer Japanese high energy physicists were extremely interested in Jason and seem well informed about its activities and membership; since 1976 when I first did fieldwork, they have asked me many detailed questions about it and they tell me that they believe no comparable organization exists in Japan. To my knowledge no high energy physicist members of Jason, in spite of their prominence, have been invited to Japan since joining Jason. The Japanese have said that they believe it would be very difficult for a high energy physicist in Japan to belong to such an organization and have the respect of one's colleagues. When I asked about other kinds of Japanese physicists, such as nuclear physicists, they said they did not know.

Even though most high energy physicists in Japan are from the more conservative universities to the east and work at the national laboratory where political activity is at a minimum, they appeared determined to avoid any association with military funding and to shun Americans who receive such funding. It would appear that Japanese high energy physicists, at least, are quite committed to having no ties to military research or funding, either direct or indirect, unlike some of their U.S. colleagues. (I have no data on whether Japanese nuclear physicists and cosmic ray physicists have any links with military funding in Japan or are concerned about foreigners who do; some certainly have had long-term collaborations with Americans who receive such funding.) The JPS long ago passed a resolution against physicists participating in projects supported by the military. That resolution was reprinted on the front page of the program for the JPS annual meeting in 1991.

Funding and Equipment

Japanese experimental scientists who are engaged in basic research are enmeshed in extensive ties to the business world. Constraints on funding to the Japanese national laboratories require that all equipment for experiments be purchased and not constructed on site, and no laboratory "shops" for building, altering, or maintaining equipment are funded. By contrast, high energy physicists in the United States highly value the capacity to make and remake, in their own shop facilities, the most complex research equipment, even if they purchase many components.

In Japan all equipment purchases from foreign suppliers over a certain very minimum cost must be approved by bureaucrats in the appropriate ministry of the government. In the United States, approximately one-third of the employees at the NSF, for example, are scientists, many of whom are on a one or two year leave-of-absence from academic positions and are quite familiar with the needs of the scientists doing research; similarly, many at the U.S. DoE are also scientists. In Japan the bureaucrats are not scientists; indeed, they are often rotated through positions within a ministry in order to learn its procedures better; the scientists at the national laboratories find dealing with these officials about purchases for scientific experiments very tiresome.

As a consequence of all these funding constraints, the Japanese experimental high energy physicists are more comfortable with another process for designing, building, operating, and maintaining research equipment. They develop very close and long-term working relationships with engineers at those Japanese firms that have the interest and capacity to provide equipment for research, usually the very large corporations with a major commitment to research and development, such as Toshiba, Fujitsu, and Mitsubishi. (Of course, these companies also have considerable influence with both politicians and agencies of the Japanese government.) In each phase of an experiment (or laboratory construction) physicists spend a significant part of their time with their suppliers' staff engineers, exploring possible specifications, modifications, and any difficulties with the equipment.

Many, many times at KEK I saw a group of three or four engineers from some company (recognizable by the similarity of their clothing—sort of like the old IBM "uniform"—white shirts, dark slacks, and ties, which most of the physicists did not wear) come to the offices of experimentalists, carrying large notebooks; they would stay at least an hour or two, talking earnestly and taking notes. All during the assembly, moving, and operation of the research equipment at the lab these companies' engineers and technicians were always there. Physicists at KEK told me that the engineers and technicians come to be quite familiar with their ways of thinking and working and vice versa. Most of the experimentalists had in their offices many large books of specifications from the companies they used. During a particularly frequent set of visits from one company, some of the younger KEK physicists called a senior scientist at the lab "Professor X" after the name of that company.

Much of the work done by company engineers and technicians in Japan is done by graduate students and postdocs in the United States—at much lower cost, at least, in the short term. (Current NSF stipends for advanced graduate students are over $14,000 and postdoctoral research associates receive about $25,000; although these amounts are quite generous compared to those in the social sciences and humanities, they are modest compared to the salaries that physics graduate students and postdocs might expect working in private businesses.) A very powerful ethos is learned by undergraduates in U.S. university laboratories about the importance of learning to work well alongside some senior esteemed technician, an ethos that extends through graduate school and into the major research labs where there is always a small cadre of exceptionally skilled technicians highly adept at transforming experimentalists' designs into prototype equipment. The machine shop facilities in which these craftspeople work are very well supplied. These facilities and the craftspeoples' salaries are funded by the laboratories' operating budgets, monies that are not available to Japanese laboratories.

This different division of research labor, due to government funding constraints —or opportunities, corresponds to a different social allocation of knowledge with different personal, social, economic, and political consequences. The work done by younger company engineers and technicians in Japan is done by physics graduate students and postdocs in the United States. In Japan these young people become, in time, senior engineers and technicians with greater opportunities and responsibilities; in the United States approximately three-quarters (the proportion has steadily increased with the scale of experiments) of the young researchers in experimental high energy physics become, at about thirty-five years old, a surplus science labor force. They learn (often with strong feelings of bitterness, anxiety, and inadequacy which seem to stay with them, sometimes accompanied by boastful stories about the past, for decades) that they must find employment elsewhere.[8]

I have found no government agency, no laboratory, and no university that keeps records on where these people find work. My own, as yet rather limited research on this topic indicates that they usually move into astrophysics, geophysics, physical oceanography, computer science, and biophysics to work on projects defined by others; some work in universities (often, apparently, on Department of Defense [DoD] funded research projects), some in government weapons labs, and some in industry. In my small number of interviews with such people, the only ones who had help from senior high energy physicists in making this transition now work on projects funded, at least in part, by the DoD. Different government policies on funding of basic scientific research benefit different sectors of society outside the laboratories: in Japan, expertise develops in the private sector from collaborating on high energy physics projects; in the United States, people trained in the very large high energy physics research teams take their skills to other projects in universities, the military (including national weapons laboratories), and, to a lesser extent, private industry.

Exceptional Elders

Particle physics research in theory and experiment was established in Japan during the 1930s by Sakata, Tomonaga, and Yukawa (Tetu 1974; Kaneseki 1974; Brown, Konuma, and Make 1980). The generation now "retiring" are the men who have been the community's organizational leaders since the postwar period; in this group are at least a half-dozen exceptionally powerful men. Many of the next generation, now in their forties and fifties, received much of their advanced education abroad and they often have continued to do collaborative research with their foreign colleagues. This group wants to work in Japan and they are eager to establish truly international laboratories in Japan. The younger people in their twenties and thirties seem to have made use of their elders' international networks to take some training abroad, but fewer have the experience of either long-term collaboration with foreigners in Japan or sustained work abroad. These generational differences are seen by the physicists as substantial, weakening communication and trust. This complicates the crucial process of succession in leadership and exacerbates the educational and linguistic differences I discussed earlier.

All these "retiring" leaders have, in addition to their forceful personalities, extensive international networks in particle physics and they have considerable influence over the formation of science policies in the Japanese government. Their institutional bases have ranged from Kyoto, Osaka, and Nagoya to Tokyo, Tsukuba, and Sendai; they include experimentalists and theorists. Laws in Japan force retirement at specific ages, but it is quite possible to influence the selection of one's successor, unlike the United States (Traweek 1989, 1992a). Some of these men will continue to wield power with or through those successors, as is the custom in Japanese society.

This partnership in leadership of two men, separated in age by ten or twenty years, is not merely a prolonged passing of the baton as in some athletic track events. It is rather more like the division of labor in U.S. universities between presidents and provosts, or chancellors and vice-chancellors: the junior ranked person attends to the day-to-day leadership of the institution while the senior person focuses on extramural relations, such as raising funds and expanding the resources of the institution, enhancing its reputation and visibility, and protecting it from any outsiders who might seek its resources. Since the war removed many of the previous generation in Japanese science, some of the retiring leaders have done both these jobs for years. They are now obliged, by their retirement, to chose their successors and some have already done so. This transition will free those exceptional elders to become science statesmen for particle physics in Japan. They already have extensive ties to the government. It remains to be seen what they will make of their new role in the 1990s and how their designated successors will wield their new authority.

Power and the Infrastructure of Big Science

Japanese high energy physicists have managed with *gaiatsu* to take advantage of the new prosperity in Japan to establish a research laboratory with state-of-the-art equipment; to modify graduate education requirements, which enables them to train young physicists in contemporary research strategies; and to help establish a university (Tsukuba) that admits some students solely on the basis of exceptional grades in one area, rather that the more customary requirement of very high averages on standardized tests in a variety of fields. At least temporarily they gained massive levels of funding for basic research in science and they learned to do experiments in collaboration with engineers from private industry. They reversed some of Japan's postwar brain drain by creating a good research environment in Japan; they brought researchers from around the world to work in Japan, making it possible for young Japanese high energy physicists to have the benefits of foreign collaboration without having to leave their own country, a privilege many North Americans and Europeans enjoy. They launched the idea of science cities in Japan; perhaps in the future their ideal of frequent, informal interdisciplinary interaction will be realized, but as yet it has not. Finally, through mass media reports about KEK they gained, at least temporarily, a public audience for basic science research in Japan.

These high energy physicists inaugurated big science in Japan and built a skeletal infrastructure for big science research: a "critical mass" of "world class" scientists, "world class" research equipment, massive levels of government funding and public support for that funding, high quality education for advanced research, and high levels of communication among researchers. This is an extraordinary accomplishment. Their problem now is how to sustain these activities. High energy physicists in Japan are only beginning to have the local or national power that their colleagues in nuclear physics have had, power that seems to me crucial. That group, along with their own international network, may indeed launch big nuclear science in Japan, perhaps at the expense of the high energy physicists who gathered the pieces together and put them in place, over the objections of some of the nuclear physicists. However, I think there is a chance that the *bachigai* outsider high energy physicists and the insider cosmic ray physicists will form a coalition and retain KEK as a laboratory for basic research in particle physics, but that is another story.

The high energy physicists in Japan work at the margins of two empires: the international scientific community, which is based in North America and Europe, and the Japanese scientific community, which is based in the universities and the Ministry of Education. A few high energy physicists used that *bachigai* position to build Tsukuba Science City, Tsukuba University, and KEK, the Japanese national laboratory for high energy physics. They accomplished this through strategic use of *gaiatsu*, among other strategies, and made the most of the Japanese public's and government's concerns about *kokusaika*. By building a national laboratory with state-of-the-art research equipment they vastly increased their status in the international scientific community, but that community is still centered somewhere else. Japanese scientists now wield some power in the international high energy physics community, but few U.S. or European physicists have yet bothered to learn about the political economy of big science in Japan. I am certain that the next generation of high energy physicists in Europe and North America will.

In this chapter I have described the strategies that various groups of Japanese physicists, bureaucrats, and politicians invent, learn, and use to make sense and to make a difference among themselves and with each other in their day-to-day negotiations. I have foregrounded the strategies of the physicists who are in the laboratories and the elders of the scientific community. I have argued that to understand the relations within and among the Japanese physics community, other national physics communities, the Japanese government, and other governments, it is necessary to understand the Japanese physicists' career strategies, gender and generational differences, their politics, their laboratories, and their funding ecology. I have frequently contrasted the Japanese to the U.S. physicists; the sharpest difference is with respect to the role of military funding, even though both communities appear to eschew any contact with the military. I briefly explored how all these issues are at play in the formation of factions about the future of high energy physics in Japan. In all of these situations the negotiation of borders and the construction of power is crucial to the production of knowledge in physics.

Notes

1. I have been studying the multinational high energy physics community (especially in the United States and Japan) since the early 1970s. I have spent about two years at the Stanford Linear Accelerator Center in California; two years at the National High Energy Physics Laboratory in Tsukuba, Japan; six months at the Fermi National Accelerator Laboratory in Illinois; and shorter visits to CERN in Switzerland, the German Electron Synchrotron (*Deutsches Elektronen-Synchrotron*—DESY) in Germany, and Saclay in France. I have also visited physics departments at universities throughout the United States and Japan and attended innumerable particle physics colloquia, workshops, and conferences. Altogether I have conducted about five years of research at some of the major national high energy physics laboratories where this community gathers.

2. The most visible of these anthropologists included Tadao Umesao, Director of the Senri Ethnological Museum. He is best known for his theory of world civilizations; see his article in *Senri Ethnological Studies* (1986:1–8), one of two special issues of that journal devoted to his theory.

3. Massive research literatures exist on the following distinct, but related topics: 1) women in science, engineering, and medicine; 2) discriminatory practices in science, engineering, and medical education, hiring, and promotion; 3) gendered ethos in science, engineering, and medicine; and 4) gender assumptions in scientific and engineering knowledge. See Sharon Traweek (1993:3–25).

4. Monica Strauss has found a similar process at work in Japanese computer science research. See her unpublished master's thesis (1982), available from the Massachusetts Institute of Technology Laboratory for Computer Science as Publication TR-383.

5. See Lillian Hoddeson (1983) for a history of the decision making about the design of the two accelerators. See Satio Hayakawa and Morris F. Low (1991:207–229) for a history of the interactions between Japanese government agencies and scientific organizations concerning high energy physics from 1957 to 1970. See Traweek (1989) for a discussion of how high energy physicists at KEK coped with the limitations on that design imposed by last-minute funding cuts. The two primary histories in English are meticulously based upon diverse formal written documents: circulars, minutes, memoranda, bulletins, reports, letters, requests, and proposals that emphasize the roles of the very few actors who had official responsibilities. Hoddeson's history emphasizes decisions about the parameters of the first accelerator at KEK; Hayakawa's and Low's jointly written history stresses the long-term factional, institutional, and political features of the conflicts among senior academic scientists and bureaucrats in decisions leading to the establishment of KEK as a national laboratory unaffiliated with any existing institutions. These histories are invaluable for the carefully documented information they contain and I recommend that my account be read in conjunction with them. However, I would caution the unwary to remember that many activities crucial to contested decisions never leave a "paper trail." One major figure in many of these decisions said to me that he found those histories "too hard" and that he preferred my "soft" accounts. When I asked what he meant, he replied that mentioning names and quoting papers not only embarrassed people who were still dealing with each other but also gave a misleading impression of certain events; he

thought that describing processes rather than chronicling statements gave a more realistic account. I am not asking that we only accept representations that are pleasing to the actors; I am arguing that we try to describe all the ways communities make and unmake decisions. I think this is particularly important in a community where everyone most certainly notes what is *not* said, written, or done at least as much as what is. In addition, timely private suggestions can lead to important public actions. It is not only in Japan that private comments or the absence of a public gesture, especially from someone who is powerful, can make a huge difference; nonetheless, in Japan this way of exercising leadership has been especially valued.

6. Because Jason members work on "classified" (secret) projects and the members choose not to identify themselves, the group is rarely mentioned in media available to the public. One exception was a newspaper article by David Graham entitled "When U.S. has a science question, it asks Jason" (1990:B1, B4), which included an interview with Professor Curtis Callan, a theoretical physicist at Princeton University and Jason's chair at that time. Graham mentioned that "most of the 56 Jason members are physicists, although there are oceanographers, mathematicians and electrical engineers, too. Their credentials are among the most impressive in science. Members include six Nobel laureates. Two Jasons are women." Graham also stated that "at least 10 UCSD faculty members have been listed as part of the Jason group, including Edward Frieman, director of Scripps Institution of Oceanography, and Herbert York, a former UCSD chancellor who was a scientist on the Manhattan Project, which built the first American atomic bomb." Photographs of York and Frieman accompany the article. He added that "the Jason group was founded in 1959 with a core of scientists from the Manhattan Project who wanted to help guide development of new technology for defense uses, as they had done more than a decade earlier by translating nuclear theory into an awesome bomb."

7. One member of a colloquium I attended chided me for identifying La Jolla as a "resort." I feel obliged to make my point more specific. In the 1940s and 1950s, La Jolla had the reputation of being home to the Navy "brass" based in San Diego; it was also notorious then in Southern California for its restrictive covenants against selling housing to Jews and many other minorities. It has long been home to the Scripps Institution of Oceanography, and during the 1960s the University of California at San Diego, which now enrolls nearly 20,000 students, was established there on land between the beachfront areas and Miramar Naval Air Base farther to the east; private enterprise laboratories, especially in biotechnology, followed. The land immediately north and especially east of the university, largely unoccupied until the mid-1970s, has become the site of phenomenal real estate development, commercial and residential, and is now called "the Golden Triangle" (parts of zip codes 92121, 92122, and 92037), at least by realtors. Up-scale tourism continues to be an important industry in beachfront La Jolla; the shopping district includes many small expensive shops and several expensive restaurants, short-term summer housing is also quite expensive. The average price of the seventy-three single family houses *and* condominia sold during March 1990 in the prized 92037 zip code was $476,629 ("Southland Home Prices," *Los Angeles Times*, May 27, 1990, p. K17); summer rentals are priced accordingly. Even with deflation in the housing market the prices remain very high. My point is that this older part of La Jolla, while certainly very pleasant, is an exceedingly expensive place for academics (and their families) to meet for six to eight weeks every summer and to do so

represents a substantial economic incentive for academics to become Jason members. It is widely believed in academia that salaries and benefits for DoD research are more lucrative than those for the DoE and far more so than those for the NSF; I have no data to confirm or refute this, although the high energy physics conferences and workshops sponsored by NSF that I have attended over the last eighteen years have never been nearly so well endowed, although quite comfortable. Accommodations and amenities at the Japanese high energy physics conferences and workshops sponsored by the Japanese Society for the Promotion of Science and by the Ministry of Education which I have attended have been comparable to those sponsored by the NSF.

8. For a study of the postdoctoral career stage and the evaluation of these post-docs see Traweek, (1989) chapter 3. I have had conversations with about twenty-five midcareer scientists who left high energy physics after completing a postdoc and I have had detailed discussions with seven.

PART 3

Conflicting Knowledge Systems

CHAPTER 11

Public Policy, Sciencing, and Managing the Future

M. Estellie Smith

Introduction

In the current climate of public policy-making, the public, business, government, and scientists are being encouraged to meet in common management arenas to address and reach consensus on the optimum approach to issues of concern to society at large. This process is opposed to top-down decision making (by, say, government agencies) and is a difficult (some say even Utopian) way to proceed. It is frequently marked by two stumbling blocks that seem bound to increase in time. First, even though such groups use scientific findings to formulate positions, scientists and their findings often disagree. That is, there is no single scientific world, even within a field of study, let alone across disciplines. Second, individuals in groups inevitably coalesce into subgroups, rendering them neither as single-minded nor as homogeneous as suggested by such phrases as "the public," "scientists," or even "the government."

One result of this interaction between science and the public is that the line between pure and applied science—never as sharply delineated as many practitioners have claimed—is increasingly blurred as scientists are called upon to focus research on issues viewed as critical to the general good. Similarly, not only do scientists constitute part of the public, but they also consist of diverse interest groups that credit opposing (or at least disconsonant) scientific positions. This extends debate and lengthens the time needed to establish common ground for discussions on consensual management design. By then, the problem and its dimensions have altered so much that the design no longer suffices. In part, this problem is inherent; the essence of science is skepticism: scientists are always moving boundaries and changing the received wisdom of yesterday. Finally, as scientists constantly bemoan, their findings are seldom the pivotal point of public policy-making. Rather, it is, increasingly, a triage process; decisions are made according to such practical concerns as economic feasibility, realpolitik, and public emotion.

This article will give a thumbnail sketch of the various faces of science in policy-making, focusing on a particular issue—the management of fishery stocks in the

United States. Such management came to the fore in the United States in 1976 when Congress passed the Magnuson Act mandating that resources of the continental shelf would be managed by eight regional management councils.[1] Council membership was to be drawn from social and natural scientists in both the public and private sectors, governmental administrators, representatives of the commercial and recreational fishing industries, conservationists, lawyers, coastal developers, the oil and tourist industries, and consumers. The Act charged the Councils to produce management regimes for fishery resources on the basis of "the best [biological, economic, and sociological] scientific data available." In the New England region especially (where my research has focused), it is widely held that the process has not been successful; both the stocks and the fishing industry are in crisis as resources have continued to decline and ever more draconian measures to protect dwindling stocks threaten traditional livelihoods in dozens of communities and thousands of people who rely on the sea. In this context, I explore the different faces of science and the problems of cross-cultural communication between two significant segments of the council: fishery managers (e.g., biologists, economists, and administrators) and members of the commercial fishing industry.

Pure and Applied Science

Whatever the outcome of human attempts to manage, it is important to recognize that science[2] not only varies in research methodology but also in scientists' cognitive environment of subtle, covert, and deeply rooted axioms; overarching institutional dynamics; and external sociocultural dynamics. A growing body of research is revealing that, no matter how those engaged in such activities attempt to disengage themselves from the broader context, the world beyond is an omnipresent factor in their work. Despite claims to the contrary, much scientific research is encouraged or constrained by the ways in which their fields relate to public issues. At the least, funding for research comes from governments, foundations, and private corporations that have agendas determining which research is viewed favorably and receives financial support and which is not.

This has certainly been true in the fishery sciences over the last several decades. The explosion of fisheries technology enabled and encouraged the commercial sector to shift from an essentially labor-intensive work pattern (where a day's catch was at most several thousand pounds) to a capital-intensive industrial fishing pattern. In the latter fleets of several dozen factory vessels move from locale to locale and, operating round the clock, employ fishing techniques that enable each vessel to haul in several thousand tons of fish per hour. Using sophisticated fish-finding equipment such as radar, sonar, and helicopters to spot schools of fish, such fleets (often coordinated by their governments or by private corporations) are able to roam the world, fishing out one locale and then moving on to another. It was the national need for food and exports, coupled with the appearance of a whole new technology of fishing gear, that gave impetus to this industrial transformation of a resource search that for millenia was akin to ancient modes of hunting and gathering.

Thus, fishing—like agriculture, stock raising, and a host of other human activities—became "scientific." Those who practiced it for a living had their work patterns altered by applied and theoretical investigations into all the components connected with those patterns—for example, net designs that permitted the escape of juveniles in order to maintain future breeding populations; the economic efficiency of vessel engines relative to trip length and average catch landed; and the decision-making process of a boat captain as to when, where, and how long to fish before returning to port. In the dramatically short time of little more than one generation, commercial fishing ceased to be something done mostly by family firms, with children acquiring the necessary skills during an extended apprenticeship, usually with kin. Today, it is not uncommon for young people who plan to enter the fisheries to attend college in order to obtain a degree in fisheries science.

I demonstrate the conflicts inherent in scientific work in the public context by examining a single publication: a recent issue of *Fishing News International* (January 1995),[3] the leading industry newspaper, which reports on all aspects of modern global fishing. Selecting one source for such a content analysis is deliberate; the data have not been culled to prove a point. For instance, a number of articles note areas of the world that have closed fishing grounds or have imposed catch limits—including Norway where the new "get tough" attitude of the Norwegian Coast Guard is aimed at protecting the shrinking stocks of mackerel in their waters (1995:4). The same issue, however, includes an article on an addition to the Irish fleet, a new 59.6-meter supertrawler (cost US$10.5 million, hold capacity 1,000 British tonnes) to be used to target that same North Sea mackerel (30). The vessel has state-of-the-art fish-finding gear and navigational equipment: The "centerpiece of the wheelhouse equipment" is "the most powerful fishing sonar available with a proven detection range of [schools of fish] up to 500 meters distant (over three miles) ... down to 1,200 meters and the seabed down to 3,000 meters" (32). A few pages later, another article headlined "120-Meter Trawling Giant!" (the second of its type to join the Dutch fleet) describes the catch of the vessel's "very successful" maiden voyage in the same northern waters. It landed 100,000 cartons of horse mackerel and herring—not the least because the ship employed five nets with fishing-circle circumferences ranging from 1,100 to 1,700 meters (33).

Still other articles call attention to scientific research on: 1) new and bigger purse seines and midwater nets that permit a vessel "to aim easily for new target resources ... with appreciable high commercial values" and at costs so low that the equipment "can pay its way even when the fish concentrations are reduced in size and density" (10–11); 2) a new design for dropping and hauling in nets that has speeded up the process and increased hauls per trip "up to 80 per cent" (9). 3) Readers are also advised that the National Marine Fisheries Service (NMFS)—

the same agency of the U.S. Department of Commerce that oversees fishing and has consistently urged the fishing councils to take more stringent measures to control fishing effort—has recently concluded studies on new, "improved" types of plastic traps that catch "about three times as many spiny lobsters and nearly four times as many slipper lobsters than the wire traps" (15).

Yet another article discusses the growing role of science in the fishing industry, reviewing the history of one of the oldest fisheries degree programs in the United States (43). The University of Washington, in Seattle, established its program in 1920, and *Fishing News International* emphasizes that the program has changed considerably since its inception. The first students were trained primarily for work in the canning and fishing industries. This "reflected the needs of fisheries at the time. The accent was ... on how to catch and process more fish." When initiated, other scientific faculty viewed the program skeptically: had the graduates been trained in "real science"? As early as 1931, the program's focus had shifted to the acquisition of "basic information about the life cycles and biology of various fish," and scientific data was deemed necessary "if wise decisions were to be made about harvesting."

The usefulness of the scientific research conducted in the program was so obvious that, in the 1940s,

> the Alaska salmon industry contributed funds to establish and support a Fisheries Research Institute at the School and a long term research program was set up to learn more about the basic biology of salmon so runs could be predicted and regulations imposed before fishing seasons even started. (1995:43)

Today, the university is quoted as underwriting research decisions that center around the perspective that

> fisheries are about much more than extracting and using a resource. Fishermen, citizens, policy makers and managers ... are struggling with the great balancing act of the '90's: how to protect species and habitat in the face of demands for more fish, more timber, more water, more power, more of everything. (1995:43)

The school now receives US$6 million a year in outside grants and contracts for research, [and also] provides technical expertise to state, regional and federal agencies, fisheries councils and the courts.

Thus, research is conducted in a bicameral arena where practitioners work both to protect the resource and support the technology that exploits the resource ever more "efficiently," that is, more intensely and at lower cost.

It is this dual functioning and function of science that not only confuses but angers many, especially the primary producers who go to sea to catch fish. Their view of the work of science and scientists is summed up in an article titled "Modern Boats Fish 'Non-stop,'" in which the owner of a marine-supply firm notes:

> Fishermen with bigger, more modern boats fish virtually non-stop; that's what it takes now to make a living.... They unload their fish, take a shower, go home and kiss the wife, then it's bye-bye again! (1995:21)

And what of those fishermen who "just want to make a living and get by without raping the stocks so that my kids will still be able to fish" (as one New Englander put it to me)? The marine-supply owner addresses them when he notes philosophically that, "Guys with the 50 to 55 footers ... take what fish they can get and grin and bear it."

In sum, scientific research both giveth and taketh away, depending on who employs the scientists and to what purpose their job descriptions direct their activities.

A Case Study: A Problem in Consensual Management

When the Magnuson Act was passed by Congress with the intent of protecting the resources of the continental shelf, it aimed to have a management regime that reflected the diverse knowledge and needs of all those with an interest in protecting the viability of marine resources. Despite regional variation, however, members of several of the councils are predominantly drawn from, on the one hand, public-sector personnel (that is, administrators, scientists, and technicians) and, on the other hand, from members of the commercial fishing industry.

Proponents of the comanagement theory of public policy argue that, through joint participation and by shared special knowledge, fishery councils can produce fishery management plans supported by the diverse groups affected.

The process, however, has proved slow, painful, and, for most, unsatisfactory. The context within which the councils operate requires members to recognize the political, economic, sociocultural, and biological forces in play. During the past three decades, market forces of consumer popularity combined with improved technology and the expansion of the human predator population have increased demand on fishing stocks. This has led not just to lower landings but, more critically from the biologists' perspective, to annual catch statistics that indicate an ominous emphasis on single (and younger) year classes. The difficulties of both management and stock maintenance in this context are compounded by additional factors: 1) the expansion in the number of targeted species, which results in increasing pressure on the entire food chain and also increases natural predation opportunities for some species and removes food resources for others; 2) the increase in pollution, which may be affecting the reproductive cycle; 3) the alteration, even disappearance of marshes, rivers, beaches, and other locales necessary for the reproductive cycle of some species—and by extension to all in the chain—forcing the species dependent on such locales to move elsewhere or die; 4) the targeting of new species, about whose particular life cycle or food chain niche we know little or nothing; 5) the volatility of market forces; 6) the rapid change of commercial fisheries technology (and rising capitalization costs); 7) the overlapping demands of multiple users; and 8) the competing lifestyles of coastal dwellers,[4] especially those in fishing communities, who see their livelihood and their way of life under attack from all directions.

The comanagement approach involves a delicate juggling of economic and political interests, conservation and predation, and the public and private good, which accounts for the dissatisfaction and anger of so many with more than seventeen years of regional council management. Charges of overmanagement are matched with counterclaims of inadequate regulation while concern grows for the present and future condition of the stocks.

Among the long-term results is the acrimony that marks debates as members in each group increasingly see other members as their adversaries, who lack the good faith and good will and necessary understanding to come to grips with a crisis that seems to have only two solutions. Either fishermen must be allowed to continue fishing (and, with their ever-more sophisticated technology, risk wiping out the stocks on which their livelihood ultimately depends) or there must be a limit on fishing. By limiting total allowable catches (TAC), or banning fishing altogether, stocks rebuild. In the indefinite interval before permitting fishing to resume, however, the infrastructure of the domestic industry would be dismantled as people switch to other occupations. Charged to "do something," those on the councils, in advisory groups, and at public hearings know they must be open to new input but not so open as to be labeled sell-outs. They must be prepared to negotiate something that affected populations would willingly abide by. Unfortunately, all but a few plans, once actually implemented, have been labeled by most as insufficient, inappropriate, probably incorrect, and certainly out of date.

My own extended observations of the New England council (since its inception), indicate that all participants identify with one or another interest group and, not surprisingly, each subgroup has a view of what constitutes relevant and robust scientific data.[5] That is, how one goes about verifying or testing the validity of data, what to do in the face of uncertainty, how best to analyze data, and how to translate such analyses into programs for management. In short, conservationists, biologists, economists, industry personnel, state and federal administrators, state and federal legislators, fishery managers, industry members, and public interest groups each possess a cultural belief system. Group membership is maintained by articulating the axioms—first principles—that constitute the basic truths of the group. Axioms are the basic formulas that, if questioned or challenged, bring frowns, pitying smirks, or charges that the questioner is "naive," "uninformed," or "just plain nuts." Such first principles support belief systems and are taken for granted as right, rational, and "going without saying." It is expected that such axioms need not be discussed because they are commonly held by all who use common sense and are knowledgeable.

A majority, if not all of the participants in the management process, fall into one of two camps relative to the nature of Nature. The two camps 1) analyze the same information differently; 2) gather, aggregate, and juxtapose different information differently, which also leads to different analyses; 3) diverge in their perceptions of how, what, where, when, and whom to manage; 4) disagree on whether it is even possible to produce long-term management plans, which many see as fundamental to the production of a stable environment for economic investment and decision making (and lack of agreement on what constitutes long- versus short-term is a major stumbling block). Both camps believe that decisions are urgently needed and that decisions should not have unintended consequences. But many of the delays, shifts, reassessments, fine-tuning, and continual stream of plan modifications have taken place because the participants are speaking uncommon languages.

The two paradigms of Nature provide yardsticks that individuals and groups use when measuring "good" versus "bad" management. Unfortunately, fundamental differences between the two paradigms, rarely made explicit, also create a chasm across which dialogue is difficult, if not impossible. Each camp feels that its position (whether derived from scientific research or practical experience) is grounded in common sense and will surely generate sound plans. In contrast, the others' position is general (or narrow), complex (or simple), rigid (or ad hoc), costly (or inadequately funded for proper implementation), or it ignores the human element (or is too vulnerable to manipulation in the political and economic arenas). If or when management programs finally are implemented, such deeply held but rarely stated worldviews—or cognitive modes—are liable to lead members of, especially, the targeted user group(s) enthusiastically to endorse, comply, or resist the rules and regulations.

In the case of the members of the commercial fishing industry, the extent to which the council produces workable management plans determines the degree to which industry members cooperate in making the plans work. The final compromise plan is often seen, however, as having gone through so many alterations that it emerges fatally flawed. Groups will respond by arguing that the plan violates federal mandates or needs fine-tuning when being implemented. The latter has led to continued modification as well as charges of ad hoc crisis management (for instance, a series of emergency actions to deal with "unexpected" overfishing or shortfalls). What user groups see—sometimes correctly—is a bewildering and contradictory flow of rules and regulations.

The distinction between the two views of Nature is as follows.[6] On the one hand are those who view Nature in classic Newtonian terms; on the other hand are those who view Nature in the terms strikingly parallel to the model suggested by the newly emerging science of chaos.[7] Adherents to the first position model the world in terms of linear relationships; adherents to the second model it in nonlinear interweavings. Chaos theory is useful for studying phenomena as diverse as weather patterns and climatic movements, stock market cycles, and irregular fluctuations in the rhythm of the heart.[8]

This paper is not a claim that chaos theory is the best model for understanding the fisheries or engaging in fisheries management. My concern is to address the apparent failure of consensual fisheries management. Why have participants in fisheries management failed to negotiate a successful program? Why have such acrimonious charges of "rigid, dogmatic thinking" and "muddleheaded opportunism" bedeviled those involved in the process?

The first group, consisting of the majority of biologists, economists, statisticians, and ecologists—those scientists and technical experts who are the lead members in marine research studies and in the agencies concerned with fisheries—see Nature as, first, a system and, secondly, as a system in which there is periodic order. The study of such a system depends on studying various species, year classes, subregions within the marine econiche, the various ports and their landings, and so on. It involves defining perimeters and parameters, identifying relevant variables, and using differential equations to describe processes that change "smoothly" over time. Therefore, fishery managers tend to speak of fish reproductive processes as if there were no interactions among overlapping generations and no unique environmental events affecting generations differently. That is, they directly identify the relationship between the number of, say, herring or cod at Time X_1 and the number of for Time X_2—they express stock dynamics as if the Time X_2 population is a simple function of the Time X_1 population.

Perhaps the classic expression of the linear view of population dynamics, the view that the nature of Nature is ordered, balanced, and in dynamic equilibrium, was given by J. Maynard Smith (1968), whose position was that populations either remain relatively constant or regularly vary around some presumed equilibrium point.[9] As James Gleick put it:

> In a real world system an observer would see just the vertical slice corresponding to one parameter at a time. He would see only one kind of behavior—possibly a steady state, possibly a seven-year cycle, possibly apparent randomness. He would have no way of knowing that the same system, with some slight change in some parameter, could display patterns of a completely different kind. (1987:73)

In the case of commercial fisheries, biologists frequently assume that fishing primarily accounts for deviations of natural populations, and, in the last decade especially, have moved to sustain the stocks by attempting to regulate human predation.[10] The complexity of the universe being modeled is recognized, however, and scientists are aware that predicting future states for natural systems is no more certain for managed than for unmanaged systems—and perhaps even less. As one manager put it, "If we can't manage the fish, we'll have to manage the fishermen."[11]

Members of the second group are mostly fishermen (and their families), buyers, and processors. They do not see Nature as random ("Things don't just happen—there's always got to be a reason") but they recognize that they operate within an unpredictable universe. ("If I knew everything that was going to make one fishing trip a winner and another a loser, I'd be God.") Natural processes are perceived as complicated, dynamic, and sensitive to small initial perturbations. Modeling of the processes suggests causal relations. Sequential patterns can stretch over so long a period that they appear aperiodic or are so frequent that they appear even, without irregularities (much as the daily ups and downs of the volatile stock market appear in a smooth curve over the period of a year).

Attempts to understand such phenomena as population dynamics (whether of humans or fish) in terms of linear relationships that can be captured on a graph can be counterproductive. Yet, those responsible for final plan production—in translating committee work, discussions, and public hearings into plans submitted for federal approval—are, for the most part, linear modelers.

Chaos argues that system dynamics can unfold in a nonrandom but unpredictable fashion. They are labeled "unpredictable" only because those studying them cannot take into account all perturbations when modeling the system. Small or initial perturbations are elements that, although often ignored, determine both calculated and real outcomes. We all learned the principle as school children:

For want of a nail, a shoe was lost; for want of a shoe, a horse was lost; for want of horse, a rider was lost; for want of a rider, a message was lost; for want of a message, a battle was lost; for want of a battle, a war was lost; for want of a war, a kingdom was lost—and all for want of a nail.

As the lost nail analogy shows, in nonlinear systems every movement has a meaning. More important, as the meaning moves through extensive networks, its significance snowballs and its potential to alter events, or systems, intensifies. This effect is nonrandom but unpredictable.

The distinction between those arguing for the existence of linear or nonlinear views of Nature is blurred. Industry members can now argue that their position is not reflective of a difference, say, between scientists and entrepreneurs, pure theory and dirty practice, intelligence and stupidity, objective managers and greedy fishermen. In short, it opens up the possibility of a more effective dialogue, one in which participants listen to, rather than talk at, each other.

Given that the chaos model argues that 1) any open system is susceptible to new input and 2) any initial condition can generate consequences that intensify up systemic levels, the model also suggests that it is difficult to have full rationality of action.[12] One of the early proponents of the chaos paradigm labeled this characteristic the "butterfly effect." As Edward Lorenz put it in an important and early statement of this approach (1979), the fluttering of a butterfly's wing in Rio de Janeiro, amplified by atmospheric currents, could cause a tornado in Texas two weeks later. Examples of this abound in the real world. Thus, it has been suggested that the crisis at the Three Mile Island nuclear plant, which led to the entire U.S. nuclear energy program being indefinitely constrained, resulted from one worker who neglected his monitoring of one gauge because it was obscured by his overgenerous belly.

For some, especially those who operate within the cognitive context of Max Weber's "rational bureaucracy" (its raison d'être is planning the future), all this is anathema. "How," they ask, "can we operationalize such a model?"

As a final note to the discussion of chaos theory—it is often difficult to distinguish between two forms of chaos. The one results from not including critical but periodic components. The second, under discussion here, is that, in a majority of natural, open systems, including new information can create a new order. Perturbed by randomness, real systems permit new variants to emerge that cannot be made to disappear by future noise.[13]

The fishermen who have participated in the management process—those who have challenged the scientific data, have resisted the imposition of management regimes, and are seen by managers, especially scientists, as naive and self-serving—have legitimated their stance by arguing in ways that seem consonant with the scientific theory of chaos. A small perturbation—say, a one-degree drop in ocean current temperature—can result in changes in fish distributions. Commercial fishermen frequently reject management plans because they believe the plans are insensitive to such realities. As one fisherman once said to me:

> By god, those people are stupid! Year after year they come out here with their charts and graphs and measuring tools and go to the same spot at the same time and try to catch fish so they can compare this year's stock with last year's and 10 years ago and so on. And they mumble about "replicability" and "sampling procedures," and like that. Jeeesus! Don't they understand that fish swim?

What the fisherman is saying is that a small change in the local water temperature, a marsh that gets drained somewhere along the coast, some vessel dumping waste overboard, can be the disturbance that leads to a change that ripples along the food chain, amplifying in scale as it moves up in scale. From this perspective, a management plan can be doubly damned—imposing overkill responses on normative "abnormalities,"[14] while ignoring the extent to which Nature is vulnerable to small natural perturbations with large consequences.

A recent example of the chasm between fishery managers and industry personnel was a dispute between an industry group and Dr. William W. Fox, then head of the NMFS. Industry personnel filed a civil suit in the U.S. District Court against Fox[15] because he had approved a ban on drift gillnetting in their fishery. The leading trade journal interviewed Fox and opined that, given that some scientists maintained there were inadequate data to justify the ban, Fox might have reserved judgment. Fox responded,

> It depends upon what you do with uncertain data. You can say, "Well, this doesn't prove there is a problem even though it may imply it. Therefore, we aren't going to take any action until we can prove it." My view is ... to react in a conservative manner in the face of uncertainty. (quoted in Fee 1990:15)

Public Policy, Sciencing, and Managing the Future 211

Fishermen, however, did not see eliminating their means of earning a livelihood as a conservative action. Said one:

> It's just one more uncertainty in a fisherman's life. Weather, the fish, the boat, and now the fishcrats—and they're the most changeable of all. Hell's bells! We could learn to live with anything if the damn Feds would just put something in place and then leave it alone for a while. But you never know what they're going to do and they're always tinkering, "fine tuning" they call it. But you know, I know, and they know you can't "fine tune" fish or the weather or the water. Only thing that's left is how I go fishing.

From another fisherman:

> It took me more than a decade to learn something about how it is out there, how to be a good fisherman—and then I'm only right some of the time because any little thing can make a big difference.
>
> But regulations change everything, all the time. It used to be that you worked out (with a little of margin of risk) when you needed a new net. Nowadays, if I buy new nets this week, there's likely to be a new net regulation next week—and mine will have an illegal size mesh, the wrong cod end, or something. And I'm stuck with thousands of dollars of useless net. But if I hear talk and decide to wait, then things just drag on, nothing gets decided, and I lose a whole trip because the damn net goes.

A third fisherman:

> I'll tell you, they may save the fish but what for? You can't be a fisherman if you can't make a living at it. If the stocks come back, there won't be anybody left to fish. You don't just stick somebody on a boat and tell him "go fish." The world's too complicated. We've been trying to manage the fisheries since the late '70s, here, Canada, Europe. Nobody's gotten it right yet. You'd think people would learn that you can't use nature and manage it at the same time. The only time I know that the stocks have gotten better was during both world wars, when fishing effort was down. So, you want to catch fish, the stocks will go down; it's as simple as that. You want stocks to go up, you have to stop fishing completely—fish farming and aquaculture too—for 10 years. That's the only way. Then learn fishing all over.[16]

Such "passionate rejection" is not limited to fishermen. It also marked responses from fisheries managers to my suggestion that discussion of the linear and nonlinear views of Nature might provide common ground for more productive dialogues among managers and industry. Only two of almost two dozen scientists were even mildly receptive ("I'd try anything if it seemed able to help things along"), despite my repeated emphasis that chaos theory was "a scientific model" that could bridge some critical gaps in the communication process. The three responses given below are representative and, significantly, none saw the use of chaos modeling as having a potential for improving their understanding of the biological systems with which they were concerned.

The first fisheries manager (biologist) said:

> If management were left up to us [marine biologists], we could do something. But in the final analysis, scientific evidence doesn't matter; it's what plays in the political sector. In any case, I don't buy chaos as a useful tool in biology. But where it would be useful would be in understanding how a biologically sound fisheries plan gets completely transformed as it moves through the political corridors.

The second fisheries manager (biologist) said:

> Nonlinear models don't tell us anything we don't already know about natural systems. But when you have to gather data, analyze it, and produce projections for the conditions of the stocks that will help write management plans, what do you suggest we do? The people up the line expect our analyses to follow a certain format. You have to do it by the book or they just send it back down. Besides, if chaos theory is correct, then the world is so full of such trivialities that nonlinear theorists must find it as difficult as linear-based theorists to map a trajectory accurately. So what profit is there in the exercise? I don't think any of us have time to worry about such philosophical issues as "the nature of Nature."

The third fisheries manager (statistician) said:

> Chaos is New Age pseudo-science crap. I don't know anyone who is seriously concerned with fisheries management that bothers with it. Once you've used it descriptively what real utility does it have for management regimes? You don't really believe fishermen when they talk like that, do you? They're just putting you on.

Conclusion

This review of the problematics of fisheries management has highlighted an issue that will increasingly bedevil modern society. Given that scientists are as much a product of their sociocultures as those who engage in other occupations, scientific processes and products are subject to most of the same influences that shape other activities. Axioms are the building blocks with which each of us constructs our basic views of the world. The variability in these views accounts for many of the scientific disputes that divide both "the public" and "scientists" on issues such as abortion, environmental pollution, global warming, or conservation concerns. The heterogeneous and multiplex world of science will also continue to confuse and divide those who look to science and scientists for ways to address the problems the world faces.

An old Irish curse is, "May all your wishes come true." All scientists have sought to have the world recognize the validity, significance, and crucial necessity of incorporating scientific research into everyday life. As this becomes more and more a reality in such areas as policy-making, public action, and the managing of our global system, scientists may regret the extent to which they are meeting with success in their quest.

Notes

1. An extended study of these councils is in Smith (1982). Smith (1988) explores the view that fishermen now see the primary risk in fishing not as the traditional dangers at sea but the as effects of fishery management.

2. I'm hard put to define "science." This is always true of the fundamental concepts in a discipline (for instance, "culture," "learning," or "life"), for the more one has delved into a topic, the more one is aware of the subtleties, the caveats, the ambiguities, the constraints, and the specificities of the subject's basic building blocks. Whatever science is and however it is defined, its doing seems to require the art of curiosity, the gift of observational tenacity, the capacity to organize and systemize, and an omnipresent sense of skepticism. From this perspective, there are fishermen who do sciencing (see White 1938) just as there are scientists who do fishing.

3. The following articles appeared in *Fishing News International* 34(1)(January 1995):

   Grounds shut in Canada. p. 3.

   Why Norwegian Coast Guard Is Getting Tougher—'Mackerel Stock at Risk.' pp. 4–5.

   Fishing Gear and Deck Machinery—12-Page Report from Around the World. pp. 6–17.

   1700 m Trawl for Dutch Supership. p. 8.

   Rise 'Up to 80 Per Cent'—But Fuel Costs Grow. p. 9.

   Forty Pairs Trawl for Tuna: French Mid-Water Technique Can Be Used Elsewhere. pp. 10–11.

   Modern Boats Fish 'Non-Stop.' p. 21.

   Second Supertrawler Joins Irish Fleet. pp. 30–32.

   120 Meter Trawling Giant! p. 33.

   Ship Sale and Purchasing Segment. All Change as Irish Rebuild Fleet. p. 41.

   University of Washington Celebrates Changes in 75 Years of Fisheries Study. p. 43.

4. Coastal dwellers include those living in urban ports, recreational coastal communities, retirement villages, or historically embedded rural fishing communities.

5. There are, of course, variations within subgroups. Canadian and American biologists disagree on a number of issues concerning North Atlantic stocks—although they use

the same statistical data. Biologists from Norway and Japan support continued commercial whaling of some species while the rest of the world sees such actions as (quoting one French biologist with whom I recently spoke) "an unconscionable rape of a resource."

6. Other views of nature also exist. For example, Tsuneo Katayama—one of the organizers of the Japan/U.S. workshop on urban earthquake hazards reduction, which was scheduled to open in Osaka just three hours before the catastrophic January 1995 earthquake struck Kobe only twenty-five kilometers away—responded to a reporter's query as to why the quake had had such disastrous results, "We were not polite enough to nature" (*The Economist*, January 21, 1995:35).

7. One of the founders of chaos theory, Robert May of Oxford University, has said that it is not surprising that chaos theory became a science but rather that it took so long to come together. It may be noted that social scientists have been receptive to the model because it has produced a paradigm for axioms they have long held.

8. Since I indicated that the chaos model has been formulated by scientists, in one sense the dichotomy being presented is simplistic; some of those concerned with fisheries management are working with the chaos model (a recent notable example in the late 1980s was the analysis of the Nova Scotia fisheries produced by Allen and McGlade [n.d.]). The methodology aims to simulate the world rather than explain it. This approach to management might be even more threatening to industry members than that currently in place. The categories of "linear" or "chaotic" thinker are ideal types; in real life, all of us use both models depending on what task we're attempting.

9. Compare this assumption with Gleick's comments on the results of research conducted by W. E. Ricker, who used the logistic difference equation to study fisheries in Australia and discovered that the growth rate parameter, X, was a nonlinear, "messy quantity" in the modeling of a stock. In modeling fisheries, he says, it must be viewed as playing a role similar to that of, say, friction in a hockey game. Gleick (1987:24) put it:

> the act of playing the game has a way of changing the rules. You cannot assign a constant importance to friction because its importance depends on speed. Speed in turn depends on friction. The twisted changeability ... asks: if you don't know the measurements, what can you say about the overall structure?
>
> "Messy" quantities in the growth rate of fishing stock are the market, aquatic pollution, changes in the biomass of every species in the food chain, and the cumulative effect of fickle consumer patterns.

10. For an interesting critique of the economists' use of Newtonian thinking, see Philip Mirowski (1989).

11. When I suggested considering managing market economics (since, after all, fishing effort is driven by market prices and demand), he was aghast.

12. For one thing, we cannot tell until after the event what inputs may have been critical. For another thing, even were they known, including all such variables would make for too complicated a model design. Finally, it might not be possible to produce a timely analysis—even with the most powerful of computers. It is for this reason that critical variables are *a priori* evaluated as "givens" or set to one side in the category of "all things being equal."

13. For an interesting discussion of this, see West and Shlesinger (1990:40–45).

14. Various recent studies are exploring the thesis that, as Seth Reice (1994) has recently put it, "biological communities are always recovering from the last disturbance" and "nonequilibrium theories of biological community structure . . . emphasize environmental disturbance and spatial heterogeneity as factors that encourage colonization and species diversity." In short, "equilibrium is an unusual state for natural ecosystems" (1994:424, 427).
15. The lawsuit was filed in Washington, D.C. in 1990 (see Fee 1990:14). The judge dismissed the plaintiffs' argument as without merit.
16. It should be noted however that industry personnel are just as guilty of calling for changes as the "fishcrats." One also hears from fishermen that if plans were instituted and left alone, the main reason one could "learn to live with them" is because creative minds would find loopholes and ways around the rules—in short, they would find ways to manipulate and cheat the system.

CHAPTER 12

Inuit Indigenous Knowledge and Science in the Arctic

Ellen Bielawski

The process of opening Western knowledge to traditional rationalities has hardly yet begun.

(Salmond 1985:260)

I'm telling you about myself. You didn't even bother telling me about yourself, you just wanted me to write about myself. I don't think that's fair. I would like to know about your parents and I would like to know about other things. I am an old man now and I am curious.

(Akuliaq 1967)

The vast and particular knowledge of the Eskimo, garnered from hundreds of years of their patient interrogation of the landscape, was starting to slip away.

(Lopez 1986:6)

Introduction

In this chapter we compare Inuit indigenous knowledge with Arctic science through the philosophy of science. Using a realist approach to Inuit knowledge and Arctic science, we allow that each contributes to understanding the natural world. One important contrast between the two is that Inuit do not separate people from nature, but Arctic scientists do. This contrast has resulted in incomplete science, management detrimental to Inuit, and conflict between Inuit and scientists in the Arctic.

Our research seeks to make Inuit knowledge more comprehensible to Arctic scientists who have, until recently, rarely recognized indigenous knowledge. We discuss two case studies: how Inuit constructed knowledge when the Government of Canada relocated them to different environments and an ethnography of Arctic science. We also examine methods for integrating indigenous knowledge and science. Integrating Inuit and scientific knowledge meaningfully and effectively is only a beginning, but the results will likely serve both Inuit knowledge and science well.

Research Context

Two groups of people—informants—are currently enriching my knowledge of the Arctic. One group is Inuit elders, who speak little if any English. The others are scientists who work in the Arctic. Two such different groups, my colleagues comment: Inuit elders born on the land, spending most of their lives as subsistence hunters, and scientists—people as passionate about their right to pursue pure knowledge as Inuit are about the rich taste of caribou marrow. I respond to my colleagues lightly: "I work with elders who speak no English, and it's quite similar to working with scientists, because they don't speak any English either."

I am, of course, trained in the Western scientific tradition, and am not fluent in Inuktitut. But I find much in common among the informants in the two case studies in progress. Both have difficulty communicating their knowledge to those who use it. Both are isolated from much of the knowledge held by the other. Both, historically, have seen little use for each other's knowledge.

The research I am doing is about the difference between Inuit indigenous knowledge and Arctic science. Conflict between ways of knowing began when Inuit (whose ancestors first occupied the Arctic nearly 4,000 years ago) and European explorers first met. The conflict has sometimes been subtle, quietly as well as savagely devastating Inuit, who nevertheless endure. The conflict continues in the form of negotiations for land, sea, and resources; for political power; for housing and health care; for culture. The difference between Inuit and Western knowledge underlies conflict in all realms. The conflict is detrimental to both cultures. (For an especially lucid discussion about damage to both colonizer and colonized, see Maracle 1992a and 1992b) Recent thinking recommends that Arctic scientists and those who use their work (managers and policymakers) resolve the conflict by recognizing the continuing existence and value of indigenous knowledge (also called traditional or local knowledge). Canada's Minister of the Environment wrote, "our task is to integrate traditional knowledge and science" (de Cotret 1991:8). The Traditional Knowledge Working Group, in Canada's Government of the Northwest Territories, strives for legislative and policy changes that will integrate traditional knowledge into policy about wildlife, health, justice, and social problems. The working group debated the meaning of traditional knowledge for over two years before reporting:

> The lack of common understanding about the meaning of traditional knowledge is frustrating for those who advocate or attempt in practical ways to recognize and use traditional knowledge. For some, traditional knowledge is simply information which aboriginal peoples have about the land and animals with which they have a special relationship. But for aboriginal people, traditional knowledge is much more. One elder calls it "a common

understanding of what life is about." Knowledge is the condition of knowing something with familiarity gained through experience or association. The traditional knowledge of northern aboriginal peoples has roots based firmly in the northern landscape and a land-based life experience of thousands of years. Traditional knowledge offers a view of the world, aspirations, and an avenue to "truth," different from those held by non-aboriginal people whose knowledge is based largely on European philosophies. (Government of the Northwest Territories 1991:11)

Surveys of Arctic and Antarctic scientists in 1986 and again in 1992 show that interest in indigenous knowledge grew significantly over that six-year period:

university researchers are now more interested in "aboriginal science." This is a matter of very special cultural significance in terms of the involvement of native northerners in research. It is a matter about which there is a feeling of urgency in the North, as many feel that the generation that has the distinctive aboriginal view of the universe and that has the local ecological knowledge is passing. (Adams 1992:iv)

Anne Salmond, in her work describing Maori epistemologies, writes that "the process of opening Western knowledge to traditional rationalities has hardly yet begun" (Salmond 1985:260). This is the problem: people now seek to integrate indigenous knowledge and science in the Arctic, but we have only begun to learn how.

Realism, Indigenous Knowledge, and Science

Research comparing Inuit knowledge and Arctic science is an attempt to interpret both in ways mutually comprehensible to Inuit and scientists. I have approached the problem of integrating indigenous knowledge and science through philosophy of science precisely because scientists need to, and some want to, grasp indigenous knowledge in terms they can understand and respect within their explanatory frameworks. I emphasize that validity in the terms of Western science and philosophy is not required to acknowledge or validate indigenous knowledge in and of itself.

Arguments about rationality and relativism are the philosophical context for comparing Inuit knowledge with Western science. Inuit knowledge did not develop in a context requiring comparison with the parameters of science, but compares well when challenged with these parameters. Inuit knowledge is consensual, replicable, generalizable, incorporating, and to some extent experimental and predictive (Bielawski 1992a; Denny 1986). Conversely, I have not yet seen evidence that Inuit controlled conditions for experiments. Nor did they, over time, increase accuracy in measurement.

Realism allows that both science and Inuit knowledge contribute to understanding the Arctic. A realist approach requires that one accept the natural world as real and holds that the objects of nature exist in and of themselves, were here before science, and will remain regardless of the activities of inquiry directed toward them. This position takes science seriously as a special form of knowledge different from the indigenous knowledge of societies without science; and it allows that indigenous knowledge also contributes to understanding the world. Hence, as stated above, there is no need to seek validation of indigenous knowledge in and of itself, only in comparison with Western science and the dominance its results and applications exert over indigenous cultures.

Anthropology is ideally suited for realist inquiry about science and other forms of knowledge (Jarvie 1986:162–171). A realist view contrasts with relativist interpretations more commonly invoked in describing and validating the indigenous knowledge of oral societies. From the realist position I am examining the difference between Western culture, which possesses science, and Inuit culture, without Western science (see Jarvie 1986; Gellner 1985). What happens to knowledge when one culture rapidly imports the products of science? On this question, philosophy of science remains essentially silent.

I write here about the Inuit of the Central and Eastern Canadian Arctic, referring to closely related peoples (for example, the Yup'ik Esdimo) in cited work from Alaska. The Governments of the Northwest Territories and Canada administer most of the study area. Its southern reaches are administered by the province of Quebec. The research problem crosses modern boundaries. I hope that this research may be applied to Inuit life in the contemporary Arctic and to the conduct of science there.

Case Study 1: How Inuit Construct Knowledge

This case study is set in an historical process where people were relocated to an environment unfamiliar to them. In 1953 and 1955, the Government of Canada resettled Inuit from the area around Inukjuak, on the east coast of Hudson Bay, to two locations in the High Arctic, at Resolute Bay and Grise Fiord. The High Arctic locations are approximately 1,700 kilometers, and about 16 degrees latitude, north of Inukjuak. The first group of Inuit were, literally, dropped off at these locations after a month-long journey, by hospital ship, from their homeland. They were expected to find food in an environment very different from their home. Informants say there are fifty-three species of animals, birds, fish, shellfish, and plants available for subsistence around Inukjuak, compared with seven species near Grise Fiord (Patsauq, January 15, 1991, interview in Inukjuak). One of these, muskoxen, was protected by law from hunting for sixteen years after the Inuit were moved to the area. Near Inukjuak, daylight and darkness occur every day all year; in the High Arctic, the sun does not rise from October to February, nor set from April to August. Hunters who had not hunted in darkness had to learn to do so. Birds, berries, fish, and shellfish that women and men gathered and hunted in Arctic Quebec were absent in the High Arctic. Women had to identify the right kind of sea ice (which loses salt content as it ages) to collect and melt for fresh water. They did this by the light of seal oil lamps and candles. The darkness, the barren land, the sea ice that endured through summer, and the cold required difficult adaptations, especially in that Inuit who were relocated were not equipped with appropriate traditional or government-supplied clothing and housing in the first years after the relocation (interviews with relocatees; Government of Canada 1990; Marcus 1991). Perhaps most significant, Inuit extended families were disrupted, their emotional, economic, and social support lost at the time and in subsequent generations.

After twenty to thirty years in the High Arctic, many of these people have returned to Inukjuak. They are negotiating an apology and compensation from Canada; the federal government has acknowledged that the relocation was a mistake. To document their case, the Inuit have given many structured interviews, quite unlike their usual long, nonlinear narratives, to researchers, consultants, journalists, lawyers, and police officers. I have conducted unstructured interviews (see Bernard 1988:204) with Inuit in Inukjuak, Grise Fiord, and Resolute Bay, in addition to having worked with Resolute Bay Inuit since 1977. My task was to draw out how they solved the problems of living in an environment where they had no previous experience—how they constructed knowledge necessary to live there practicing a subsistence lifestyle.

Prior to this field work I reviewed Nunavik oral histories for data pertaining to Inuit knowledge. From Inuit today and from oral accounts of historical Inuit life, I am trying to derive Inuit epistemology, as others have derived epistemology for several non-Western "schemes of human nature and reality" (Overing 1985:17; Borofsky 1987). The case study is intended to yield data on how Inuit know what they know.

Nunavik oral histories are nearly silent on matters of epistemology except when the context of specific knowledge—the story that is supposed to contain it—has been lost. When people say that they cannot remember stories and instead "just say it" (Tuniq 1985), they are most direct about how they know something. When they feel that they have lost the context for their knowledge, they speak about how they know it. One hunter, asked how he worked out the location and movements of caribou herds, said "because we are Inuit, we can do that" (Iqaluk, January 18, 1991, interview in Inukjuak). What does "because we are Inuit" imply?

In her oral history, recorded in 1985, Martha Tuniq (born 1924) refers to four ways of knowing. In this, her account is unusual. First, Tuniq describes hunting with her father, stating that "she knew about it because he told her." Second, her account implies that she learned by doing it (hunting) with him:

> I use to go out hunting with my father and he would tell stories. He use to tell me "if you can remember what I have taught you, do not forget them." Mostly about hunting skills.... Whenever my father went hunting I would follow. He could not leave me behind because he had no helper beside me that is why I know his hunting skills. In the wintertime when we were hunting he taught me how dangerous the ice and snow can be. I use

to be very cold when he was teaching but I would wait until he finish. (Tuniq 1985:4)

The third way of knowing that Tuniq describes is knowing something because she heard about it:

If the ice is not supported at the bottom it can be very dangerous.... If the weather was very stormy you won't be able to tell which part of the ice is dangerous. If it's nice weather you can tell by looking at the snow and ice.... The place where the ice or snow has no support is mainly beside the mountains, there is a few on land too dangerous too but not as dangerous as the ones on ice and snow. I know about it since I've heard about it. (Tuniq 1985:4)

Fourth, Tuniq describes knowing something because she saw it.

One time I watch a man dieding [dying] from kidney which was a beluga whale. Before the starvation he had eaten some kidney but he did not take all the blubber off of the kidney. He was walking and met some people on dogteams so he ate some kidney which was not really clean, which had some blubber. After he had reach his home, all day he was just eating and feeling the kidney he had ate when the night fall he died and his stomach bloated after. It was cost [caused] by the fat which was on the kidney.... What I have to say is to those people who are going to eat or store the kidney for a few days, what they have to do is take all the fat and the musels [muscles] off of the kidney even a small piece of fat. Never eat the part where the fat has a lot of meat it can be poisonous. The reason why I know about it is because I watched a man die from it. (Tuniq 1985:6–7)

In Western philosophy, these latter two kinds of knowledge are known as knowledge by description (she heard about it) and knowledge by acquaintance (she saw it).

Tuniq also describes the validation of the knowledge she gained from her father. "Everything what my father taught are true up to now," she says (Tuniq 1985:4).

Markusi Iyaituk (born 1906; interviewed in 1985) distinguishes between concrete knowledge of artifacts and the spoken, discussed knowledge of his elders. He states,

Old people in the old days refused to tell us about things when we were young. That's why we don't know as much as we should.... I only know about things other people know. Sometimes I ask other old people if they know stories about the past but they usually answer that they weren't allowed to listen or ask questions. We weren't allowed to listen to elders talking.... I can only talk about the things we owned, like sleds. (Iyaituk 1985:9–10)

Elsewhere, Iyaituk, like Tuniq, describes learning things through doing and experience. Iyaituk also recounts his detailed knowledge of seal anatomy, astronomy, species of algae and their uses, and midwifery, such as complications that result when the umbilical cord stretches.

Another elder, Salmonie Alayco (born 1906, interviewed in 1985) describes learning astronomy that would be put to use in navigation.

Every single morning, I was asked to go and check if the constellation "Quturtuuk" were in view before the constellation "Attuuk" appeared. My father asked me to go and check whether the constellations "Quturtuuk" "Attuuk" were in sight at dawn before the first light of the day. That's how we used stars to tell time in the morning. I was also asked to check from which direction the wind was from and whether there were any clouds. I didn't think he was teaching me but he was. (Alayco 1985:11)

These accounts and others show that Inuit ways of knowing include knowing through doing and experience, knowing through direct instruction, and knowing through stories told in order to convey knowledge. Inuit oral histories and classic Inuit ethnography document in depth how Inuit knew the environment and practiced a worldview that enabled their survival. The current question is not so much how Inuit adapted to the Arctic (an environment characterized by outsiders as harsh and empty, but rich for those with appropriate resource exploitation strategies, social relationships, and spiritual relationships with the environment) but how their worldview and some uniquely Inuit knowledge survive under modern conditions.

The interviews with Inuit of Inukjuak, Resolute Bay, and Grise Fiord today support the conclusions I drew from the oral histories: Inuit knowledge resides less in what Inuit say than in how they say it and in what they do. Inuit knowledge contrasts with science in that "pure knowledge is never separated from moral or practical knowledge" (Overing 1985:17; see also Sayer 1984:16-40). Arctic ethnography fortunately records a great deal of practical knowledge. However, as Fienup-Riordan comments, we have paid much less attention to *why* Eskimos do what they do (1990a:4). The difficulty of answering questions of "why" and "how" is emphasized in studies attempting it. Salmond, Overing, and others emphasize that an openness to other rationalities, a critical understanding of assumptions built in to Western epistemology, and linguistic comprehension beyond the norm even for anthropology are required (Salmond 1985:245; Overing 1985:17; Borofsky 1987).

Case Study 2: An Ethnography of Arctic Science

The second case study begins with a significant contrast between Inuit knowledge and Western science. Inuit do not separate people from nature. Arctic scientists do. As one scientist said to me, "it's a little bit different in archaeology and anthropology, but in geology or biology, people are overburdened" (background interview, June 1990, Resolute Bay). The bifurcation between culture and nature, between social and natural sciences (see Margolis 1987:xv) in Arctic research denies the unity of people and nature in Inuit interpretations of the world and strategies for living in it.

Fienup-Riordan takes this contrast further in her work on Yup'ik Eskimo ideology. She writes that

> Western science assumes an inherent differentiation between humans and animals and focuses on the explanation of the relationship between originally independent parts; the Yup'ik Eskimos stand this basic assumption on its head and assume, for instance, that men and animals are analogically related as human and nonhuman persons.... The focus of explanation shifts to the creation of difference out of an original unity. (Fienup-Riordan 1990a:4; see also 1990b:167–91)

This contrast led to such things as Canada relocating Inuit to the High Arctic without grasping any of the inseparable social and natural implications for them. For example, Inuit subsistence depends both on local, detailed environmental knowledge and on sharing practices that are dictated by kin relationships. The relocated Inuit lost both reliance on local knowledge and kin on whom to rely and with whom to hunt and share. The contrast between Western and Inuit worldviews also led to many of the Western strategies for community development, health care, education, and justice that have isolated Inuit from at least half of their reality, the natural world. It also led me to my second case study in progress, an ethnography of Arctic science.

Firmly grounded in the Western intellectual tradition, Arctic science nevertheless differs in specifiable ways from science that is geographically, financially, and culturally closer to its southern support bases. Arctic science is shaped by environment (e.g., low species diversity, polar extremes), history (e.g., four nations conducting science and applying technology amidst Inuit culture), politics (e.g., sovereignty in the Arctic), and humanism (e.g., the Arctic in the southern imagination; wilderness conservation). Arctic science attracts some who prefer Arctic vastness and extremes for its own sake; others who find it free of the "overburden" of a western-educated public, who pay taxes for research. Some see it as a "playground," others as the key to global research questions (background interview, Geological Survey of Canada, November 1990, Ottawa).

The ethnography of Arctic science involves documenting the patterns of research behaviors, beliefs, and attitudes of a sample of Canadian Arctic scientists. It differs from the usual ethnography because the community of Arctic scientists is not local, even in a particular research laboratory, in contrast to much social research on scientists (for example, Traweek 1988). Scientists who do research in and on the Arctic are dispersed among universities, research institutes, and government departments, bound only by a common interest in the place. Most of the time, scientific research has been conducted in the Arctic, but most scientists live and work far to the south. The expansion of government bureaucracy into the Arctic requires more, albeit relatively few, scientists to live in the Arctic.

The sample of scientists for the ethnography is drawn primarily from those receiving logistical support from the Polar Continental Shelf Project (PCSP), Department of Energy, Mines, and Resources, Government of Canada, in 1989, 1990, and 1991. PCSP is an agency that provides aircraft charters to field sites via off-strip aircraft; research and camping equipment; coordination of field research activities; and accommodation at base camps on Canada's polar continental shelf. Scientists apply to the project for airtime and other support, and this support is key to effective fieldwork on Canada's Arctic coast and in the Arctic Islands.

Research has proceeded through questionnaires to 198 scientists, 147 of whom responded, for a response rate of 74.24 percent. The questionnaire comprised nineteen pages of fifty-eight qualitative and quantitative questions. Topics included arctic experience; education and training for the Arctic by discipline; methods; arctic interest; contact with Inuit; research influences (mentors and teachers, colleagues, and students); research topics; and funding, logistics, and personnel. Data on respondents' gender, age, citizenship, affiliations, education, and other qualifications are included. Following analysis of this data, interviews will be conducted with selected respondents at their research institutes. Participant observation, supplementing my own fieldwork in the Arctic since 1975, will also be conducted at scientists' field sites.

The questionnaire survey was completed in June 1992. It is not uncommon to read of social research based on a response rate of less than 30 percent to questionnaires (Bernard 1988:259). Given the high response rate in relationship to the length and complexity of the questionnaire, it is evident that scientists currently active in the Arctic care deeply about their research and about being accurately represented to other people. Preliminary analysis of Arctic scientists' responses to questions about contact with Inuit suggest extreme disparities in attitudes.

Many respondents recognize that Inuit indigenous knowledge is relevant to their research, but few have significant contact with Inuit. Many note reading explorers' accounts of the Arctic, but few other than anthropologists note reading Arctic ethnography as part of their introduction to the Arctic. Scientists are mixed in their views of the relevance of Inuit knowledge. The data have not yielded a clear correlation with the quantity and quality of scientists' contact with Inuit. Several scientists who work on oceanography, glaciology, and ice physics see Inuit knowledge as significant to their research. Most geologists regard Inuit knowledge as irrelevant to their research. One scientist researching indicators of climate change writes that "aboriginal peoples the world over have little understanding of the history of the earth" but indicates no contact with Inuit. Several scientists studying climate change note that they do not have contact with Inuit because they do not work near an Inuit community.

The Polar Continental Shelf Project Base Camps are located at Tuktoyuktak, an Inuvialuit (western Canadian Inuit community, and three miles from Resolute Bay, a High Arctic Inuit community. Nearly all of the scientists in the sample work out of or travel through one of these bases, in one of these communities, each field season. In these and other contemporary Arctic communities, the contrast between scientists' and Inuit worldviews is invisible on the physical landscape. While Inuit are not settled in permanent communities, their occupancy and perception of land are not confined to physical settlements, nor is their land use (land use is a Western concept that has little meaning for Inuit). Inuit perceive any activity on the land as significant to them, including the field research activities of scientists. If not conducted properly, activities on the land may interfere with the appropriate relationships within the environment, among animals, plants, rocks, spirits, and people. Not to have contact with Inuit while traveling through, if not working near, their communities, indicates the narrow focus of the scientist, and ignorance of the Inuit worldview of any activity on the land.

The sample of scientists includes, as well, a small number of people who study birds or mammals that are important to Inuit subsistence. These scientists are uncertain that Inuit knowledge has any relevance to their research but suggest that their research is important for Inuit. Younger scientists, or those trained most recently, are attuned to Inuit reaction to their research. This group describes contact with Inuit through community negotiations, information sessions, and very occasionally through research partnerships. Scientists who have worked in the Arctic for more than two decades describe either no contact with Inuit or employing Inuit as laborers or field guides. Some exceptions to these patterns exist. The change in scientists' attitudes toward, and relationships with, Inuit over time reflects the political gains Inuit have made toward self-determination since 1970.

Interpretation of correlations between these and other variables awaits completion of the analysis. As with all ethnographies, categorized data collected through questionnaires will be integrated into the context of results from interviews and participant observation.

Integrating Indigenous Knowledge with Science in the North

Arctic science is being influenced by the social context in which it is conducted, although the attitudes described earlier have been entrenched:

> Many, including, I suspect, most residents of the North, would say
> that researchers view northern Canada and Antarctica in much the same
> way. They see both as laboratories for research, principally natural
> science research, that is only incidentally of relevance to human kind. I
> believe that there is more validity to this image than many of us in research
> care to admit. However, I also believe that there are, at last, signs of
> change in attitude inside and outside the research community.
> (Adams 1992:iii)

These changes are coming about in part because of relatively recent but extensive research on indigenous knowledge, especially traditional environmental knowledge (for example, Andrews 1988; Berkes 1988; Bielawski 1984, 1992a, 1992b; Bielawski and Masuzumi 1992; Colorado 1988; Cruikshank 1981, 1984; Feit 1988; Freeman and Carbyn 1988; Johnson 1987, 1992; Merculieff 1990; Saunders 1992; Sioui 1992; Waldram 1986).

The key intellectual problem for research integrating indigenous knowledge and science is discovering categories for data collection that match the aboriginal and scientific worldviews. Research utilizing traditional knowledge thus far relies primarily on extracting aboriginal knowledge from its context so that it matches categories of information determined by the needs of scientists. This approach serves the needs of science. Indigenous knowledge, however, is both context-embedded and implies correct, spiritually based relationships within the environment. When scientists attempt to extract bits of data from the indigenous context, meaning may be altered or lost.

Indigenous knowledge is increasingly influential in research using approaches grouped under the term "participatory action research." Ryan and Robinson (1990) describe participatory action research, and its intellectual heritage under other names, in a case study from Gwich'in, Northwest Territories, Canada. Key features of participatory action research include:

> —a commitment to the community controlling the process, from setting the research agenda, through consultant and trainee selection and project development, to budgeting and annual project review;
>
> —a commitment to community ownership and control of all research products and their use. This meant that copyright was retained by the community;
>
> —a strong and continuing reliance on the capability of community adults as trainee researchers, teachers, writers, and project advisors;
>
> —a shared commitment to advocacy on behalf of the community on issues of its choosing;
>
> —a commitment to a group dynamic and consensual process of decision-making and a feminist interrelational approach; and
>
> —a commitment to working oneself out of a job within a specified time. (1990:59)

In the Canadian North, projects based on some or all of these features are not uncommon among native communities, social scientists, and community development workers in various fields. Results rarely enter the larger body of scientific literature (Ryan and Robinson 1990:58–59) although they are achieved with processes that fit within the parameters of science (Bielawski 1984, 1992a). Natural science research projects using participatory action methodology are a more recent phenomenon (Fleming 1992; Johnson 1992; Bielawski and Masuzumi 1992).

Comanagement of Arctic species and other resources attempts to combine the expertise of scientists and native land users. The model and process for comanagement, however, is primarily Western, scientific, and bureaucratic, as Cizek describes in his case study of the Beverly-Kaminuriak Caribou Management Board (BKCMB):

> Althouqh the BKCMB does incorporate aboriginal users into decision-making, it cannot be considered a model of complete integration, since government biologists do not appear to have accepted the indigenous system of wildlife managment.... On the whole, it is possible to say that the BKCMB does incorporate some indigenous knowledge into the management of caribou, but it should not be considered a model of integration, since the balance of authority rests with the state system." (Cizek 1990:16)

Cizek also demonstrates how the BKCMB emphasizes Western epistemology and bureaucratic practice (1990:14–19).

Conclusion

The case studies should illuminate how Inuit knowledge and Arctic science both contribute to understanding the Arctic. Realist theory supports the argument that Inuit knowledge and Western science each contribute to understanding the Arctic. Each is successful in specific realms. Inuit knowledge has both a moral and practical place in the contemporary Arctic (for comparative studies, see Overing 1985). But no Inuk will abandon her or his antibiotics, electric sewing machine, or snowmobile. Methods for integrating indigenous knowledge and science in research and application are being proposed, tried, and refined. Results of this work will likely serve Inuit and science more appropriately than has been the case since contact between Eurocanadians and Inuit began.

Notes

As always, I thank Inuit with whom I work, past and present. I thank the scientists who devoted considerable time to completing questionnaires. The Inuit Relocation Case Study was supported by Multiculturalism Canada, Department of the Secretary of State. The Ethnography of Arctic Science was supported by the Social Sciences and Humanities Research Council of Canada, Science, Technology, and Policy Program. I am very grateful to the Documentation Centre, Avataq Cultural Institute, for use of the Inuit Archive, and especially to Sylvie Cote Chew for her invaluable assistance there.

CHAPTER 13

Popular Delusions and Scientific Models
Conflicting Beliefs of Scientists
and Nonscientist Administrators in the Creation
of a Secret Nuclear Surveillance System

David Jacobson and Charles A. Ziegler

In the decade following World War II, scientists such as J. Robert Oppenheimer and Vannevar Bush wielded unprecedented power as advisors to government administrators. Thus, the examination of decision-making processes involving both scientists and nonscientist administrators during this period is essential for understanding the relationship between science and government at a time considered by many historians to have been crucial in determining the nature of the American scientific enterprise for the second half of this century.

The goal of our research is to contribute to an understanding of this relationship by studying an aspect of decision making within this period that has remained relatively unexplored—namely, the role played by the science-oriented beliefs of the participants in the decision-making processes. For this purpose, we have selected as a case study a sequence of events and decisions that occurred between the end of World War II and 1949 and that led to the creation of a secret surveillance system for monitoring foreign atomic bomb tests. This was the so-called Long Range Detection System that in 1949 identified the explosion of the Soviets' first atomic bomb despite their efforts to keep this proof-of-design test hidden.

We chose this case for several reasons. First, the involved scientists and administrators held distinctly different science-based views about the need for and the feasibility of a monitoring system, a situation that produced a considerable amount of documentary evidence about the nature of their relevant beliefs on these issues. Second, the determination of the influence of beliefs on decision making was simplified by the secrecy surrounding the development of the monitoring system—that is, legislative and public pressures on participants that could otherwise confuse the analysis were minimal. A third reason for selecting this case material is to fill a gap in the historical record. For, despite the immense strategic importance of the Long Range Detection System, the decision-making processes involved in its creation have never been the subject of serious inquiry: until recently many of the relevant documents were classified and the interpretation of some of the declassified documents requires considerable technical expertise.

We have dealt with the latter problem by using a multidisciplinary team in carrying out our research—a team that included a chemist, physicist, and science historian, as well as anthropologists. Our experiences with this approach suggest that the collaboration has been very effective in ensuring that our application of the interpretive methods of anthropology to the material of this case has produced an analysis that is consonant with science as it was understood and practiced in the period under consideration.

Our analysis deals with a rather complex organizational system that ultimately involved more than eleven governmental agencies, three industrial firms, a university, and one foreign government. Our complete description of this system has recently become available (Ziegler and Jacobson 1995). Therefore, rather than touch upon all of the descriptive material, we will, on this occasion, present one facet of our analysis that can be related to some general assumptions and propositions about scientific beliefs that have long characterized the work of historians and sociologists of science. We will first present some of these assumptions and propositions and then go on to describe their implications for understanding the science in this case and the formulation of the science policy related to it.

In the past two decades a number of social scientists have studied the interplay of scientific beliefs and other elements of our culture. These investigations indicate that beliefs about science held by the general public are usually a distorted rendering of scientists' scientific beliefs. For example, one commentator states,

> The people who gladly vote billions for scientific research are far from understanding its inner character; and the points of view associated with it have never been altogether assimilated to the culture even of the West. The "popular delusions" which the scientist encounters with surprise upon his occasional forays outside the laboratory are the normal beliefs of a world which uses, but does not understand, the learning he develops. (Handlin 1972:253)

Implicit in this statement is an assumption that is commonly found in studies of scientific beliefs, namely, that scientists' scientific beliefs are "truer"—in the sense of being a better basis for predicting outcomes—than the distorted public versions of these beliefs. Other studies, for example, on water fluoridation (Mausner and Mausner 1955) and on cigarette smoking (Reiser 1966), suggest that when such "popular delusions" influence government decision making the resulting policies are retrogressive. Two questions stem from such studies: is it possible for "popular delusions" to be truer than the scientific beliefs on which they are based, and, if so, can they have a benign influence on related science-based policy decisions?

In examining portions of our material that seem relevant to these questions we were forewarned by the caveats of several social scientists, most notably Barry Barnes, who have written extensively on the subject of scientific beliefs. In pointing out the inadequacies of some of the studies in this area, Barnes has noted that the flaws stem chiefly from what he terms "an exaggerated stress of the special status of scientific beliefs" (1972:269). And, indeed, our research indicates the answer to both of the questions just posed is "yes"—a result that seems to reinforce Barnes' implicit admonition to avoid canonizing scientific beliefs. However, as we will try to demonstrate, we believe that this reinforcement is more apparent than real and that the relativism advocated by Barnes must be tempered by realism.

We will begin by describing two public beliefs about atomic bombs that were common currency in this country in the immediate post-World War II period, when the idea of monitoring foreign atomic activities, such as bomb tests, was first seriously considered in government circles. We will then sketch the role these public beliefs played in the decision to develop a monitoring system. The two beliefs were: first, that some other country had, or would soon have, the bomb; and, second, that atomic explosions were always accompanied by airborne radioactivity or "fallout" that was readily detectable at great distances from the explosion.

Public Beliefs (or "Popular Delusions")

More specifically, the first public belief was that America's atomic monopoly would be ephemeral because the so-called "secret" of the bomb, in terms of the physics of how to make a bomb, was known by the scientists of all nations. This message, drawn from popularized scientific writings of the time (Marshak et al. 1946) and amplified by the media, made a deep impression on the public, as attested to by surveys made by social scientists in 1946 and 1947. For example, one such study (Eberhart 1947) indicated that at least 75 percent of the people who held opinions on the subject believed that another country already had the bomb or would have one within three years. Moreover, the great majority of respondents believed the other country to be the Soviet Union. These surveys also revealed a close relationship between media reports that often quoted scientists' statements about atomic matters and public beliefs about the bomb.

The second public belief, involving the concept of fallout, also reflected the writings of scientists and was further promulgated by media descriptions of atomic explosions. Moreover, the popular press drew public attention to unclassified reports in scientific journals showing that the fallout from U.S. bomb tests was easily detected at distances of several thousand miles by nongovernment scientists using standard equipment in college laboratories (e.g., Coven 1945, Fearson et al. 1946). This view that a foreign atomic bomb test would be betrayed by airborne radioactivity was also expressed on popular radio programs. For example, William Laurence of the *New York Times*, in a radio interview aired in December 1947, answered a question about the possibility of detecting a Soviet bomb test radiologically by pontificating that this was so easy to do that "the possibility of anyone testing atomic bombs in secret from now on is practically nil as far as we know."

Influence of Public Beliefs on Policy

Government administrators, including Atomic Energy Commissioner Strauss and Navy Secretary Forrestal (later to become Defense Secretary), shared these two public beliefs about the imminence of a Soviet bomb test and the efficacy of detecting it at great distances radiologically, that is, by monitoring fallout. For example, these beliefs made both the urgency and the feasibility of monitoring so obvious to Forrestal that when told that no monitoring was yet being carried out he replied, "Hell, we must be doing it!" (Strauss 1962:213). Given this mindset, it is not surprising to find that when an interagency committee was set up to evaluate radiological monitoring as a primary method of intelligence gathering no scientists were included.

The committee took the urgency of the need for such monitoring and its technical feasibility as givens. The chief problem was seen as organizational, and debate centered on deciding which agency should be assigned to carry out the monitoring task. Again, the belief that fallout would be the primary indicator of a foreign test made the Air Force the obvious choice since it controlled the aircraft suitable for a worldwide, high altitude air-sampling program. Committee recommendations were accepted by the Joint Chiefs of Staff and in the fall of 1947 the U.S. Air Force was directed to establish a secret nuclear surveillance system to monitor foreign atomic explosions.

The documented deliberations of the interagency committee (Ziegler and Jacobson 1995:79–83) leave little doubt that the public belief about the imminence of a Soviet bomb espoused by some committee members created a climate of urgency in which the policy decision to establish a monitoring system was made, and the public belief about fallout affected the decision to assign responsibility for monitoring to the Air Force. Since, in retrospect, both decisions can be seen to have facilitated the timely establishment of an effective monitoring system, the influence of these public beliefs on the government policy on monitoring can be considered benign.

Scientific Beliefs

The implementation of the decision to create a monitoring system, which administrators expected to be smooth, began to falter when the need for urgency in monitoring was brought into question by estimates of the Soviet atomic timetable. These estimates were developed by scientists who were well aware that the "secret" did not lie in the physics of the bomb, which by 1947 was, indeed, well known, but rather in two areas in which the Soviets were thought deficient. The first area was in the knowledge of a complex mass of technological detail often having nothing to do with nuclear physics, such as the techniques for making superpure graphite. The second, and more important, area was in the knowledge of means to rapidly convert low-grade uranium and thorium ores into fissionable material.

Such speedy methods of processing low-grade ores had yet to be developed by the United States. U.S. scientists had been able to produce an atomic bomb in a relatively short time only because they had access to uranium ore of an extraordinarily high grade. This ore could be processed, using existing techniques, in a fraction of the time needed for the low-grade ore typically found in deposits around the world. Indeed, there was then only a single known source of such high-grade ore—the Shinkolobwe mine in South Africa. To ensure the prolongation of its atomic monopoly, the United States had initiated, during World War II, a super-secret program called Murray Hill Area (Ziegler and Jacobson 1995:21-33). This program, carried out at enormous cost, involved open-ended agreements to purchase the entire output of Shinkolobwe and all other known or suspected sources of uranium and thorium in allied or friendly countries throughout the world. After the war, Murray Hill Area scientists, who were skilled in geology and ore processing, made careful estimates of the probability that the Soviets would find high-grade ore in their own country or in the few areas that remained accessible to them. On a statistical basis, this probability was considered very low.

Based on the results of Murray Hill Area, a scientific model for predicting the duration of the U.S. atomic monopoly was constructed. Using this model, complicated scenarios for the projected Soviet atomic program were formulated, including all known parameters such as the rate of production of fissionable material from low-grade ores. Most of these scenarios suggested that the technological infrastructure of the Soviet Union had been so damaged by the war that, even if the low-grade ores known to be available to them were fully utilized, the Soviets would not produce a bomb until after 1952. This view, which might be called the belief of "insider" scientists, placed the Soviet bomb considerably further in the future than did the public belief.

It is noteworthy that, because of the supersecret status of Murray Hill Area and the strict policy of compartmentalizing information, very few scientists and administrators were privy to the scientific model for predicting the duration of the atomic monopoly. Hence, apart from these "insiders," many scientists and most administrators espoused the public belief that the Soviets would soon have the bomb. But among top leaders and those scientists who were aware of the rationale for a long-enduring atomic monopoly, a monitoring system was not considered an immediate requirement. One result, according to the recollections of a contemporary observer, was a "general disinclination to engage in monitoring based on the feeling it would be a waste of time, personnel, and money" (Strauss 1962:213). Thus, although top government officials and scientists who were "insiders" accepted the need for eventually creating a monitoring system, they questioned the urgency of the need. Hence, there was no "top down" pressure to establish such a system.

A much more serious setback arose when high-level scientific advisors entered the picture. This occurred when the U.S. Air Force sought funding to develop the monitoring system, an action that brought the matter to the attention of the Research and Development Board. The board was a body of scientists that had been established to oversee interagency cooperation in scientific research. The board's Committee on Atomic Energy, which included prestigious scientists such as Oppenheimer, raised serious doubts about the technical feasibility of the proposed monitoring system. These doubts stemmed from the fact that the scientists' belief about fallout was quite different from the public belief espoused by administrators. Most important, the scientists' frame of interpretation of the term "fallout" included concepts that led them to believe that airborne radioactivity, measurable at great distances, was not an inevitable concomitant of atomic explosions.

In order to understand their position, it was necessary for us to establish the scientists' model of a high-altitude atomic bomb explosion as it existed at that time (see Ziegler 1988:205; Ziegler and Jacobson 1995:111–13). A detailed description of this model falls outside the scope of the present paper, but, in brief, its salient feature was that it made a clear distinction between vaporized radioactive fission products produced during the explosion and the dust particles sucked up from the earth on which such fission products were deposited and which thus became radioactive in proportion to their surface area—a distinction that was blurred in the public belief about fallout. Dust was only present if the fireball created by the explosion touched the ground—something that happened if the bomb was exploded at altitudes below about 500 feet. But if detonation occurred at high altitudes, the model indicated that the atomized material produced by the bomb would consist chiefly of radioactive fission products in the form of extremely minute, gas-like particles dispersed in the atmosphere where radioactive decay and geometric dilution would cause their concentration—at distances greater than a few hundred miles—to fall to levels undistinguishable from fluctuations in the natural radioactivity always present in the air.

Moreover, at great distances from the explosion the collection of sufficient bomb debris to allow fission products to be chemically analyzed was highly problematic. And such chemical analysis was essential if atomic bomb explosions were to be reliably distinguished from other possible events that could produce airborne manmade radioactivity such as reactor accidents or the purposeful simulation of an atomic bomb explosion by detonating a mixture of TNT and radioactive material. Differentiation between these possibilities was essential, since, to intelligence analysts, each implied drastically different scenarios for the Soviet effort to develop an atomic bomb.

The gloomy prognostications of the scientists about the usefulness of radiological monitoring via air-sampling brought into question the rationale for selecting the U.S. Air Force as the monitoring agency, Even more disturbing to military and civilian Air Force administrators, the technical feasibility of creating an effective monitoring system, which had been taken as a given, suddenly seemed doubtful.

Thus, late in 1947, when scientists and administrators met to plan the development of the nuclear surveillance system, the scientists' models relevant to the advent of a Soviet bomb and to the nature of an atomic explosion led them to conclude that the need for monitoring was not urgent (since a Soviet bomb was thought to be at least five years away) and that existing methods for identifying the atmospheric radioactivity produced at long range by an atomic explosion were not adequate for monitoring (although they might be made so through further research). The scientists' scientific beliefs were thus at odds with the public beliefs espoused by administrators, which were that monitoring was urgently needed (since a Soviet bomb test was thought to be imminent) and that atomic explosions could be readily identified at long range by monitoring airborne radioactivity (since all such explosions were thought to produce fallout that would betray an atomic explosion occurring anywhere in the world).

Nor could their differences be resolved. Among the reasons for this, two appear primary. First, and most important, because of secrecy and the strict application of the compartmentalization policy, administrators were not given details of why the scientists believed as they did. Today it is difficult to imagine the almost surrealistic degree of secrecy and compartmentalization regarding all atomic matters that prevailed in the immediate post–World War II years, but the following example is a dramatic illustration. In 1946, the Joint Chiefs of Staff completed an elaborate contingency plan for a possible war with the Soviet Union, code-named Pincher, that was based on completely unrealistic assumptions about the atomic capability of the United States. According to historian Gregg Herken (1982:219), this gaffe occurred because the Joint Chiefs were not allowed access to the necessary atomic information, including the number and production rate of atomic bombs!

Second, to the degree that the rationale for the scientists' beliefs had been explained to them in briefings, administrators lacked the scientific training to correctly interpret this information. Instead, they perceived the scientists' views as motivated by an excessive concern to iron out every last technical detail before getting on with the monitoring task. This belief that the scientists' position was merely an expression of technical preciosity appeared reasonable against the background of media reports that emphasized both the fragility of the U.S. atomic monopoly and the ease with which foreign atomic activities, such as the test of a bomb, could be detected radiologically at long range.

Nevertheless, despite their reservations about monitoring, the scientists agreed that, with further research, techniques for reliably monitoring a Soviet atomic bomb test might be developed. Hence, the program to establish a nuclear surveillance system proceeded. Although the "popular delusions" of administrators had a positive influence on the timely decision to initiate this program under the aegis of the U.S. Air Force, there is no evidence that their beliefs, which the administrators continued to espouse, accelerated the program after it began. But, at least, such beliefs engendered a feeling of urgency among administrators that had the effect of mitigating bureaucratic delays that might otherwise have retarded the program.

So the question of whether it is possible for public beliefs about science to have a benign influence on policy decisions can, in this instance, be answered affirmatively. But can such public beliefs be "truer," as measured by their predictive accuracy, than the corresponding scientific beliefs of scientists? This question also must be answered affirmatively in this case since, as we will next describe, nature had some surprises in store for the scientists.

Nature Adjudicates

The Soviet atomic bomb project, which began in 1942 and went into high gear in 1945, was plagued in its early years by difficulties in producing fissionable material from the supplies of low-grade ore available from mines in the Soviet Union and, after 1945, in allied countries. But no evidence has surfaced to suggest that the Soviets overcame these difficulties by finding ore of a grade equal to that of the Shinkolobwe mines. There is, however, evidence that they made discoveries that resulted in speedier methods of processing ore, methods that, according to a 1950 Central Intelligence Agency report, "were unexpectedly successful" (Herken 1982:332). These technical advances allowed the Soviets to produce enough material for a bomb in about half the time estimated by U.S. experts, whose projections were based on the Murray Hill Area project and on standard ore treatment methods. As a result, the "popular delusion" that a Soviet atomic bomb test could occur as early as 1949 proved to be correct.

As if to demonstrate a kind of cosmic evenhandedness, nature also provided a surprise for U.S. scientists, who made an unanticipated discovery that ensured that when the Soviet bomb test occurred, the United States would have an effective means of detecting it. This discovery occurred shortly after the 1948 U.S. atomic bomb tests in the Pacific, code-named Sandstone. The primary purpose of Sandstone was to assess new bomb designs, but a secondary aim was to evaluate the effectiveness of long-range monitoring methods.

The Sandstone bombs were detonated atop towers. Since the explosions were near the ground, they produced large amounts of fallout that, not unexpectedly, provided fission products (radioactive elements) that were collectible via air-sampling at long range in quantities sufficient for chemical analysis, a finding that confirmed that a Soviet bomb test could be detected and identified as such if the explosion was at or near the surface of the earth.

But it was feared that the Soviets might opt to conduct their first bomb test as an airburst, by dropping the bomb from an airplane or supporting it with a captive balloon, where detonation would occur above about 1,800 feet. The fireball of such an airburst would not contain dust from the earth below, a necessary ingredient of fallout. Instead, only the bomb casing and its contents would be vaporized and dispersed as submicron-sized particles of bomb debris. According to the then extant model of an atomic explosion (which had not been disconfirmed by the five atomic explosions prior to Sandstone), the subsequent agglomeration of bomb debris particles would be retarded by ionization to the extent that quantities of fission particles sufficient for chemical analysis would not be collectible at long range—that is, at distances needed to monitor a bomb test deep within the Soviet Union. Indeed, the inability to detect airbursts was at the heart of the scientists' doubts about the technical feasibility of creating a reliable monitoring system.

But the unanticipated occurred. During a routine microscopic examination of bomb debris from Sandstone, a few minute, but perfect, spheres were discovered amid the jagged fallout particles. According to a scientist who participated in the discovery, "it was considered quite a revelation to find not only particulate matter, but matter in terms of shiny, metallic-looking spheres that were just beautiful" (Ziegler and Jacobson 1995:137). The spheres could be explained only on the basis that they were created by the coalescing of vaporized material of the bomb itself, free from any association with the dust sucked up by the explosion. This meant such spheres would also be present in a high-altitude airburst. Moreover, they had been collected with existing methods at long range and, most important, they were chemically analyzable.

Suddenly and unexpectedly, the possibility of radiologically detecting any above-ground Soviet atomic bomb test became a reality, a possibility that was rapidly exploited by administrators who felt that their initial view about the "obvious" feasibility of this method had been right all along, a view that stemmed from the "popular delusion" about fallout. By April 1949, the nuclear surveillance system was in routine operation and early in September of that year the system successfully identified the first Soviet bomb test.

Implications for Science and Science Policy

This case study demonstrates that it is possible for public beliefs to be better predictors of outcomes than the corresponding scientific beliefs of scientists and that these "popular delusions" may have a positive influence on related science-based policy-making. Do these findings support the view that science is merely another form of discourse, based on knowledge that is no better or worse than other types of knowledge? We think not.

Our findings seem anomalous precisely because they are counter to the trend established by other cases involving the interplay of scientific and public beliefs such as those cited earlier, a trend that indicates that the track record of scientific beliefs in making predictions is far better than that produced by other kinds of knowledge such as "commonsense," untested (or untestable) speculation, and political or ideological beliefs. To perceive the predictions based on the scientific models described in this case as wrong is to fail to understand that the conclusions drawn from such models are expressed in terms of probabilities, not epistemological certitude. What this case study shows is the improbable sometimes happens.

It is true that administrators in powerful positions, like Forrestal and Strauss, supported the development of monitoring on the basis of public beliefs that were garbled versions of scientific beliefs. But because the process by which scientific beliefs become garbled is unsystematic and random, the public belief on this topic might well have been different and its influence on policy retrogressive. Thus, although in this case the influence of "popular delusions" on policy-making was benign, this anomalous finding points to the contingent nature of such influence, not to the efficacy of a superficial and garbled version of science in guiding policy.

The focus on beliefs in our analysis further indicates that once the program to develop the monitoring system had been initiated, it was the scientists' scientific beliefs, modified through a reiterative process of hypothesis testing and experimentation, that were crucial to its success—not the "popular delusions" that administrators continued to embrace (Ziegler and Jacobson 1995:37–58, 99–140). This reiterative process is highlighted in this case study by the unanticipated discoveries that mandated changes in the scientists' models related to the atomic monopoly and to fallout, discoveries that illustrate one way in which scientific paradigms are changed.

Those who espouse the position of radical relativism (as summarized by Webster 1991:15–32 and Lynch 1993:39–116) argue that such alterations in scientific paradigms are produced solely by a dialogic process that reflects power shifts within the scientific community and changes in the ambient culture. In other words, they contend that scientists delude themselves in thinking that nature adjudicates their questions. On the other hand, scientists such as Cromer (1993) and Gross and Levitt (1994) and philosophers such as Siegal (1987) have presented arguments for rejecting this view.

We believe that this case study can be best analyzed by fully understanding both the relevant science and the scientists' viewpoint since to do otherwise is to ignore the culture of the group being studied. From an anthropological perspective, it seems important to note that the involved scientists acted as if they believed that nature exercised a degree of constraint over what they could say about it. As this case demonstrates, such constraints are especially dramatic when they come as a complete surprise to all parties in the "dialogic process" by which science is constructed. In sum, we believe that the anomalies in this case study do not support the decanonization of science as a special kind of knowledge. Instead, they highlight the need to recognize that the issue of the authoritative nature of scientists' scientific beliefs is one that must be addressed by practitioners of an anthropology of science.

Notes

The research on which this article is based was supported by a grant (RH-20893-89) from the National Endowment for the Humanities.

We wish to thank Stanley Goldberg, Robert P. Multhauf, and Lloyd R. Zumwalt for their help and collaboration in carrying out this research.

This view of the problematic status of scientific beliefs is more fully presented in other statements of the "strong program in the sociology of knowledge" (see, for example, Barnes 1974, Barnes 1977, and Bloor 1976).

CHAPTER 14

Japanese Science and Western Hegemonies
Primatology and the Limits Set to Questions

Pamela J. Asquith

Introduction

Cultural studies[1] of science include at least three kinds of research, two of which have been applied to Japan at one time or another. One approach is to question whether other ways of practicing science, or alternative epistemologies, can be considered as science at all. That is, in comparing the effects of culture on science, can we speak of more than one kind of science? Science, after all, can be defined so that nothing outside a Western tradition will be admissible. Yet, if science is considered as an epistemology in the sense of a coherent, internally logical, and systematic way of ordering knowledge about the world, and which has predictive and testable capabilities, other rationalities may count as sciences. Pioneering classics dealing with specifically "eastern" approaches to knowledge (Capra 1975; Holbrook 1981; Needham 1956), foreshadowed the study of other comparative epistemologies of pre- and post-technological societies around the world (Selin 1992; Tambiah 1990).

If science is viewed, however, as a specific methodology, based on a particular set of underlying assumptions, limits are then set to what can legitimately be investigated. When restricted in this way, these "other" ways of gaining and ordering knowledge may not be considered as science. Yet, in several quarters, a reexamination of what we call science is underway. An increasing list of sociological and anthropological studies of mainstream scientists and science institutions have revealed that ideals about the scientific enterprise show a less than perfect isomorphism with actual practice.

A second genre of cultural studies of science is represented by this research on the sociological and cultural underpinnings of knowledge production. On the microlevel, social scientists have examined the effects on science of how "facts" are produced in a scientific laboratory, from the discussions at the water fountain, to lobbying for laboratory space and funding, to the influence of personalities and of contemporary concerns of society. Latour and Woolgar's (1986) *Laboratory Life*, which described their study of the Salk Institute in California in the mid 1970s, is an early example of this. Their study has been followed by others, such as Charlesworth's et al. (1989) study of an Australian scientific community, and Traweek's (1988) research on Japanese and American physicists (expanded in this volume). At the macrolevel, historians, sociologists, and feminist scholars[2] of science have considered the effects of wider societal forces and social mores on the pursuit of science as, for instance, Fox Keller (1982) on cytogenetics, Latour (1987) on scientists and engineers in society, and Haraway (1989) on primatology.

A third thrust of metascience or studies of science is discourse analysis, which, in its various forms and disciplines, examines how our use of language delineates our studies and findings. Metaphor, especially, is important in this understanding of the scientific enterprise (Paprotte and Dirven 1985; Ortony 1979), as are semiotics and rhetoric. Most discourse analysis has been done on English as the lingua franca of science.

One area in which other traditions of thought have been considered to provide alternatives to our atomistic reductionist approach to knowledge of the world is so-called environmental philosophy. Callicott and Ames (1989), for instance, look to East Asia for both a different relationship to nature and for a new vocabulary, in order to replace the arid formulas typical of Western science that have become, they maintain, exceedingly abstract and remote from ordinary Western experience. Although a closer scrutiny of actual behavior toward the environment on the part of some East Asians, both historically and presently, has revealed that this is something of a romantic fiction (see Bruun and Kalland 1995), Asian and other cultures continue to attract those in the endeavor to articulate an alternative epistemology.

In the field of primatology examined in this paper, an alternative approach has done more than simply provide a different expression to the findings of science. It has also generated different insights, questions, and hence new avenues of research. Japanese, European, and U.S. studies of primate behavior began within a decade of one other in the 1930s and 1940s, and yet have had fundamental differences from the outset. Japanese reports of naturalistic primate behavior were more personal, anthropomorphic, and richly detailed as to individual animal's life histories. The questions that arose from their studies yielded far-reaching insights that the Western approach did not. However, potential directions for research generated by Japanese findings were largely ignored by Western researchers, even though the work was published in English in a readily accessible journal produced by the Japanese. Only after many years were those insights, that "other" knowledge, subjected to mainstream Euroamerican methodology for testing and found to be valid.

This paper thus uses both the first and second genre of cultural studies of science mentioned above in describing the cultural and intellectual foundations of the Euroamerican and Japanese studies of primatology. They are also used in seeking, in the history of intercultural debate in this discipline, the hegemonic effects of mainstream science on limits set to questions (and answers) in field studies of primate behavior. This paper will give a brief overview of the development of the Japanese and Western traditions in field studies of primate behavior[3] and dwell in more detail on the factors affecting a more biological approach in the West as compared with a more sociological and anthropomorphic approach in Japan. Limits to interpretations of primate social lives, particularly in the Western studies, as well as delays of from ten to twenty years in recognizing some Japanese insights are shown to have resulted from culturally embedded scientific predispositions. At the same time, limitations to dissemination of findings derived from the Japanese tradition are shown to have resulted from their apparently unfashionable views of science in this field, and their way of presenting findings, resulting in nonacceptance in international fora.

The Study of Primates

Primatology is the study of the behavior, evolution, and psychobiology of our closest living relatives in the animal kingdom, the apes, monkeys, and prosimians. Studies of these species have been used to test what might reasonably be speculated about the evolution and behavior of our own species. As such, it is a highly self-reflective field of study, with descriptions of apes and monkeys based on and reflecting prevailing views about human society, and about humankind generally. Primate studies offer a view of our presuppositions about humans through what scientists attribute to similar, but different species, and through the theories they choose to explain the animals' behavior.

Primatology is also one of the few modern fields of research that was initiated independently in Japan at approximately the same time as in Western countries, with minimal exchange between the cultures during the formative stages. It thus provides a highly visible example of cultural effects on a science and of the effect of the presentation of data on hypothesis formation.

Development of European and U.S. Primate Field Studies

Other than a few anatomical dissections of a limited number of animals, systematic study of nonhuman primates did not get underway until the twentieth century. In the United States before 1920 there were no scientific studies of the natural behavior of primates. Concentration had been on studies of anatomy, taxonomy and, later, comparative psychology. In 1924 the Yale University Laboratory was set up under U.S. comparative psychologist Robert Mearns Yerkes. He saw the need for field studies to complement the experimental laboratory studies of the psychobiology of primates that were characteristic of comparative psychology at the time. He thus coupled the study of animals in their natural environment with comparative psychology. Naturalistic studies of primates began in earnest with Clarence Ray Carpenter (1934), one of Yerkes' assistants, who published a study of howling monkeys (*Alouatta palliata*) on Barro Colorado Island in the Panama Canal Zone, followed by several studies on other primate species. In 1938, he also trapped over 500 rhesus monkeys (*Macaca mulatta*) near Lucknow, India, and shipped them to Cayo Santiago, a small islet of about fifteen hectares off the southeastern coast of Puerto Rico. Just over 400 monkeys survived the trip and the descendants of the colony remain as a field site for primatologists today.

Carpenter's studies were largely of factors affecting the coordination and cohesion of the social group, looking at the effects of sex and sex ratios, rank order, and interactions within and between groups. Carpenter viewed sexual activity as the unifying locus of the individual organism and of organic society. He believed it should be possible to construct a formula for the central grouping tendency of each species. This tendency he considered to be the result of fundamental physiological and psychobiological forces, not of ecological or environmental conditions. This outlook was to change radically in the 1960s.

At approximately the same time as field studies began in the United States, Sir Solly Zuckerman, a South African anatomist, made a detailed behavioral study of a colony of hamadryas baboons (*Papio hamadryas*) at the London Zoo in England. The 1932 book resulting from this study, *The Social Life of Monkeys and Apes*, was considered a milestone in primatology. Unfortunately, it was not recognized at the time that the behavior displayed by the captive and overcrowded hamadryas baboons was highly aberrant. Although his work attracted considerable attention, including that of Carpenter in America, his incorrect conclusion was that the main determining factor of social grouping was sexual attraction.

During World War II there was a lull in primate field studies everywhere, but in the 1950s, the number of studies increased dramatically. They seemed to ignite almost simultaneously in Europe and the United States. *Primatologia* (subtitled the *Handbook of Primatology*) was begun in 1956 in order to publish studies on the morphology of primates. *Bibliotheca Primatologica*, first published in 1962, was devoted to comparative biology of primates, including behavior, physiology, taxonomy, and morphology. The journal *Folia Primatologica* appeared in 1963, providing a forum for behavioral studies of primates, among others. Articles on primates were published in other journals before these specialist journals appeared, such as in the *Zeitschrift für Tierpsychologie*, begun in the 1940s, which included articles in German, French, and English; *Animal Behaviour*, begun in 1953; and *Behaviour*, begun in 1947. Primate scientists attending three major conferences held in 1962 in London, Germany, and New York, represented the institutional and teaching nodes from whose lineages most practicing Euroamerican primatologists have come. In 1964 the International Primatological Society was formed. Since there were no naturally occurring primates in Europe or North America, these researchers were dependent on African institutions and goodwill to carry out their studies. Uganda, Tanzania, Kenya, Ethiopia, and South Africa were the field source of much early data on primates.

Development of Japanese Primate Field Studies

The Japanese macaque (*Macaca fuscata*) inhabits forested areas throughout Japan except for the northernmost island of Hokkaido and islands south of Yakushima off the southern tip of Kyushu. Thus, it is not surprising that the first studies of primates by Japanese scientists were made on this local species. However, while this proximity meant that they did not have to mount expensive international expeditions, at first it proved to be very difficult to follow and to observe the monkeys. Primatology had its beginnings in 1948, at Toimisaki in southeastern Kyushu, when Kinji Imanishi and two of his students, Shunzo Kawamura and Junichiro Itani, started a sociological study of semiwild horses. While there, they spotted a troop of monkeys with mothers and infants, adult males and juveniles who were crossing a sunny ridge above the ocean. The monkeys made an indelible impression on the small group of researchers.

The Primates Research Group was formed at Kyoto University in 1951 by Imanishi and Denzaburo Miyadi. Included in the group were eight young researchers, who conducted a series of preliminary surveys of the Japanese macaque habitat throughout the country, from Yakushima in the south to Shimokita, the northernmost extension of the monkeys' range on the main island of Honshu. Despite their efforts, it was four years before they could begin systematic observations of the monkeys. This was because the animals were shy of humans; many groups had been subjected to hunting pressures. They could be glimpsed only as they crossed paths in the forest or fed in the trees. In order to observe the monkeys for longer periods of time, attempts were made on the island of Koshima to provision them with food along their travel routes in the forest. The monkeys accepted the food, and the breakthrough came in 1952 when researchers lured them with sweet potatoes to a sandy beach that was the one open area on Koshima offering an unobstructed view of the monkey group. Finally, troop composition and interactions among all the troop members could be observed. At the same time, the monkeys at Takasakiyama in northeastern Kyushu were provisioned. Takasakiyama grew into and remains today a prime site both for tourists coming to view the monkeys at their provisioning site in front of the temple, and for continuing primate behavior research. By 1961, more than twenty troops had been provisionized throughout the country (Itani 1961). Research was conducted on their ecology and social behavior in a wide variety of habitats, from subtropical to subfrigid zones in Japan.

After successful provisioning, long-term studies based on individual identification of all troop members resulted in pioneering work on dominance rank order (Kawai 1958; Kawamura 1958), parental care (Itani 1959), cultural behavior (Kawamura 1959), vocal communication (Itani 1963), and sexual behavior (Tokuda 1961–1962). Early Japanese primate studies were not restricted to the Japanese macaque and in 1957 research trips to Southeast Asia and Africa were mounted.

Kyoto University produced the bulk of primate behavior research in centers developed between 1962 and 1985. These included the Laboratory of Human Evolution Studies (formerly the Primates Research Group); the Primate Research Institute (PRI) situated in Inuyama city about 300 kilometers from Kyoto; the Center for African Area Studies; and the Koshima field laboratory. Recently, a field laboratory has been added at Yakushima. A further, private facility was begun in

1956 called the Japan Monkey Center (JMC), owned by a private railway company (Meitetsu Railway). Monkeys were purchased from Yakushima to release for purposes of tourism by the JMC. In addition to an amusement park and a free-ranging troop of monkeys attracted to the facility by regular provisioning, the JMC has an impressive zoo collection of primates, a natural history museum, and research facilities where primatologists also work. The JMC also produces two journals: *Monkey*, a Japanese language popular journal about primates and the center's activities, and the international journal *Primates*, begun in 1957, which has been published in English since 1959.

Osaka University researchers also had an early interest in primate studies. After visiting Konrad Lorenz (founder, with Niko Tinbergen, of modern ethology) in Austria in 1953, Yoshiaki Maeda introduced ethology to his laboratory. Ethological, or naturalistic studies of animal behavior, provided an enormous contrast to the behaviorism current in U.S. psychology at that time. Members of the Osaka school are oriented to comparative psychological studies, while Kyoto University members are more oriented to sociological and ecological studies of primates. Today, Japanese primatologists are affiliated with universities and research facilities throughout Japan, although most received their training at Kyoto or Osaka.

Initial Contacts Between Japanese and Western Primatologists

Japanese scientists initiated primate field studies in 1948, but their work remained largely unknown in the West until 1958. About that time, despite large geographic, cultural, and linguistic distances, the pioneers of primatology met sporadically. In 1958 Kinji Imanishi visited several European and U.S. centers of primate research, and met such pioneers as Adolph Shultz, Michael Chance, Clarence Ray Carpenter, Sherwood Washburn, and Harry Coolidge. In 1961 Imanishi and colleagues met Jane Goodall at her field site just as she was beginning chimpanzee studies at Gombe Stream, Tanzania.

Besides personal contacts, a few lectures were given by Japanese primatologists at international conferences (Imanishi 1957; Miyadi 1959). Perhaps the most fortuitous occurrence for the future cross-fertilization of primatology was the presence of Jean Frisch, a graduate student at the University of Chicago in the 1950s. He discovered papers, books, and the journal *Primates*, all published in Japanese about primates. Frisch is a Jesuit priest who was learning Japanese and who intended to teach in Japan. In an article submitted to *American Anthropologist*, Frisch (1959) described the JMC and outlined what he considered to be original features of Japanese studies. Imanishi (1960) described the Japanese method of study. One Japanese, Hiroki Mizuhara, participated in the Stanford Primate Year in 1962–1963, although he did not contribute to the publication that arose from it. Stuart Altmann had some of the early Japanese papers translated into English and collected in a book, *Japanese Monkeys*, edited by Imanishi and Altmann (1965). Carpenter sustained collaboration with Japanese primatologists, making three trips to Japan between 1964 and 1974, even giving advice on how to manage and develop the Takasakiyama colony and contributing to the general development of field methodologies (Carpenter and Nishimura 1969). Since the third issue, published in 1959, the Japanese journal *Primates* has been published in English even though articles were almost all written by Japanese primatologists until the 1980s. Despite these initial contacts, deep-seated differences between the two traditions became apparent during the 1960s in the way that published reports of behavior were presented.

Theoretical and Cultural Perspectives

In the West, after Yerkes' and Carpenter's early work in primatology, people from several fields began to converge on primate studies in the 1950s. Physical anthropologists, psychologists, and zoologists initiated field studies. Scientists in each of these disciplines pursued questions that were relevant to their own field. Primatological research conducted by physical anthropologists was, whether anatomical or behavioral, united by a common interest in human evolution. In the 1950s and 1960s the "new physical anthropology," as espoused by Sherwood Washburn, urged study of primates in their natural environment so as to see how the anatomy actually functioned in a behaving animal. This was a radical departure from the traditional static comparison of structural features of primate anatomy. Findings about how primate structures had evolved to adapt to living conditions could be compared with humans. Once unique features of humans had been identified, anthropologists could propose selective mechanisms which might explain how distinctively human traits had evolved (Gilmore 1981). Similarly, studies of naturalistic behavior generated hypotheses about the evolution of human social behavior.

The disciplinary background and rationale of Japanese primate studies was somewhat different. The founder of Japanese primatology, Kinji Imanishi, was an ecologist who had long been interested in the origins of animal societies. His interests began to overlap with those of anthropology after studies of Japanese macaques stimulated his interest in the origin of human society and of the human family in particular. He reasoned that clues to solving questions regarding the origin of the human family would be provided by two types of research. The first focused on the social structure of the great apes ("primatological approach") and the second on the social structure of current hunter-gatherers ("anthropological approach") (Nishida 1990:10). Once it was found that social structure differed significantly among the Great Apes, Japanese primatologists concluded that rather than try to reconstruct what protohominid society was like through direct comparison with ape species, they should aim to make ethological and ecological comparisons between ape and human groups, considering the differences as important as the similarities in understanding the evolution of human society.

Besides differences between the Japanese and Western traditions' understanding of humans, basic differences are also evident in religious beliefs about the human place in nature. The introduction of the idea of Darwinian evolution in 1859 was so offensive to many Westerners in part because it represented a change in the natural order from top/down (God to humans, or man made in the image of God) to bottom/up (animal to humans, or humans evolving from animals). In Japan, however, evolutionary theory, introduced in 1877, was subject to none of the religious-based criticisms that marked Western debate (Morse 1936). In the Christian worldview, only humans have souls or minds. "Proof" of mind in animals, or other humans for that matter, has been a thorny problem in the philosophy of mind and has also confronted science with the need for verifiable evidence. Thus, the assumption of mental qualities associated with complex social behavior in nonhuman primates, along with other animal species, has been assiduously avoided by most Western researchers. Certainly, the idea of soul or something like individuality in monkeys is, by definition, almost impossible to objectify. But for the Japanese, a belief in the existence of a soul and individual persona in animals allowed for the attribution of motivation, feelings, and personality in animals' behavior. These aspects are included in Japanese explanations of behavior (Asquith 1986a, 1986b).

Methodological Differences

Provisioning

Originally, three features were said to characterize Japanese primate research: provisioning, long-term studies, and individual identification of the animals. Provisioning, or feeding, drew the group into the open and allowed clear observation of behavior and identification of individual animals. However, the merits and drawbacks of provisioning have been, and continue to be, hotly debated by primatologists throughout the world (for example, Fa and Southwick 1988). Leaving aside the debates about the pros and cons of provisioning, we can examine here what it meant to Japanese and Western primatologists in cultural terms.

Whereas provisioning was part of the Japanese methodology for studying primates initially, it was never part of a traditional methodology for Western researchers. Nonetheless, two of the longest studied primate populations in the West are, or have been, provisioned in the past: the Cayo Santiago rhesus macaque colony, and the Gombe Stream chimpanzees. While Westerners debated effects of provisioning on population processes and behavior in these and other groups, Japanese talked about their relationship with the monkeys they were feeding and how it allowed experimentation to be done. (One was the famous "peanut test," in which a desirable item of food was thrown between two animals and the dominant animal was expected to retrieve it while the subordinate animal ignored it.) By providing food for the animals, the Japanese said, the researcher positively entered the group and made contact with the monkeys. At the time when most Western researchers advocated strict neutrality in the study of animals, this was indeed unique.

Social Structure and Long-Term Studies

In the West, those studying "structural features" of primate societies also sought evolutionary, or adaptive mechanisms. Physical anthropologists turned to social anthropology for data collection methods and theoretical orientation. Primate studies of the 1960s focused on social structure as an ordered arrangement of individual members. Human social relations were thought to be controlled by norms, rules, and patterns that each individual knew. Within a nonhuman primate group, rules and norms were overwhelmingly argued to be relationships of dominance and subordinance, especially among males. Ultimately, the adult male dominance system was viewed as being the stable force that held the society together. With few exceptions, studies tended to be of short duration and if many animals were individually identified, it was the males, who were considered to be the prime players in group social dynamics. Until the ground-breaking work of Thelma Rowell (1966), who noted differences among groups of forest- and savannah-dwelling baboons of the same species, it had also been thought that once you had observed a single group, you knew the "species specific" behavior for the entire species.

Japanese researchers' idea of long-term observation, on the other hand, contrasted dramatically with most early Western studies that lasted from a few weeks to a few months. Japanese researchers were seeking to understand the so-called species society (Asquith 1986a; Imanishi 1960), which included the behavior and interrelationships of no less than all members of all groups of a species. According to this framework, one began with observation of particular groups, and the study was possible only through individual identification of all the animals. Japanese have kept genealogical records of monkeys extending back to the 1950s for some troops. Perhaps the uniqueness of Japanese primatology is in part reflected by this diachronic or historical approach. They do not observe behavior only from a synchronic standpoint (that is, the study of mechanisms of behavior at a given time), but also from a diachronic one, by studying the life history of individuals and then ultimately the sociological history of the troop in terms of change over time.

An important aim of Japanese studies was to understand what position each individual held in the animal society. More than one primatologist has said that they sought parallel phenomena among primates to what occurs in Japanese society. Since status and social relationships are given much attention, it was natural to look for similar phenomena in the monkey groups. One of the first questions to be asked of monkey society was, Is there anything like rank order? This may at first appear similar to Western concerns with dominance in primate groups. However, Western researchers considered male rank order and mating access, whereas the Japanese discovered that, in Japanese macaques at least, male rank was only one factor determining social relationships and group composition. Females had a rank order too, and the stable core of the group was made up of lineages of related females, not males.

The Individual and the Group

In the West, during the 1970s, sociobiology provided an important theoretical development in which focus was placed on "individual selection." This refers to the individual acting for its own benefit or to maximize its own reproductive advantage ("individual fitness"). Apparently altruistic behavior could also be explained as ultimately "selfish" or for the individual's genetic survival (for instance, in protecting offspring, other relatives, or unrelated individuals who could be expected to reciprocate in the future). Thus, for example, it was maintained that an individual lives in a group only because group life is to its advantage. Edward Wilson's (1975) book *Sociobiology* was a landmark in focusing the direction of primate studies toward these ideas of individual fitness, and sociobiology underlies the theoretical orientation of most Western primate studies today.

Japanese researchers considered each individual of an entire species as a member of a society and as contributing to maintain the species society as a whole. Thus, intergroup relations were also important to the whole picture of group life. Early on, they recognized variability or what they called "subcultural" behavior among different groups of the same species. Likewise, the history of both the individual and the group was considered equally important, while this dual history was almost wholly ignored by Western primatologists in their shorter-term studies. To the Japanese, current variability among groups and change in individuals and groups over time had to be ascertained for a complete picture to emerge of the dynamics of a society. This differed from earlier Western views of species-specific behavior as being demonstrated by a single group of a species, and which lacked a consideration of the single group's relations with other groups, of the entire species as a whole, and of the diachronic aspect.

Thus, a major source of difference between the Japanese and Western traditions in primatology up to the early 1980s lay in the significance accorded by Japanese primatologists to both individual variability and to the "group" considered as a society. The latter concept is in strong contrast to the view sometimes expressed by sociobiologists that the social group is simply an aggregation whose form has evolved through environmental and biological pressures on each of its individual members. The Japanese approached primate studies from a cultural rather than a biological point of view. While there are certainly exceptions among Western primatologists (for instance, Goodall 1986, 1990; Smuts 1985; Strum 1987), for most, the underlying goal has been to explain behavior and social structure in terms of how an individual's reproductive success is maximized. In Japanese studies, although evolutionary theory provides an important underlying framework, proximate, nonevolutionary explanations of particular behaviors or forms of social grouping have tended to be more frequent in their reports.

In this regard, it is important to distinguish between articles written by the Japanese for English-language journals, or the international science community, and those written in Japanese for home consumption. As primatology matured, and as Japanese and Western researchers incorporated each other's findings, the Japanese had to write articles in ways acceptable to foreign reviewers. One thus finds more evolutionary explanation and, lately, sociobiological theory, in their English-language publications than in Japanese publications. The latter, in fact, include specific reference to research approaches such as *kyôkan*, implying a sympathy with or responsiveness to the animals (Kawai 1969), or *tsukiai*, a mixing or getting on with the animals (Kuroda 1987) that would be out of place in English reports. The Japanese feel there is more to primate behavior than is implied by a mechanistic explanation in sociobiological terms, and most theories proposed in the West were considered too logical and simplistic by the Japanese.

What the Japanese Saw

These theoretical, cultural, and methodological differences gave rise, with few exceptions, to a disregard of Japanese insights in international debate in the important early years of primatology's development. That is, while particular behavioral data on Japanese monkeys were sometimes cited in Western reports, their hypotheses about the underlying dynamics of the animals' societies were largely ignored. For their part, Japanese researchers made efforts to get their point across to Western colleagues, publishing their journal *Primates* in English from the third issue. They even drew on Western classical sources for some of the names they gave to individual monkeys (for example, Bacchus, Jupiter, and Titan) featured in their reports. To understand the basis for Western responses to the early reports, some passages are reproduced at some length below. These passages reveal some of the anthropomorphic reporting of behavior, but they also reveal much more. One can note the respect of the scientist for the keepers, and for the monkeys themselves. More important, they reveal the insights of early Japanese observations that were not referred to even twenty-one years after their original publication when certain primate social dynamics were "rediscovered" by Western scientists. It was, indeed, mostly those Western scientists who were criticizing other Western findings (and not finding wide support for their ideas), who drew on Japanese findings in the first decades of research.

The following discussion was published by Junichiro Itani, one of the leading figures in Japanese primatology.[4] The first passages appeared in 1961 in the *Japan Quarterly* in an article titled "The Society of Japanese Monkeys." The passages are based upon almost ten years of observations of the Takasakiyama monkeys. These are largely continuous observations as, when scientists were not there, on-site keepers kept reports of major happenings (deaths, leadership changes, and so on). Keepers, like the scientists, could distinguish among individual monkeys.

> We were responsible for christening more than 1,000 wild monkeys in all. Some of them we shall probably never forget as long as we live. One of them in particular—a monkey on Takasakiyama called Jupiter—will always be remembered by all of us. He was a monkey of great spirit and valor. Always on the alert, he was strict at times to the point of cruelty. He retained these qualities all through the eight years that we knew him. When he died, on January 16 this year, he must have been over 30....
>
> Nevertheless, throughout the eight years that we knew him, Jupiter was the number-one leader of the Takasakiyama troop. To his dying day, he remained

king of this especially large troop. Though we missed him, his death gave us a chance to study a number of exceedingly interesting problems involved in understanding the way troops of Japanese monkeys behave. The first of these concerns the life-span of the Japanese monkey; another the stability of the leader's position in the troop. It would not be possible for a single individual to retain the position of most dominant leader for as long as eight years if the troop did not have a firm framework.

Yet Jupiter in his last years was quite feeble. Physically, he was no match for the younger and more energetic males in the classes known as subleaders and peripheral males. He could hardly have maintained his position without something other than the strength of his teeth and arms. Our numerous records convinced us that it had some connection with the individual's influence and achievements, and particularly with the confidence of the females who make up the center of the troop. Jupiter's life and death afforded valuable evidence in support of this view.

After Jupiter's death Titan, the troop's second leader, took over quite smoothly without any chaos in the troop. Where Jupiter had been intrepid and adventurous, Titan was a sedate, dignified monkey. It will be interesting to study the effect on the troop as a whole of such a difference in the personality of the leader. (Itani, 1961:424)

The following passages are a continuation of the story of the Takasakiyama group's leadership, published as an editorial in the journal *Primates*, six years later:

Bacchus, the most dominant male of the Takasakiyama-A troop, has been missing since May 18, 1967. He is the oldest of the wild Japanese monkeys we have ever known, and has been known by many research workers since they succeeded in feeding the troop at the end of 1952.

From 1952 to January 1961 the most dominant male of the troop was Jupiter, and after his death Titan took the leadership of the troop. After Titan died in June 1964, Bacchus, who was then the 2nd leader, took the leadership of the troop. His disappearance probably means his death due to old age. On the day before his disappearance a keeper saw Bacchus walking toward the mountain a little later than the other monkeys, and that was the last time he was seen.

In the Spring of 1953 I named him Bacchus. Even in those days he was pretty old, being slow in motion, and his face was always red as if he were drunk. He was the 4th dominant male at that time; however, I thought that he was older than Jupiter. Recently the long hair of his haunch and back turned quite white. He was not as dauntless as Jupiter, nor as generous as Titan. Nevertheless, he possessed his own dignity, and dominated the troop. It may have been that he had, so to speak, the personality of a grandfather.

Bacchus, as well as Jupiter and Titan, must have maintained the position of leader to the last in spite of senility. In the Takasakiyama-A troop from which Bacchus has disappeared, Boor, who was the 2nd male, is leading the troop instead of Bacchus.

Bacchus, who has played an active part in our field notes must now be sleeping somewhere in Takasakiyama covered with beautiful young leaves.

When the last proofs of this issue were about to be printed, I received a call from a keeper of the Takasakiyama National Park on July 7. He told me that he had seen Bacchus after fifty days' absence on July 6.

Several days before they said that a white monkey sometimes appeared in a loquat patch at the eastern foot of the mountain. The keepers did not believe it true, saying that it was a ghost of Bacchus. But the white monkey proved to be really Bacchus.

We want to make him come back to the status of leader, if possible. But it will probably be impossible, and Bacchus himself will not desire to make his comeback. It seems that in the society of monkeys there are still too many problems that we cannot forecast. (Itani 1967:89–90)

In the next issue of *Primates*, the editorial postscript is continued:

On August 2, 1967, an unusual change occurred again in A troop of Takasakiyama; after Bacchus disappeared, Boor led the troop of over 700 head as the most dominant male; however he also suddenly disappeared. Some observations which seem to be related to Boor's disappearance have been made. On that day the monkeys of B troop were feeding at the feeding ground, while those of A troop were resting above the ground. Then Boor went down toward the feeding ground with a male of A troop to drink water. Boor must have probably forgotten that he was in the expanse of B troop. At that time he was suddenly attacked by the males of B troop and was bitten by some of them. He barely fled from there to the expanse of A troop. Mr K. Kawanaka, who observed the scene, says that Boor seemed not to be fatally wounded; nevertheless, Boor has not been found in A troop since.

If he has not died from the wounds, how should the relation between the above-mentioned occurrence and Boor's disappearance be interpreted? I suppose that a special psychological balance is necessary for a monkey who is placed in such a special status as "most dominant male," and that the loss of this balance may make it difficult for him to maintain status. Some people will say that this is a too anthropomorphic idea, but such a type of "power-loss" may occur in a society of monkeys.

The six leaders that I named in 1953 after individual identification, have all disappeared, Boor being the last. Dandy, Saruta and Aome, who were once sub-leaders, are leading the troop in place of Boor; however, monkeys of the lower classes are partially lacking because of the desertion from the troop or troop fission. Therefore it can be said, from the view of the social composition, that A troop has fallen into a very insecure state. At present, dynamic relations among A, B and C troops are very delicate, and as a result there are many sociologically interesting problems. For example, a tendency can be seen, from observations so far made that the dominance of a troop does not always depend on the size of the population, but rather the troop which keeps good social integration becomes dominant over a troop which is not well integrated. We research workers want to keep watching in order to determine how the monkeys will acquire a new equilibrium by means of their social intelligence. (Itani, 1967:187-88)

Although the *Primates* quotations are editorials, not data papers, this too was part of the Japanese style of reporting, giving an overview of findings and speculations based upon them, complete with a descriptive setting. These are, precisely, features of the Japanese approach to their study.

Western Reactions

In an important early paper, Donald Sade (1972:378) commented that "a consistent, logical, and generally accepted methodology has not yet emerged in the study of social behavior of primates.... Thus much of the work on social organization and behavior of free-ranging primates has remained speculative." Debate over the effects of different methodologies coincided with what Japanese researchers had supported from the beginning of their studies. For instance, Sade noted that critics of longitudinal studies maintained that such studies simply produced a larger amount of data, producing a diminishing return for effort. Sade disagreed with his Western colleagues, noting that longitudinal study produces new information and new conclusions. For example, researchers such as Koford (1963) and Kaufmann (1967) said that dominance hierarchies in female rhesus monkeys (*Macaca mulatta*), which belong to the same genus as the Japanese macaque, could not be determined. On the contrary, long-term observations (by Japanese researchers as well as by Sade) in both Japanese and rhesus macaques showed that linear dominance hierarchies exist among the adult females of the group, and that they are highly stable, unlike the male hierarchies. Furthermore short-term research such as that of Washburn and DeVore (1961) on baboons (*Papio species*) were impressionistic and emphasized a few obvious or spectacular aspects of behavior, often centered on the more visible males of the group. Such observations were then used as a description of the social organization of the group. Sade noted that this produced a model (which influenced primatology for a decade until shown to be wrong), but was not tested in other, short-term studies. Sade further questioned Hall and DeVore's (1965) report on the absence of female hierarchies in baboons, based on the kind and duration of their observations.

A second feature of Japanese studies was identification and familiarity with individual members of the primate group. Sade noted that, for instance, Kawai (1958) could make use of genealogical information (based on others' long-term studies of a group) for a ten-day study but could conclude something about interaction—in this case that the successful monkey in a paired competition was usually the offspring of the dominant mother.

Besides some of the methodological aspects, the Japanese observations quoted above contain two points that revolutionized Western studies of primates when finally recognized in long-term research by Westerners. The Japanese findings were not cited even then. One point was that, with reference to Jupiter of the Takasakiyama troop, the maintenance of the position of alpha male was not dependent upon strength, but upon "the individual's influence and achievements, and particularly with the confidence of the females who make up the center of the troop" (Itani 1961:424). It was not until 1982 that Frans de Waal reported in his influential and (to Western researchers) groundbreaking book *Chimpanzee Politics* that "physical strength is only one factor and almost certainly not the most important one in determining (male) dominance relationships" (1982:95). In the same work, he also suggested that the alpha male must protect females if he is to receive their support in return (125). A rival male could undermine female support of another male by "punishing" them, or could attack females to show them that the alpha could not protect them. De Waal also referenced prior work on female support in a Western study by Ron Nadler (1976) which described captive female gorillas helping a male become a leader at the Yerkes Primate Center. No mention is made of Japanese reports. *Yerkes Newsletter*, in which Nadler's report appeared, is also less accessible than the international journal, *Primates*, where the Japanese findings were published.

De Waal also noted that, "the influence on group processes does not always correspond to rank. It depends too on personality, age, experience, and connections (1982:210). De Waal regarded the oldest male and oldest female as the most influential members of the group. Granted, he was writing about chimpanzees, and not a macaque species, but it was precisely his point that such advanced primates as chimpanzees could have subtle social strategies. Such things might have been sought sooner in apes, had they been recognized as occurring in monkey species. So too, female roles in primate societies might have been recognized earlier rather than the long-held predominant Western view of male-centric troops.[5]

Differences in orientation affected many aspects of data obtained on primate behavior in the two traditions. To put it simply, the Japanese methodology allowed them to "see more" of primate behavior and its proximate causes. With their long-term diachronic and synchronic studies, and willingness to attribute more complex motives to animal behavior, the Japanese discovered much about primates that Westerners have only lately come to realize. Several pioneering Japanese insights besides those mentioned above have been discussed elsewhere (Fedigan and Asquith 1992): for instance, the importance of female kinship to understanding dominance, affinitive relations, and group fission; "incest" avoidance, particularly between mothers and their male offspring; nonsexual "friendships" between unrelated male and females; the inverse relationship between male dominance and mating access. Reynolds (1992) noted that the Japanese discovery of kinship in nonhuman primates was not taken up by Western anthropologists or primatologists due to their regarding kinship as a purely cultural (human) phenomenon. Yet kinship became a fundamental concept for understanding non-human primate society. Japanese had felt freer to use this concept, even with its anthropocentric semantical load.

Thus, reaction to Japanese primate studies can be summarized as, initially, suspicion of and disregard for Japanese data because of their anthropomorphic flavor (Frisch 1963:239–241); then admission by Western scientists that the Japanese data are accurate (Reynolds, 1976:137), but with reservations about their method (Carpenter 1960). A slight alteration of a common saying—"They find the right answer in the wrong manner"—might summarize the second attitude to the Japanese work. Although differences remain, as is evident especially in Japanese-language journals and books, both traditions, since the 1980s, have come closer. Many Western primatologists have come to agree with the importance of long-term research, despite the practical difficulties of accomplishing this in foreign countries. Reportage of findings in international journals are now virtually indistinguishable. Both freely draw on the findings of the other, and cooperative projects are successfully completed. Yet, because they are aware that papers submitted for review to English-language science journals would be rejected if they did not conform to a nonanthropomorphic and less speculative format, many of the ideas debated among Japanese primatologists have been and continue to be buried. What was thought to be a less rigorous approach in Japanese primate studies was, in fact, a carefully considered methodology.

Conclusion: The Decolonization of Science

A number of somewhat naive suggestions based upon one cultural variable or another have been made about the foundations of "Japanese-style" science, especially as regards their relationship to nature. These range from relating the aesthetics of Japanese gardens and tea houses to methodologies in science (Watanabe 1974); to suggestions that the use of Chinese characters in writing caused Japanese to tend to nonlinear modes of thought, with more emphasis on pattern recognition, as contrasted with Western linear thought processes (Kikuchi 1981); to suggesting physical bases for different ways of processing in Japanese and non-Japanese brains (Tsunoda 1975, 1978). Others suggest, with Motokawa (1989), that Western science is hypothesis-oriented and Eastern science is fact-oriented.

The fact remains that the greatest single problem confronting Japanese scientists has been their continuing isolation from the world science community (Bartholomew 1989). This refers to more than the problem of language, large as that problem is. To give a personal example, before beginning my studies of Japanese primatologists in 1981, the primatologist Yukimaru Sugiyama wrote to a Western primatologist, Vernon Reynolds, with regard to my proposed study: "The biggest problem which we can't write in our scientific articles in English is the basic way of thinking by the language barrier" (letter, March 21, 1980). I may not know precisely what Sugiyama meant by this, but I can hazard a guess based upon fourteen years of interaction with the language and scientific tradition.

In primatology, Japanese reports were, among other things, too anthropomorphic to be taken seriously by most Western researchers. They appeared to be collecting vast amounts of detail and not putting it into any theoretical framework. Yet, even the short examples given above reveal how primatology could have developed more quickly if Japanese material had been taken more seriously. More generally, in terms of language and presentation of findings in English-language journals or at international conferences, Japanese scientists may seem atheoretical, uncommitted, or even illogical. This can be related both to practical conditions and cultural predispositions. First, their focus in certain fields and at certain times in the history of a discipline may be directed more toward a building up of details to support one position or another. This may be seen as necessary before stating any specific position. What may have contributed to the notion that Japanese scientists tend to build on others' theories, rather than develop their own, is the way in which they write their reports. Scientific reports written in Japanese do not typically state conclusions (see Leggett 1966; Motokawa 1989). Instead, they try to describe one fact from various points of view. These points may be connected by imagery to other points, rather than by strict logic. Why do Japanese scientists not state a firm conclusion? It would, they say, close their world. A typical example is found in biologist Motokawa's (1989) observation that a conclusion means that "this statement is the truth." Such a statement, he maintains, is definitely false because our words can never be absolutely true. Anthony Leggett (1966), a physicist, wrote a paper for a Japanese journal of physics in which he suggested to Japanese physicists how to write scientific papers in English. He told them not to use ambiguous terms but to be definite and commit themselves. It may often be the case that what is in fact a commitment to one idea or another is simply not understood as such by the foreigner. There is another aspect to this. That is, despite many successes in science, an apparent lack of self-confidence exists in presenting Japanese findings to the world community. For example, Traweek (1988) has noted that among the Japanese physicists she studied, the scientists feel there is some suspicion of their work on the part of Westerners. She reports that they have asked foreign scientists into one of their premier physics research laboratories in Tsukuba City, near Tokyo, in order to gain credibility.

Ultimately, the scientific hegemonies affect not only our knowledge base, but the ability of others to contribute to it. In this paper, reflection upon the development of primate studies is not intended as an exercise in science-bashing. The methods of Western science obviously contribute much to our understanding of the world around us. The point is simply that Western science will remain at best an impoverished view if we continue to impose the Western cultural experience of the world on the questions we pose.

Notes

I am indebted to Ellen Bielawski, Linda Fedigan, Laura Nader, and an anonymous reviewer for helpful comments on this paper, and to Masao Kawai, Junichiro Itani, Yukimaru Sugiyama, and many primatologists in Japan who have facilitated my study of Japanese primatology. The research has been supported by Monbusho, the Social Sciences and Humanities Research Council of Canada, the Izaak Walton Killam Trust, and the Japan Foundation. I am grateful to all of these granting bodies.

1. For a consideration of theoretical underpinnings to cultural studies of science, see Rouse (1992).
2. See Haraway (1994) for further comment and a bibliography on interlocking cultural, feminist, and science studies.
3. The passages on the historical development and methodology of the primatology traditions have appeared in Asquith (1994).
4. Recognition from the international community came when Itani was awarded the Thomas Henry Huxley Memorial Award for his contributions to anthropology in 1984.
5. An exception to this was, again, someone who used findings from long-term studies by the Japanese and the few Western researchers who were conducting them (Lancaster 1973).

Epilogue

CHAPTER 15

The Three-Cornered Constellation
Magic, Science, and Religion Revisited

Laura Nader

> There are no peoples however primitive without religion and magic. Nor are there, it must be added at once, any savage races lacking either in the scientific attitude or in science, though this lack has been frequently attributed to them.... On the one hand there are the traditional acts and observations, regarded by the natives as sacred, carried out with reverence and awe, hedged around with prohibitions and special rules of behavior. Such acts and observances are always associated with beliefs in supernatural forces, especially those of magic, or with ideas about beings, spirits, ghosts, dead ancestors, or gods. On the other hand ... no art or craft however primitive could have been invented or maintained, no organized form of hunting, fishing, tilling, or search for food could be carried out without the careful observation of natural process and a firm belief in its regularity, without the power of reasoning and without confidence in the power of reason; that is, without the rudiments of science.
>
> (Malinowski 1925:17)

With these words Bronislaw Malinowski pursued the polemics of his book, *Magic, Science and Religion*. "The problem of primitive knowledge has been singularly neglected by anthropology" (1925:25), he noted. Some myths needed to be dispelled by examining "the three-cornered constellation." Challenged by the dogma that primitive peoples were categorically irrational, Malinowski set about distinguishing magic, science, and religion from each other, and outlining their respective cultural functions. Although he was not first to make these distinctions, he was probably the first anthropologist to pursue these distinctions in fieldwork.

Science is guided by reason, he argued, corrected by observation, and open to all, a common good of the whole community. Magic is made by tradition, impervious to reason and observation. It is occult, not easily available except by mysterious initiations, and it serves to bridge the unknown. Religion lifts man above magic to humility and the realization "of human impotency in certain matters" (1925:19). It "establishes, fixes, and enhances all valuable mental attitudes" (89); focuses on the crises of life, especially death; and thereby ritualizes man's optimism.

Malinowski posed the question: "Every primitive community is in possession of a considerable store of knowledge based on experience and fashioned by reason. Can this primitive knowledge be regarded as a rudimentary form of science ... ?" (1925:26) He continued: "Science, of course, does not exist in any uncivilized community as a driving power, criticizing, renewing, constructing. Science is never consciously made. But on this criterion, neither is there law, nor religion, nor government among savages" (35). Malinowski argued that whether we are dealing with pre- or postliterate societies, all peoples operate within the domain of the Sacred and the Profane, within the areas of magic and religion, and of science.

A further finding of the Trobriand study was that magic flourishes in settings where man cannot rely completely upon his knowledge and skill. Malinowski's illustration is often quoted: "In the lagoon fishing, where man can rely completely upon his knowledge and skill, magic does not exist, while in the open-sea fishing, full of danger and uncertainty, there is extensive magical ritual to secure safety and good results" (1925:31). His study also reported that the Trobriand Islanders were able to distinguish magic from science, an observation of potential importance. If they were able to distinguish between science and magical activity, between empirical knowledge and ideas operating under the influence of desire, their entire society shared an understanding of risk and uncertainties of life. Because all were aware of these uncertainties, if they failed, they failed together. For Malinowski, knowledge was related to biological advantage and knowledge was the reason for survival in primitive communities. But there is more to knowledge than biological advantage or utilitarian purpose. In "modern science," no distinction is made between magic and religion by the main actors. Publics are not aware of the inherent risks and uncertainties of life because magic and religion are not marked as separate or different activities. It is worth hypothesizing that modern scientists practice more magical thinking in their highly experimental and dangerous research work than do the Trobrianders in the rough waters of the Western Pacific.

Few anthropologists have pursued the challenge of Malinowski's provocative essay in the decades that followed its publication.[1] In an evaluation of the work of Malinowski (Firth 1957), neither the word "knowledge" nor the word "science" appears in the index, although kinship, law, and magic do. Indeed, S. F. Nadel (1957:189–208) wrote his essay on "Malinowski on Magic and Religion" and made no reference to Malinowski's work on "Magic, Science and Religion." Nadel is silent on the subject of science; his own constellation is "magic, religion, and mythology" (1957:208). In the same volume E. R. Leach writes about "The Epistemological Background in Malinowski's Empiricism" (1957:128–129). Leach takes Malinowski to task for attempting

> to impose "rationality" upon his savages.... Malinowski maintained, no doubt rightly, that Trobrianders are at least as rational as twentieth-century Europeans. He stressed that "civilized" as well as "savage" life is packed with

magical practices.... Where he seems to err is in maintaining that the ordinary man distinguishes consistently between the magical and the nonmagical.... In seeking to break down the dichotomy between savagery and civilization Malinowski argued that primitives were just as capable as Europeans of making such distinctions.... He would have had a much better case if he had insisted that Europeans are ordinarily *just as incapable* as Trobrianders of distinguishing the two categories. (emphasis added)

Since Malinowski, anthropologists who have investigated native science or indigenous taxonomic systems began to systematize native knowledge and others to collate such knowledge. Although their contributions are important, the lessons they have drawn have not always been persuasive. Authors still assume that systematic classification is restricted to societies that have reached a sophisticated level (Morris 1976). For C. R. Hallpike (1979), primitives are impoverished and it is the impoverished environment that accounts for what he sees as their intellectual retardation.

In this paper I examine, as Leach suggested, whether Euroamericans are more capable than indigenous peoples of distinguishing "science" from "magic" and "religion." The three-cornered constellation—magic, science, and religion (or desire, reason, and humility)—is a likely universal, and therefore observable among scientists and technologists in the most developed of contemporary science and technology.

Experiencing the Problem

The closer historians of science look at the great achievements of science, the more difficulty they find in distinguishing science from pseudoscience and from the political, economic and ideological contexts. Scientists' philosophical views about nature, man and society appear to play a very important part in formulations of the substance of major scientific ideas.... Science is much more like the messy world of social and political intercourse than working scientists care to believe, or are willing to concede. (Young 1972:103–104)

In the late 1970s I began work on government-supported research dealing with domestic energy policies in the United States. The work was first conducted at the federal level as part of the National Academy of Science's Committee on Nuclear and Alternative Energy Systems (CONAES), and resulted in several studies, one of which I oversaw—*Energy Choices in a Democratic Society* (Nader et al. 1980). A second study, *Distributed Energy Systems and California's Future* was conducted at the University of California, and resulted in several additional reports and publications (Nader et al. 1977; Nader and Milleron 1979). Both experiences stimulated me to scrutinize and question some basic assumptions of scientists and engineers working on energy questions.

It was during participation in this energy work that I became interested in the predominant ideologies of energy experts, the social groups and networks of which they were a part, and the equity conceptions that developed as a consequence of their work experiences. In the following pages, I report my observations and the observations on my observations as received from a wide spectrum of physical scientists. I then return to the three-cornered constellation alluded to earlier in order to address the question, "When is science scientific?" My data is taken from the more than one hundred responses from physicists and engineers in the United States and Europe to an article that I published in *Physics Today* summarizing my findings (Nader 1981) and the difficulties of "distinguishing science from pseudo-science and from the political, economic and ideological contexts" (Young 1972:103).

The development of commercial energy through scientific enterprises is a recent phenomenon, one associated with the observation that fossil fuels are being used up at a much greater speed than nature is able to replenish. At the turn of the century the president of the American Chemical Society predicted in his national address that the United States would be running on solar energy by the 1970s: "The sun directly and indirectly will monopolize the power of the country" (Wiley 1902:163). Yet the energy shortage took on crisis proportions only after the oil embargo of 1973. Since that time public debate on energy supply has stimulated questions about dominant paradigms in scientific and methodological thought, in which energy sources such as nuclear, solar, coal, or others are enmeshed, in order to understand entrenched manners of thinking about and controlling energy sources. The structure of scientific and technological debates over the capability of large and now planetary systems to adapt to changing resource circumstances were (and are) the subject matter in energy research.

As part of the CONAES work I chaired a resource group mandated to consider lifestyle changes that would accompany changes in per capita consumption of energy by the year 2010. What would happen to consumer tastes if energy levels remained the same or increased two-to-threefold over several decades? *Energy Choices in a Democratic Society* (Nader et al. 1980) described two futures. One was characterized by high efficiency, the other by low energy use and high technology. The high-efficiency scenario was designed to explore energy use in a society much like our present society projected into the future, without major changes in attitudes, but with improvements in amenities roughly consistent with those in recent decades. In the second instance, we explored a society in which attitudes toward resources change. It is a comfortable society that has decentralized land use and work; its members value thrift and self-reliance and are less vulnerable to terrorism and violence. Our exercise in decoupling tightly coupled beliefs (such as "low energy use indicates aversion to high technology") illustrated the possibility for a society to have both high technology and low energy use.

In the process of doing this work, we unselfconsciously questioned hitherto unquestioned assumptions by providing answers to "problems," and by questioning presumed relationships between the quality of life, lifestyles, and the amount of energy produced and consumed in a given society. First, we observed that there are many, although not limitless, ways to use any amount of energy. Ireland and New Zealand expend about the same amount of energy each year per capita—in very different ways. The Republic of Ireland is frequently used as an example of social and economic stagnation and even misery, while New Zealand is typically seen as a society in which everything runs so smoothly and progressively and equitably that its only fault is dullness. Within any one level of total energy use there are widely varying fluctuations in quality of life. Individuals vary in their energy use, regardless of the energy available. Next, we learned that supply technologies such as oil, gas, and solar may be less intrusive than others, such as nuclear. We also learned that an abrupt rise or decline in energy use will have disruptive effects on society, that changes incorporating grass roots input could have effects and results different from those imposed by business or government. Technologies that render citizens' civil rights, liberties, health, and safety vulnerable require a focus on the supply form and an emphasis upon the necessity for consumers rather than producers of energy to define needs. Furthermore, we found that there was no basis for believing that the continued growth of technology would not dramatically affect changes in life-style (rather than preserve it as some believed). Growth changes and will continue to change the fabric of life in all industrialized countries. Contrary to accepted belief, considerations of low energy, high-technology societies expand the range of choice and increase the time within which we can plan for the age after fossil fuels. Because many of the questions are ecological, attentiveness to context is critical. We were fighting isolating or context-stripping devices of scientific models.

Initial reaction to these "discoveries" was denial; we were describing impossible futures. We were planning low-energy futures that "go against the grain of human nature"; we were pursuing "a return to caves and candles." Further criticism impugned our methodology. Although our resource team represented a wide range of disciplines, many of which rely heavily on numbers, we were told our work was not quantitative enough, and that we were unscientific. I was repeatedly encouraged to turn out "more tables, less prose" because energy experts, it was alleged, do not read. The belief that numbers in themselves are useful was rampant and was attached to a peculiar importance given to prediction.

Anthropologist E. Colson (1973) has observed the role of divination as a decision-making procedure that legitimates the basis for choice. People, "primitive" or "civilized," wish to minimize risk-taking where the future is uncertain, or at least to legitimate decisions should there be a need for accounting. But while numbers may indeed provide, as she says, "Tranquility for the Decision-Maker" in energy affairs, they sustain the illusion of concrete reality, and avoid important issues not amenable to modeling processes.

Eventually this use of numbers and modeling sabotages public confidence in the "scientific." Questions of civil rights, of freedom, of social structure, of democracy, of quality of life, or of equality are not easily discussed numerically or through modeling. Numbers also dilute the dangerous and the unthinkable: the distancing function. Losing 2 percent of the population makes a possible disaster easier to bear than losing thousands or millions of people; thus percentages are preferred to actual numbers of people when discussing nuclear accidents. Scientists and engineers are expected to bring their special skills to a problem that may include leaving behind their human understanding of a situation; like professionals in law, for example, they are trained to rise above sympathy.

A summary statement of my observations would include the predominance of group-think; people who thought differently than the group were told they were off the track. Part of group-think was a lack of respect for diverse solutions and for diverse kinds of intelligence, an avoidance of technologies other than nuclear or coal, and a preference for abstract rather than concrete thinking (as if figuring out how to dispose of waste was equivalent to actually doing it). Memos discussed nuclear, coal, and nonnuclear energy. Nonnuclear meant solar, which was described as "an orphan child," "not very intellectually challenging," and "just a bunch of mirrors." My notebook was full of observations on unexamined assumptions such as: change means maintaining the status quo, progress is equal to growth, more energy expenditure doesn't change lifestyle while less does, societies only change from the top down, technological fixes can solve human problems and forestall crisis. In addition, I tried to understand the concept of the Big Toy. In answer to my question, "Why did you go ahead with nuclear energy without having first solved the nuclear waste problem?" I heard "because it is interesting, it is fascinating, it is fun" even though the same people would unself-consciously agree that such a direction takes us into unknown and dangerous waters.

What was striking about the science work I observed was that assumptions were often made on the basis of faith (see Kuhn 1970:167 on Western sciences' rejection of critical attitudes as an element crucial to success). For instance, the connection between rate of increase of energy use and rate of growth of gross national product was assumed (see Nader and Beckerman 1978). In the United States from 1935 to 1955, both energy use and quality-of-life social indicators were positively correlated, but in the period between 1955 and 1975, energy use generally increased while social indicator research shows a decline in quality of life, a negative correlation. Neither correlation shows the true relationship because per capita energy use and quality of life are related to each other in very complex ways, as we indicate. Human affairs do not break down in terms of scientific disciplines, which after all evolved for research and teaching convenience. For purposes of making policy, the boundaries of academic disciplines are limited.

Magic, Science, and Religion Revisited

My observations on social and cultural "Barriers to Thinking New About Energy" were published in *Physics Today* (Nader 1981) and stimulated rich response from scientists and engineers. Letters came from older physicists indicating awareness that the behavior observed is selected for, learned, and transmitted from generation to generation. Although they all had explanations, one blamed the situation on physics education:

Physics texts present the subject matter in an orderly manner but usually fail to indicate how the present concepts developed. Few theories were eagerly adopted when the theories were first announced.... Physics students are given problems with known answers. They are not trained to be original ... physicists work in groups and often for large organizations.

A scientist from a major research laboratory commented generally about groupthink and the dilemmas of a modern science interested in "results." He was describing cultural patterns that play themselves out in a Western style of doing science:

> I have a deep personal and professional concern about the use of human ingenuity in the improvement of the human condition. It's clear that when a group of people think alike, most of their brains are redundant. Group thinking has become (maybe it's always been, I don't know) so common in the "educated" parts of our society that it's almost funny.... As far as I can tell, the educational process does little to enhance original thinking and a great deal to stop it.... As a manager of a research enterprise, I try to help create an environment in which creativity can flourish.

A teacher and observer of scientists remarked on the change in scientists as science and technology have become more and more intermeshed in their quest for absolute certainty:

> As I. I. Rabi said ... physics used to be done by people, people with first, middle and last names. Now it is done by laboratories—CERN, Brookhaven, Fermi, etc.—and the individual people are largely faceless, hence not specifically responsible. Having come from a family of physicists, of Rabi's vintage, and having known many of the men whose names were legend in their time even as they are remembered now in the history of science accounts, I sometimes suggest in full seriousness that most of the practitioners whose attitudes (and blindnesses) you properly decry simply are not physicists at all; they are technicians, perhaps very good ones but technicians none the less. And so I wonder, as Rabi wondered in his talk, whether the age of physicists is over. If it is, then maybe we can understand the kinds of myopia the profession seems to perpetuate.... Changes ... will come only as we change our methods, our own behaviors and our own attitudes toward science....

A scientist at a major national laboratory spoke about the process of becoming a scientist, the role of power and money, and the practices that are intricately tied to dominant institutions and their interests:

> The experience in most universities is that many students enter stating a preference to study hard science yet only a few succeed. Those that leave

science do so because they fail to survive the implied and actual rigors of the technical curriculum. Thus, a strong selection process for narrow academic brilliance is in effect. The most creative students often resist the scientific dogma they are taught and fail to survive to graduation. This happened to Einstein and is still true today. The best of the survivors, judged entirely by their academic performance, go to graduate school where they evolve complex yet very narrow research projects, ultimately incomprehensible to everyone but the student, for which they are awarded a Ph.D. At this point they have become incompetent to do anything of practical value so they 1) go into industry where they are retrained; 2) join a research group in a national laboratory; 3) teach; or 4) continue to work on the same narrow research project. At this point purely social factors take control as the scientist realizes he will not generate any new, earth-shaking ideas. Thus, he 1) enters management; 2) becomes a research group leader; 3) becomes department chairman; or 4) writes a gigantic research grant proposal to the government. Thus in the final analysis money and power become the end of scientific enquiry. The construction of a large accelerator becomes a highly visible project which can generate great respect for the scientists involved irrespective of the scientific worthiness of the venture. What people do not realize is that very little scientific consideration goes into the generation of most large research projects.... The people you met were not concerned with the energy problem so much as getting a piece of the action.

Yet another physics professor comments on the religious nature of growth assumptions and the mission to modernize American society and the world:

Growth is the American Religion and a religion is something you accept without subjecting it to analytical scrutiny. The growth religion is rare among religions because with the growth religion one can predict the consequences of courses of action. But the true believers don't wish to see the predictions made. When someone does make the study of the consequences of the growth religion the high priests don't want to hear heresy.

A Nobel Laureate old enough to have experienced the era before the militarization of science reported that the intellectual climate has changed in ways that have deep cultural implications for science and democracy:

Yes, that's the way it is, and well guided by an unexpressed system of rewards and punishment.... In the first half of this century we had a generation of monumental physicists—Einstein, Bohr, Heisenberg, Pauli, Schrodinger, and on and on, all of whom knew that what physics is about is reality; and that physics (science) can explore only part of reality and by far the smaller part. That kind of thought is now virtually forbidden in scientific literature.

In fact, I was surprised at the number of responses that mentioned censorship, absence of dissent, or coercion. Some indicated that they were amazed that my comments were published in a science journal. Many of their additional remarks are not peculiar to science, however, for as many of us know they could be generalized to other professional groups, and possibly to our society writ large. The comments are particularly interesting in the context of discussion about science because it has been argued by so many that the very nature of the scientific enterprise requires freedom of inquiry. Secrecy is the enemy of science. The chair of a physics department is unusually frank about the daily controls in science work:

> There are many of us (even males) who agree with your point of view and share your puzzlement at the attitudes and approaches followed and fostered by those who participate in the planning process. Of course, many individuals ... are bullied into effective silence by those who prevail in the upper echelons of planning—whether it be think tank, government, or academy based. This is even true to a disappointing degree in the university community. The price of questioning or dissent (or asking questions like "Why not solar?") can result in professional ostracism. And, of course, for those who aren't quite sure which way to go, it is clear that the best thing to do is to be tougher than the next tough guy. And toughness doesn't mean wimpy solutions like less energy use or such "soft" scenarios, but rather showing that we are indeed tough guys who can "overcome nature," build a bigger bridge, pump water faster, construct a safe structure on a fault zone, build a safe nuclear power plant, walk on the moon, climb the mountain because it's there, shoot down the other guy's missiles, build a better accelerator, dump our garbage where it pleases us, keep women and other lesser beings in their place, and do it "my way." ... Of course, not all of the undertakings or concepts I have expressed above are all bad. Some of them stem from the same basic drives that make humankind achieve so many brilliant and wonderful things. But somehow achievement has become linked with production and consumption. And those who are best at these two things are those who determine what the rest of us should do, think, and have.

Numbers of scientists speak about a bureaucratized science. A staff scientist at a major government laboratory demystifies the experience of science work by describing scientists as bureaucrats largely intent on building or maintaining empires. The following story by a retired physicist indicates what can happen to science work in a bureaucratic setting. It is a story that I have heard over and over from highly placed scientists speaking offstage:

> At one time in my career as a physicist I worked with five other physicists. One day when I was in the office of the head of the project I noticed that he and one of my colleagues were using a foolish and time consuming method of converting the tracings on a chart to numbers for a table. Immediately my

five colleagues recognized the truth of my observation. Three of them laughed but the project head did not think that it was amusing. He stated that the reason that I noticed the foolishness of the procedure was because I always wanted to take time to think. He said that he was too busy doing research.... For the first three years of our study there was a lively controversy between our leader and theoretical physicists from other organizations. Our measurements did not fit the theory. One day when I was bored doing what I was supposed to be doing I decided to determine the cause of the difference between theory and measurement. I decided to go over every step of our procedure. At step one I found that a coefficient that we used was obtained from a table published long ago. In our great rush to do research the wrong value was read from this table and no one had checked the reading of the table. When the correct number from the table was used our experimental results agreed with the theory. In the great rush to do research three years of heated argument was found to be useless. I was not awarded a gold star for this little piece of work....

Both hierarchy and bureaucracy seem to get in the way of frank and open expression. Science as organized rationality is not what most of these scientists experience.

A chemical consultant, in reference to my "delightful account of the experiences of a lady anthropologist in the creative deliberations of a macho bureaucracy," also commented on bureaucracy:

It tells more about what is wrong with American creativity than most of the articles I've read, for the reason that only an outsider—in this case, a token female liberal—can open her mouth. If you are a member of the conference sponsor's organization and depend on it for such things as food and shelter, you had better go along with the conference's purpose and vote Ja! ... The sad thing about bureaucracies, of course, is that the communication becomes "top-down" very quickly as the bureaucracy grows, because no one dares not to pass an idea down, while many a brown-nosing middle manager considers that his sole task.

A distinguished physicist and engineer expresses relief at my depiction of scientists and engineers, and raises the issue of desire and accusations of irrationality:

For years I have felt like the ultimate minority. I have been characterized by my peers as illogical, a kook.... I am glad that you have brought up the pathological blinders of the technocratic establishment. I now feel that I am not irrational. I merely refuse to include human beings as numbers in sets of equations.... I have often compared the attitude of physicists and engineers to cowboys; the only difference being that cowboys get drunk and run over people in their jacked-up pickup trucks. Scientists get drunk on technology and may run over us all with their dangerous toys.

The letters are filled with thoughtful observations that belie any attempt to talk about scientists as a homogeneous class, and indicate scientists' views about the dominant culture in science and the context in which science work is conducted. The writers of these letters have a capacity for sharp social assessment. The following observation indicates a concern with blinders that result from education and work settings, value preferences, and arrogance. One needs only to remember the record of failure of endless projections of energy futures:

> I think there are two characteristics of the scientist which are quite pertinent to your commentary. The first unites the scientist with the rest of humanity: cupidity—all other things being equal, there is greater opportunity for personal gain working on billion-dollar projects than on thousand dollar projects. (I'd wager that the version of solar sometimes called "the power tower" has fairly staunch defenders in scientific or at least engineering circles.) The second distinguishes the scientist from the rest of humanity: specialization—scientific education is an extended process of focusing, learning more and more about less and less.... (In effect, so long as the scientists with whom you worked stuck to central station power, nuclear, hydro, coal, what-have-you, they were the experts—as soon as they had to confront a 4m^2 solar collector on a residence, they had no greater skill than the random home owner.)

Science is an omnibus word connoting a body of scientifically validated knowledge, an organized rationality, or an attitude toward knowing. In *How Natives Think* Lucien Levy-Bruhl (1910) contrasts the native (unable to reason to conclusions) with the modern mind (surrounded by an atmosphere of logical potential). The contexts in which scientists work indicate room for the opposite kind of behavior, however. How scientists set research agendas and how they construct "facts" are all profoundly social and cultural processes (Latour and Woolgar 1979). As a senior physicist observed:

> I have noticed over the years an apparent irrationality on the part of many colleagues when it comes to discussing matters related to energy. All the data in the world cannot bring about a change in energy strategies, once the person has decided on a particular course. This is all the more strange as many of these people are competent and intelligent scientists who are usually willing to alter their view in the light of new data.
>
> What seems to be happening is that many scientists who work in energy policy and energy technology have different social and personal values and these values are rationalized by appeal to numbers and data. Many of the disagreements I have witnessed about energy policy or one technology vs. another, is really a disagreement over values. Normally, a disagreement over values would be stated as such, but I think physical scientists are uncomfortable in the subjective realm of values. This inability to recognize the true

source of conflict leads to the nonsensical and ridiculous statements you mention in your comment.

A science editor lamenting that scientists are not taught about concern for the consequences of their work writes:

> Science has two faces. The actual experimentation into which value judgements should not enter and the technological applications of our experimentation into which value judgements certainly should enter.

In a letter to the editor of *Physics Today*, a reader examines the myths that are propagated, indicating that much more than the scientific method is at work and underscoring the technological dominance of science:

> Most pure and applied scientists have become protagonists for the myth that what is technologically feasible is also scientifically desirable.... The momentum for rationalizing and propagating this myth has been the economic role of the high technology industries in contemporary society and the intellectual attitudes of scientists who work in these industries. I believe that Nader has correctly identified the source of these attitudes as intellectual arrogance untempered by humanism, an unquestioning acceptance of the value and effectiveness of big aggregated business or scientific "from the top down" projects and contempt for disaggregated "from the bottom up" projects....
>
> The big is good and the bigger is better attitude seems to be in contradiction with the accepted scientific philosophy that Occams razor is used to arbitrate between alternate explanations or solutions to problems. It has always been my approach as both a physicist and engineer that the most reliable and safest engineering solutions are usually the simplest: complex solutions should always be anathema to a good pure or applied scientist.

A critical examination of unwarranted assumptions and cultural superstitions leads back to questions of practice particularly in the case of engineers working on "real" problems. One engineer expresses optimism:

> It is some dozen years ago since I first questioned the general energy growth assumption before audiences of students, engineers and the public. Naturally, I was ignored or laughed at but my general thesis that "the growing demand for energy reflects lack of ingenuity and imagination rather than a real need" is now listened to politely. Even engineering myths and taboos change with time.

Those who locate the problem of paradigm in bureaucracies seek solutions there also:

There are two fundamental answers to the political, economic, and environmental ills which beset our society. Both require "grass roots" thought, which can come from any intelligent and unbiased individual who will accept the discipline of the scientist rather than the discipline of the bureaucracy. They are "creative conservation,"... and "cooperative decentralization," which is the process by which bureaucracies become less bureaucratic, and by which those at the bottom assume a greater responsibility for creativity and improvement in the real quality of life.

There are pleas for obvious solutions:

Over the last ten years it has fascinated me that obvious solutions could provide such threats to the technical minds of this country. Even solar has its elite which are disturbed by simple systems. We should hope that those with whom Dr. Nader worked on CONAES continued to ignore simple alternative energy systems because the last thing we need is a gold plated, over engineered, extremely costly government designed system.... While Department of Energy experts were stating that alcohol as a fuel was just too expensive, farmers were putting up their own stills and making fuels at a cost much less than that which the DOE had estimated.... While those at the top argue, those at the bottom build.

Another writer observes that engineers in their pursuit of the dangerous, challenging, and costly have betrayed the ethos of engineering. The general theme for those commenting on solutions is one that calls for breadth, integrated or ecological thinking, what one author calls a category of solution in the true humanizing spirit of "Renaissance Men":

Realistic solutions demand a breadth of understanding and wisdom. The development of breeder reactors, enrichment processes, fusion reactors, solar power satellites, etc., may involve the solution of fascinating problems of great interest to many physicists but they should not forget that, fundamentally, they are only complex solutions to the energy problem that undoubtedly have simpler, holistic and environmentally and societally benign solutions.

Many scientists might disagree with "Barriers to Thinking New About Energy" and the responses to it, but few wrote about their disagreement. Among those who did write, only one defended standardized thinking as well as nuclear power, and unwittingly recalls Levy-Bruhl's (1910) contrast between the "savage" and the rational scientific worldview. The author, who does not identify his affiliation, put the argument this way:

The question is, whose bias was based on fact and reality and whose was based on myth and fear? More importantly, which bias was purposefully selected by consideration of scientific data involving technology and economics, and which bias was unrecognized. I submit that 1) when a person sees everyone else out of step it is time for them to reexamine their own position, and that 2) physical laws, their measurement and expression in numbers, and economics are valid constraints on human thought where engineering is concerned.... Among people who understand nuclear power and the present need for energy independence in this country it is obvious that there is no other viable option. No other energy source can equal nuclear in the areas of safety, economics, and acceptable environmental impact, regardless of the biases one might hold.... I submit that it is available technology, economics, and human need that have determined that the breeder reactor is the way we must ultimately go.... Historically it is people who work with people who cause almost all human suffering. Politicians, religious leaders, ideologists, propagandists, media personnel ... it is among these ranks that we find sowers of hatred, distrust, fear, war, terror, etc., that are the dealers in subjective judgement, metaphysics and opinion. They are not physical scientists or engineers, and they do not practice scientific rationality or objectivity. An overview of human history will reveal that those who work with numbers and objects almost without exception have improved the well-being of everyone.... It is entirely proper that in a serious discussion of energy production for today, solar should not be considered. Additionally, it would be ludicrous to not consider nuclear power: the one form of energy generation that economically, socially, environmentally and technically meets the needs of the people.

With rare exception the letters indicate general agreement about machismo, professional blindness, and number running. They overwhelmingly indicate that for myriad reasons, ranging from specialization and territoriality to work-place bureaucracy and entrenched position, contemporary science suffers from self-censorship as well as censorship by others. Both European and U.S. physicists identified with the group-think problem, or what one person described as the "mind cage," and many stated that they have observed or inferred much of what I wrote about energy ideologies. It was reassuring to have my comments validated, but worrisome that distinguished scientists felt unable to speak out or to alter their conditions, except offstage. Power lies in hegemonic science, less in individual scientists.

When Is Science Scientific?

Every specialized discipline tolerates a certain number of deviants although central values prevail. Among energy experts, central values are being scrutinized and challenged because energy demand predictions were wildly off base, as were the most sophisticated assessments of energy supply. Part of what is being challenged because of the perceived crisis is the hierarchy of particular technologies. More important perhaps is the challenge to the doing of science and the creation of policy research that relates to science. Scientists get tied to particular methods. Energy research has been torn between isolationist models and the science of totalities. It is as Young said (1972), "science is much more like the messy world of social and political intercourse than working scientists care to believe." European and EuroAmerican science and technologists are unable to separate "science" work from scientific result and apparently desire overpowers reason.

In an essay called "Knowledge for Survival," Paul Feyerabend (n.d.) reminds us that the problem of who is to judge the experts was addressed by Plato and Protagoras (and probably by those before them). Plato thought that experts must be judged by super-experts. Protagoras thought that experts should be judged by all, that people must be given the power to supervise experts. Feyerabend notes further that Plato argued that while experts are good in their own fields, they lack a sense of perspective as to how different fields mesh. Protagoras (in Plato's dialogue of the same name) argues that much knowledge is acquired not by special instruction but by human contact in a rather unstructured way. Protagoras, according to Feyerabend, believed that knowledge acquired in life experiences sufficed to judge the experts:

> Like the members of a jury ... they discover that experts are liable to exaggerate the importance of their work; that different experts may have different opinions on the same matter; that they are perhaps relatively informed in a narrow domain, but quite ignorant outside it; that they rarely admit this ignorance, often are not even aware of it but bridge it with high sounding language thus deceiving both themselves and others.

The many forms of knowledge, expert and lay, may be requisite for a future.

Those who judge experts need to understand that scientific research is subject to political, financial, and entrepreneurial constraints. Because scientists are forced to adapt to the climate in which they work, a "subjective" dimension enters the picture. For these and other reasons philosophers like Feyerabend argue that science has no special properties that might justify decision-making privileges, even though an important task of the scientist is to establish facts, laws, and theories. Decision making requires more than expert knowledge, and in this Feyerabend would include love, pity, compassion, humility, a sense of personal perspective, a sense of the "sacred," and I would also add wisdom. Full humanity, he argues, must become part of the production of knowledge.

While Malinowski challenged the dogma that primitive man was characterized by irrationality, Leach was correct to direct us toward examining the widely accepted belief that science and scientists are characterized by rationality. Malinowski found magic, science, and religion demarcated among the Trobrianders; among energy experts we find reason and desire intermingled.

Today most scientists work as employees, not in universities, but in large-scale government and industrial laboratories, and yet this fact alone cannot explain a good part of why conformity is the rule. As professionals, scientists cannot be an impartial source of knowledge in conflicts between a dispersed public and centralized large-scale organizations, but scientific workplaces are useful contexts for understanding how science culture operates to legitimate the vested interests of industries, utilities, banks, mining companies, and governments. The public has paid for scientific research, but has not required the scientific community to educate them; nor has the public educated scientists with their questions. We have achieved a high level of technological development, yet, most citizens are technological illiterates. We are not the Trobrianders, who as a society shared an understanding of risk and uncertainty. In the past, discussions of equity mostly centered on who had what piece of the pie, but the sheer force of technology has moved us to expand our views. We have developed technologies that are life shaping, that have the potential for irreversible destruction, and that do not recognize national borders.

Anthropologists have been interested in how cultures are born, how they develop, how long they are viable, and how they disintegrate (Yoffee and Cowgill 1988). We have documented examples of long and continuous cultural life spans. Australian aborigines and the Great Basin tribes in the Western part of the United States have continuous histories that run into thousands of years. For cultures more like our own Western culture, the durations run shorter spans. According to A. L. Kroeber (1944), complex cultures average from one thousand to fifteen-hundred years; he also notes that Western culture took form about 1050 or 1100 A.D. Anthropologists have recorded the sudden disappearance of civilizations such as that of the Maya of Central America. The reasons for the fall are still unknown, but we do know that their high priests were not able to save the civilization. The anthropologist needs to articulate the principles that allowed the so-called simpler societies to survive when civilizations did not.

It is a humbling experience for us to know that civilizations rise and fall, flourish and disappear, and to know also that the same happens with science (White 1979). Great scientific traditions, too, after periods of long productivity, decay. Decay in science sets in with repression, which is why distinguished scientists such as James Bryant Conant (1951) emphasize the "importance of absolutely untrammeled discussion and debate." The societies that have nourished the extraordinary rise of modern Western science in the last three centuries are those having to face problems for which science has as yet no answers; problems that stem from changes often directly traceable to science and technological achievements. The scientific attitude, which both pursues and criticizes the production of knowledge, is not confined to scientists; it may even be an attitude shared with nonscientists.

Notes

During the 1980s, earlier versions of this paper were presented by invitation at the ANZAAS meetings; University of Queensland, Australia; Wellesley College; Massachusetts Institute of Technology; San Francisco meetings of the American Physical Society; and the Gordon Conference.

1. For an example of a recent popular collection of ethnoscience see Jack Weatherford's book, *Indian Givers—How the Indians of the Americas Transformed the World* (1988).

Bibliography

Abir-Am, Pnina. 1992. A Historical Ethnography of a Scientific Anniversary in Molecular Biology: The First Protein X-ray Photograph (1984, 1934). *Social Epistemology* 6 (4):323–355.

Adams, Richard. 1975. *Energy and Structure.* Austin: University of Texas Press.

Adams, W. P. 1992. Commentary: Polar Science and Social Purpose. *Arctic* 45(4):iii–iv.

Adas, Michael. 1989. *Machines as the Measure of Men—Science, Technologies, and Ideologies of Western Dominance.* Ithaca, New York: Cornell University Press.

Agnew, Harold. 1981. Interview. *Los Alamos Science Magazine* (summer/fall).

Akuliaq. 1967. Letter in the Avataq Cultural Institute. Documentation Centre, Montreal.

Akunzemann, J. 1977. Isolation and Identification of Flavonol Glycosides in *Foeniculum vulgare.* Z. Lebens, -Unters Forsch 164:194–200.

Alayco, Salomonie. 1985. Interview at Akulivik. Avataq Cultural Institute, Oral History Project, Montreal.

Alkiewicz, J. 1983. Clinical Evaluation of an Aerosol Medicinal Plant Preparation. *Herba. Pol.* 29 (3–4):281–286.

Alkire, William H. 1970. Systems of Measurement in Woleai Atoll, Caroline Islands. *Anthropos* 65:1–73.

Allen, P. M., and J. M. McGlade. n. d. *Managing Complexity: A Fisheries Example.* Report of the Global Learning Division of the United Nations University. Cranford, Bedford, U.K.: International Ecotechnology Research Center.

Altieri, Miguel. 1983. *Agroecology: The Scientific Basis of Alternative Agriculture.* Boulder, Colorado: Westview Press.

Alvares, Claude. 1988. Science, Colonialism and Violence: Aluddite View. In *Science, Hegemony and Violence: A Requiem for Modernity*, edited by A. Nandy. Delhi, India: Oxford.

American Association for the Advancement of Science. 1993. *Benchmarks for Science Literacy.* New York: Oxford University Press.

Anderson, Eugene N. 1967. The Ethnoichthyology of the Hong Kong Boat People. Ph. D. diss., University of California at Berkeley.

Anderson, G. Christopher. 1989. NIH Genome Database Breaks New Ground. *The Scientist* (September 4):3–21.

Anderson, John R. 1990. *The Adaptive Character of Thought.* Hillsdale, New Jersey: Lawrence Erlbaum Associates.

Andrews, T. 1988. Selected Bibliography of Native Resource Management Systems and Native Knowledge of the Environment. In *Traditional Knowledge and Renewable Resource Management in Northern Regions*, edited by M. M. R. Freeman and L. N. Carbyn. Boreal Institute for Northern Studies Occasional Publication 23, Edmonton, Canada: University of Alberta.

Arditti, Rita, Pat Brennan, and Steve Cavrak, eds. 1980. *Science and Liberation*. Boston: South End Press.

Arias, J. 1991. *El Mundo Numinoso de los Mayas*. Serie Antropología. Tuxtla Gutierrez, Chiapas: Talleres Gráficos del Estado.

Asquith, Pamela J. 1986a. Anthropomorphism and the Japanese and Western Traditions in Primatology. In *Primate Ontogeny, Cognition and Social Behavior*, edited by J. G. Else and P. C. Lee. Cambridge, U. K. : Cambridge University Press.

———. 1986b The Monkey Memorial Service of Japanese Primatologists. In *Japanese Culture and Behaviour: Selected Readings*, edited by T. Sugiyama Lebra and W. P. Lebra. Honolulu, Hawaii: University of Hawaii Press.

———. 1994. The Intellectual History of Field Studies in Primatology, East and West. In *Strength in Diversity: A Reader in Physical Anthropology*, edited by L. K. Chan and A. Herring. Toronto, Canada: Canadian Scholars' Press.

Austin, James S., and Aaron David McVey. 1989. The Impact of the War on Drugs. *Focus*, San Francisco, CA: The 1989 National Council of Crime and Delinquency Prison Population Forecast (December):1–7.

Avadhoot, Y., and K. C. Varma. 1978. Antimicrobial Activity of Essential Oils of Seeds of *Lantana camara*. *Indian Drugs Pharmacological Index* 13:41–42.

Avirutnant, W. , and A. Pongpan. 1983. The Antimicrobial Activity of some Thai Flowers and Plants. *Journal of Pharmacological Sciences* 10 (3):81–86.

Ayoub, S. M. H. 1983. Composition with an Algicidal and a Mollusicidal Effect and Methods for Controlling Mollusks and Algae. *Patent-Brasil* 8108 (521):1–14.

———. 1984. Polyphenolic Molluxcicides from *Acacia nilotica*. *Planta Medica* 50:532–534.

———. 1985 Algicidal Properties of Tanins. *Fitoterapia* 56:227–229.

Barinaga, Marcia. 1990. Doing a Dirty Job—the Old-Fashioned Way. *Science* 249:356–357.

Barnes, Barry. 1972. On the Reception of Scientific Beliefs. In *Sociology of Science*, edited by Barry Barnes. Harmondsworth, U.K.: Penguin Books.

———. 1973. The Comparison of Belief-Systems: Anomaly Versus Falsehood. In *Modes of Thought: Essays on Thinking in Western and Non-Western Societies*, edited by R. Horton and R. Finnegan. London: Faber and Faber.

———. 1974. *Scientific Knowledge and Sociological Theory*. London: Routledge and Kegan Paul

———. 1977. *Interests and the Growth of Knowledge*. Boston: Routledge and Kegan Paul.

Bartholomew, James. 1989. *The Formation of Science in Japan*. New York: Yale University Press.

Bartlett, F. C. 1958. *Thinking: An Experimental and Social Study*. New York: Basic Books.

Bateson, Gregory. 1979. *Mind and Nature: A Necessary Unity*. Toronto, Canada: Bantam Books.

Beecher, C. W. W., and N. R. Farnsworth. n.d. *The Organization and Function of NAPRALERT.* Program for Collaborative Research in the Pharmaceutical Sciences. Chicago, Illinois: University of Chicago, College of Pharmacy.

Benedict, Ruth. 1934. *Patterns of Culture.* Boston: Houghton Mifflin.

Bennett, David. 1983. Some Aspects of Aboriginal and Non-Aboriginal Notions of Responsibility to Non-Human Animals. *Australian Aboriginal Studies* (2):19–24.

Berkeley Scientists and Engineers for Social and Political Action (SESPA). 1972. Interview with Charles Townes. *Science Against the People—The Story of Jason.* (Berkeley, California: SESPA).

Berkes, Fikret. 1977. Fishery Resource Use in a Subarctic Indian Community. *Human Ecology* 5(4):289–307.

———. 1988. Environmental Philosophy of the Chisasibi Cree People of James Bay. In *Traditional Knowledge and Renewable Resource Management in Northern Regions,* edited by M. M. R. Freeman and L. N. Carbyn. Boreal Institute for Northern Studies Occasional Publication 23. Edmonton: Canada: University of Alberta.

Berlin, Brent. 1973 The Relation of Folk Systematics to Biological Classification and Nomenclature. *Annual Review of Ecology and Systematics* 4:259–271.

———. 1974. *Principles of Tzeltzal Plant Classification: An Introduction to the Botanical Ethnography of a Mayan-Speaking People.* New York: Academic Press.

———. 1978. Ethnobiological Classification. In *Cognition and Categorization,* edited by E. Rosch and B. Lloyd. Hillsdale, New Jersey: Lawrence Erlbaum Associates.

———. 1992. *Ethnobiological Classification: Principles of Categorization of Plants and Animals in Traditional Societies.* Princeton, New Jersey: Princeton University Press.

Berlin, B., D. E. Breedlove, and P. Raven. 1974. *Principles of Tzeltal Plant Classification.* New York and San Francisco: Academic Press.

Bernal, J. D. 1939. *The Social Function of Science.* London: Routledge.

Bernard, H. R. 1988. *Research Methods in Cultural Anthropology.* Newbury Park, California: Sage Publications.

Bielawski, Ellen. 1984. Anthropological Observations on Science in the North: The Role of the Scientist in Human Development in the Northwest Territories. *Arctic* 37 (1):1–6.

———. 1992a. Cross-Cultural Epistemology: Cultural Readaptation Through the Pursuit of Knowledge. In *Looking to the Future,* edited by M.-J. Dufour and F. Therien. Inuit Studies Occasional Papers 4. Ste-Foy, Canada: Laval University.

———. 1992b. Indigenous Knowledge in Contemporary Inuit Culture. presented at First Nations: A Current Event, at the Institute for the Humanities at Salado, Salado, Texas, October 23–25, 1992.

Bielawski, Ellen, and B. Masuzumi. 1992. Dene Knowledge on Climate: Cooperative Research in Lutsel k'e, Northwest Territories. Paper presented at Biological Implications for Global Change: Northern Perspectives, at the Royal Society of Canada Global Change Program, October 23–25, 1992.

Bishop, Jerry E., and Michael Waldholz. 1990. *Genome.* New York: Simon and Schuster.

Bittles, A. H., and D. F. Roberts, eds. 1992. *Minority Populations: Genetics, Demography and Health.* London: Macmillan.

Black, M. 1962. Metaphor. In *Models and Metaphors,* edited by M. Black. Ithaca, New York: Cornell University Press.

Bloor, David. 1976. *Knowledge and Social Imagery.* Boston: Routledge and Kegan Paul.

Blum, Deborah. 1987. Nuclear Labs: Bulwark against Test Bans. *Sacramento Bee.* (August 2).

Bohlmann, F., and A. Suwita. 1977. Naturally Occurring Terpine Derivatives. 106. On Further Alpha-Longipenine Derivatives from the Compositeae. *Chem. Ber.* 110:3572–3581.

Borofsky, Robert. 1987. *Making History: Pukapukan and Anthropological Constructions of Knowledge.* Cambridge, U.K.: Cambridge University Press.

Bourdieu, Pierre. 1977. *Outline of a Theory of Practice.* New York: Cambridge University Press.

Brecht, Bertolt. [1938] 1947. *Galileo.* New York: Grove Press. 1966.

Brightman, Robert. 1993. *Grateful Prey: Rock Cree Human-Animal Relationships.* Berkeley: University of California Press.

Broad, William. 1985. *Star Warriors: A Penetrating Look into the Lives of the Young Scientists Behind Our Space Age Weaponry.* New York: Simon and Schuster.

———. 1992. *Teller's War.* New York: Simon and Schuster.

Brokensha, David L., D. M. Warren, Oswald Werner, et al. 1980. *Indigenous Knowledge Systems and Development.* Lanham, Maryland: University Press of America.

Brown, J. S., A. Collins, and P. Duguid. 1989. Situated Cognition and the Culture of Learning. *The Educational Researcher* 18 (1):32–42.

Brown, Laurie, M. Konuma, and Z. Maki. 1980. *Particle Physics in Japan, 1930–1950,* 3 vols. Kyoto, Japan: Research Institute for Fundamental Physics of Kyoto University.

Bruun, O., and A. Kalland, eds. 1995. *Asian Perceptions of Nature: A Critical Approach.* Surrey, U. K. : Curzon Press.

Bureau of the Census. 1991. *Statistical Brief* (February).

Busch, Lawrence, William Lacy, Jeffrey Burkhart, and Laura Lacy. 1991. *Plants, Power and Profit: Social, Economic, and Ethical Consequences of the New Biotechnologies.* Oxford, U.K.: Basil Blackwell.

Calderon, J. S., and L. Quijano. 1989. Prochamazulene Sesquiterpine Lactones from *Stevia serrata. Phytochemistry* 28 (12):3526–3527.

Callicott, J. Baird, and R. T. Ames. 1989. *Nature in Asian Traditions of Thought.* New York: SUNY.

Cambrosio, Alberto, and Peter Keating. 1988. Going Monoclonal: Art, Science, and Magic in the Day-to-Day Use of Hybridoma Technology. *Social Problems* 35:244–260.

Cantor, Charles R. 1990. Orchestrating the Human Genome Project. *Science* 248 (April 6):49–51.

Capra, Fritjof. 1975. *The Tao of Physics.* Berkeley, California: Shambhala Press.

Carpenter, C. R. 1934. A Field Study of the Behavior and Social Relations of Howling Monkeys (*Alouatta palliata*). *Comparative Psychology Monographs* 10 (2):1–168.

———. 1960. Comments. *Current Anthropology* 1 (5–6):402–403.

Carpenter, C. R., and A. Nishimura. 1969. The Takasakiyama Colony of Japanese Macaques (*Macaca fuscata*). *Social Organization and Ecology*, vol. 1. Basel, Germany: Karger.

Carraher, T., D. Carraher, and A. Schlieman. 1982. Na vida dez, na escola, zero: Os contextos culturais da aprendizagem da matimatica. *Caderna da pesquisa* 42:79–86.

Carraher, T., D. Carraher, and A. Schlieman. 1983. Mathematics in the Streets and Schools. Unpublished manuscript on file at Recife, Brazil: Universidade Federal de Pernambuco.

Carraher, T., and A. Schlieman. 1982. Computation Routines Prescribed by Schools: Help or Hindrance? Paper presented at NATO conference on the acquisition of symbolic skills in Keele, U.K.

Carrithers, Joe. 1993. Naturally Gay: Celebration and Concern Greet Genetic Study. *Frontiers* (July 30):15ff.

Casalino, Larry. 1991. Decoding the Human Genome Project: An Interview with Evelyn Fox Keller. *Socialist Review* 21 (April–June):111–128.

Chapman, Malcolm, ed. 1993. *Social and Biological Aspects of Ethnicity*. New York: Oxford University Press.

Charlesworth, Max, Lyndsay Farrall, Terry Stokes, and David Turnbull 1989. *Life Among the Scientists: An Anthropological Study of an Australian Scientific Community*. Melbourne, Australia: Oxford University Press.

Chatkaeomorakot, A., P. Echeverria, D. Taylor, et al. 1987. HELA Cell-Adherent Escheria Coli in Children with Diarrhea in Thailand. *Journal of Infectious Diseases* 156 (4):669–672.

Cizek, Petr. 1990. *The Beverly-Kaminuriak Caribou Management Board: A Case Study of Aboriginal Participation in Resource Management*. Background Paper No. 1. Ottawa, Canada: Canadian Arctic Resources Committee.

Cochran, Thomas, William Arkin, and Milton Hoenig. 1987. *Nuclear Weapons Databook, Volume II*. Cambridge, Massachusetts: Ballinger.

Coleman, James W. 1993. *The Criminal Elite: The Sociology of White Collar Crime*. 3d ed. New York: St. Martin's Press.

Colliere, W. A. 1949. The Antibiotic Actions of Plants, Especially the Higher Plants, with Results with Indonesian Plants. *Chronicle of Nature* 105:8–19.

Collins, Harry. 1985. *Changing Order: Replication and Induction in Scientific Practice*. Beverly Hills, California: Sage.

Colorado, Pam. 1988. Bridging Native and Western Science. *Convergence* 21 (2–3): 49–69.

Colson, Elizabeth. 1973. Tranquility for the Decision-Maker. In *Cultural Illness and Health: Essays in Human Adaptation*, edited by L. Nader and T. Maretzki. Anthropological Studies no. 9. Washington, D.C.

Colwell, Rita R. (ed) 1989. *Biomolecular Data: A Resource in Transition*. Oxford: Oxford University Press.

Conant, James B. 1951. A Skeptical Chemist Looks into the Crystal Ball. *Chemical and Engineering News* (Sept. 17, 1951): Vol 29: 3847–9.

Conklin, Harold. 1954. The Relation of Hanunoo Culture to the Plant World. Ph. D. diss. Yale University.

Cook-Degan, Robert M. 1991. The Human Genome Project: The Formation of Federal Policies in the United States, 1986–1990. In *Biomedical Politics*, edited by Kathi E. Hanna. Washington, D. C. : National Academy Press.

Corea, Gena. 1985. *The Mother Machine: Reproductive Technologies from Artificial Insemination to Artificial Wombs*. New York: Harper and Row.

Coven, Albert W. 1945. Evidence of Increased Radioactivity of the Atmosphere. After the Atomic Bomb Test in New Mexico. *Physical Review* 68:279.

Cromer, Alan. 1993. *Uncommon Sense: The Heretical Nature of Science*. New York: Oxford University Press.

Cruikshank, Julie. 1981. Legend and Landscape: Convergence of Oral and Scientific Traditions in the Yukon Territory. *Arctic Anthropology* 18 (2):67–93.

———. 1984. Oral Tradition and Scientific Research: Approaches to Knowledge in the North. In *Social Science in the North: Communicating Northern Values*. Occasional Publication no. 9. Ottawa, Canada: Association of Canadian Universities for Northern Studies.

Currie, Elliot. 1985. *Confronting Crime: An American Challenge*. New York: Pantheon.

Damm, H., and E. Sarfert. 1935. *Inseln um Truk: Puluwat, Hok, und Satawal*. Ergebnisse der Sudsee-Expedition 1908–1910, Series 2. B, vol. 6, pt. 2. Hamburg, Germany: Friedrichsen, De Gruyter.

Darnovsky, Marcy. 1991. Overhauling the Meaning Machines: An Interview with Donna Haraway. *Socialist Review* 21 (April–June):65–84.

Davis-Floyd, Robbie. 1992. *Birth as an American Rite of Passage*. Berkeley: University of California Press.

de Cotret, R. R. 1991. Letter to the Editor. *Arctic Circle* 1 (4):8.

de la Rocha, O. 1986. Problems of Sense and Problems of Scale: An Ethnographic Study of Arithmetic in Everyday Life. Ph.D. diss., University of California at Irvine.

de Oliveira, M. M., et al. 1972. Antitumor Activity of Condensed Flavonols. *Annais Acadêmia Brasileña da Ciencias* 44:41–51.

de Waal, F. B. M. 1982. *Chimpanzee Politics*. London: Jonathan Cape.

Denny, J. Peter. 1986. Cultural Ecology of Mathematics: Ojibway and Inuit Hunters. In *Native American Mathematics*, edited by Michael P. Closs. Austin: University of Texas Press.

DeWitt, Hugh. 1986. Labs Drive the Arms Race. In *Assessing the Nuclear Age*, edited by Len Ackland and Steven McGuire. Chicago, Illinois: Educational Foundation for Nuclear Science.

Dhar, M. 1968. Screening of Indian Plants for Biological Activity. *Indian Journal of Experimental Biology* 6:232–247.

Divine, Robert. 1978. *Blowing on the Wind: The Nuclear Test Ban Debate 1954–1960*. New York: Oxford University Press.

Domhoff, G. William. 1967. *Who Rules America?* Englewood Cliffs, New Jersey: Prentice-Hall.

Doolittle, Russell F. 1987. *Of Urfs and Orfs: A Primer on How to Analyze Derived Amino Acid Sequences*. Mill Valley, California: University Science Books.

Douglas, Mary. 1970. *Purity and Danger: An Analysis of Concepts of Pollution and Taboo.* Harmondsworth, U.K.: Tenquin Books.

Downey, Gary L., Joe Dumit, and Sharon Traweek, eds. In press. *Cyborgs and Citadels: Anthropological Interventions in the Borderlands of Technoscience.* Santa Fe, New Mexico: School of America Research Press.

Drell, Sidney, Charles Townes, and John Foster. 1991. How Safe Is Safe? *The Bulletin of the Atomic Scientists* 47 (3):35–40.

Dreyfus, H. L., and S. E. Dreyfus. 1986. *Mind Over Machine: The Power of Human Intuition and Expertise in the Era of the Computer.* New York: The Free Press.

Dubinskas, Frank, ed. 1988. *Making Time: Ethnographies of High-Technology Organizations.* Philadelphia, Pennsylvania: Temple University Press.

Dunbaugh, Frank M. 1979. Racially Disproportionate Rates of Incarceration in the United States. *Prison Law Monitor* 1 (9):

Durkheim, Emile. 1954. *The Elementary Forms of the Religious Life.* New York: The Free Press.

Duster, Troy. 1990. *Backdoor to Eugenics.* Boston: Routledge and Kegan Paul.

———. 1992. Genetics, Race and Crime: Recurring Seduction to a False Precision. In *DNA and Crime: Applications of Molecular Biology in Forensics,* edited by Paul Billings. Cold Spring Harbor, New York: Cold Spring Harbor Laboratory Press.

———. 1995. Post-Industrialism and Youth Unemployment. In *Poverty, Inequality and the Future of Social Policy: Western States in the New World Order,* edited by Katherine McFate, Roger Lawson, and William Julius Wilson. New York: Russell Sage.

Easlea, Brian. 1983. *Fathering the Unthinkable: Masculinity, Scientists, and the Arms Race.* London: Pluto Press.

Eberhart, Sylvia. 1947. How the American People Feel about the Bomb. *Bulletin of the Atomic Scientists* 3:6:146–149, 168.

Elizabetsky, E. 1986. New Directions in Ethnopharmacology. *Journal of Ethnobiology* 6:121–128.

Ericsson, C. D., T. F. Patterson, and H. L. Dupont. 1987. Clinical Presentation as a Guide to Therapy for Travelers' Diarrhea. *American Journal of Medical Sciences* 294 (2):91–96.

Escobar, Arturo. 1995. *Encountering Development: The Making and Unmaking of the Third World.* Princeton, New Jersey: Princeton University Press.

Etkin, Nina L. 1994. *Eating on the Wild Side: The Pharmacologic, Ecologic, and Social Implications of Using Noncultigens.* Tucson: University of Arizona Press.

Evans-Pritchard, E. E. 1937. *Witchcraft, Oracles and Magic among the Azande.* Oxford, U.K.: Oxford University Press.

Fa, J. E., and C. H. Southwick, eds. 1988. *Ecology and Behavior of Food-Enhanced Primate Groups.* New York: Alan R. Liss, Inc.

Fabrega, H. J. 1970a. Dynamics of Medical Practice in a Folk Community. *Milbank Memorial Fund Quarterly* 48 (4):391–412.

———. 1970b. On the Specificity of Folk Illnesses. *Southwestern Journal of Anthropology* 26:305–314.

Fabrega, H. J., and D. B. Silver. 1970. Some Social and Psychological Properties of Zinacanteco Shamans. *Behavioral Science* 15:471–486.

———. 1973. *Illness and Shamanistic Curing in Zinacantán*. Stanford, California: Stanford University Press.

Farnsworth, N. R., and J. M. Pezzuto. 1983. *Rational Approaches to Development of Plant Derived Drugs*. Paper presented at Segundo Simpósio de Productos Naturais. Joao Pessoa, Brazil.

Fausto-Sterling, Anne. 1985. *Myths of Gender: Biological Theories about Women and Men*. New York: Basic Books.

Fearson, R. E., A. Wendell Engle, Jean Thayer, Gilbert Swift, and Irving Johnson. 1946. Results of Atmospheric Analyses Done at Tulsa, Oklahoma, During the Period Neighboring the Time of the Second Bikini Bomb Test. *Physical Review* 70:564.

Fedigan, L. M., and P. J. Asquith. 1992. Arashiyama Research as a Microcosm of Larger Trends in Primatology. In *Current Primatology II. Behavior, Ecology and Conservation*, edited by N. Itoigawa, et al. Tokyo: University of Tokyo Press.

Fee, Russ. 1990. NMFS Director Fox Is Criticised as Running a One-Man Shop. *National Fisherman*. October 4(15).

Feit, H. 1973. The Ethno-Ecology of the Waswanipi Cree: Or How Hunters Can Manage Their Resources. In *Cultural Ecology: Readings on Canadian Indians and Eskimos*, edited by Bruce Cox. Toronto, Canada: McClelland and Stewart.

———. 1978. Waswanipi Realities and Adaptations: Resource Management among Subarctic Hunters. Ph. D. diss. McGill University.

———. 1988 Self-Management and State-Management: Forms of Knowing and Managing Northern Wildlife. In *Traditional Knowledge and Renewable Resource Management in Northern Regions*, edited by M. M. R. Freeman and L. N. Carbyn. Boreal Institute for Northern Studies Occasional Publication 23. Edmonton, Canada: University of Alberta.

Feng, P. C., and L. J. Haynes. 1962. Pharmacological Screening of Some West Indian Medicinal Plants. *Journal of Pharmacy and Pharmacology* 14:556–561.

Fetter, Steve. 1987–1988. Stockpile Confidence under a Nuclear Test Ban. *International Security* 12 (3):132–167.

———. 1988. *Toward a Comprehensive Test Ban*. Cambridge, Massachusetts: Ballinger.

Feyerabend, Paul. 1991. *Knowledge for Survival*. Unpublished manuscript.

———. 1978. *Science in a Free Society*. London: NLB Press.

Fienup-Riordan, Ann. 1990a. *A Problem of Differentiation: Boundaries and Passages in Eskimo Ideology and Action*. Paper presented at the Seventh Inuit Studies Conference, Fairbanks, Alaska, August, 1990.

———. 1990b. *Eskimo Essays*. New Brunswick, New Jersey: Rutgers University Press, New Brunswick.

Finney, Ben R. 1979. *Hokule'a: The Way to Tahiti*. New York: Dodd, Mead.

Firth, R., ed. 1957. *Man and Culture: An Evaluation of the Work of Malinowski*. London: Routledge & Kegan Paul.

Fischer, Margit. 1971. Psychosis in the Offspring of Schizophrenic Monozygotic Twins and Their Normal Co-Twins. *British Journal of Psychiatry* 118:43–52.

Fleck, Ludwik. [1937] 1979. *Genesis and Development of a Scientific Fact.* Chicago, Illinois: University of Chicago Press.

Fleming, M. M. 1992. Reindeer Management in Canada's Belcher Islands: Documenting and Using Traditional Environmental Knowledge. In *Lore: Capturing Traditional Environmental Knowledge,* edited by Martha Johnson. Dene Cultural Institute, Hay River. Ottawa, Canada: International Development Research Centre.

Fortun, Michael. 1991. Mapping Genes and Making History: Molecular Biologists Recast Human Genetics in the 1980s. Paper presented at the Joint Atlantic Seminar for the History of Biology, Princeton, New Jersey, April 5–6.

Foster, G. M., and B. G. Anderson. 1978. *Medical Anthropology.* New York: John Wiley and Sons.

Foucault, Michel. 1980a. *Power/Knowledge: Selected Interviews and Other Writings 1972–1977.* New York: Pantheon Books.

———. 1980b. *The History of Sexuality, Volume I: An Introduction.* New York: Vintage Books.

Frake, Charles O. 1994. Dials: A Study in Physical Representations of Cognitive Systems. In *The Ancient Minds: Elements of Cognitive Archaeology,* edited by Colin Renfrew and Ezra Zubrow. Cambridge, U.K.: Cambridge University Press.

Frazer, J. G. 1911–1915. *The Golden Bough: A Study in Magic and Religion.* London: Macmillan.

Freeman, M. M. R., and L. N. Carbyn, editors. 1988. *Traditional Knowledge and Renewable Resource Management in Northern Regions.* Boreal Institute for Northern Studies Occasional Publication Number 23. Edmonton, Canada: University of Alberta.

Frisch, J. E. 1959. Research on Primate Behavior in Japan. *American Anthropologist* 61 (4):584–596.

———. 1963. Japan's Contribution to Modern Anthropology. In *Studies in Japanese Culture.* edited by J. Roggendorf. Tokyo: Sophia University Press.

Fujimura, Joan H. 1987. Constructing Doable Problems in Cancer Research: Articulating Alignment. *Social Studies of Science* 17 (May):257–293.

Gallego, F., A. Swiatpolk-Mirski, and E. Vallejo. 1965. Contribución al Estudio del *Chenopodium botrys* L. en Relación con su Aceite Essencial. *Farmacognosia* 25:69–89.

Gamble, Clive. 1994. *Timewalkers: The Prehistory of Global Colonization.* Cambridge, Massachusetts: Harvard University Press.

Gault, Robert H. 1932. *Criminology.* New York: Heath.

Gay, J., and M. Cole. 1967. *The New Mathematics and an Old Culture.* New York: Holt, Rinehart, and Winston.

Geertz, Clifford. 1973. Deep Play: Notes on the Balinese Cockfight. In *The Interpretation of Cultures,* edited by Clifford Geertz. New York: Basic Books.

———. 1983. Local Knowledge: Fact and Law in Comparative Prespective. In *Local Knowledge: Further Essays in Interpretive Anthropology,* edited by Clifford Geertz. New York: Basic Books.

Gellner, Ernest. 1985. *Relativism and the Social Sciences*. Cambridge, U.K.: Cambridge University Press.

George, M., and K. M. Pandalai. 1949. Investigations on Plant Antibiotics. *Indian Journal of Medical Research* 37:169–181.

Ghodsi, M. B. 1976. Flavonoids of *Foeniculum vulgare. Maj-Daneshagah Darusazi* (2):10–14.

Gieryn, Thomas I. 1983. Boundary-Work and the Demarcation of Science from Non-Science: Strains and Interests in Professional Ideologies of Scientists. *American Sociological Review* 48:781–96.

———.1995. *Boundaries of Science: Handbook of Science and Technology Studies*. Thousand Oaks, California: Sage Publications.

Gilbert, Walter. 1990. Molecular Biology Is Dead—Long Live Molecular Biology: A Paradigm Shift in Biology. Unpublished manuscript. (Early version of *Nature* 1991.)

———. 1991. Towards a Paradigm Shift in Biology. *Nature* 349 (January 10):99.

Gilmore, H. A. 1981. From Radcliffe-Brown to Sociobiology: Some Aspects of the Rise of Primatology within Physical Anthropology. *American Journal of Physical Anthropology* 5:387–392.

Gladwin, Thomas. 1970. *East Is a Big Bird: Navigation and Logic on Puluwat Atoll*. Cambridge, Massachusetts: Harvard University Press.

Gleick, James. 1987. *Chaos: Making a New Science*. New York: Viking Penguin.

Gluckman, Max. 1954. *Rituals of Rebellion in South-East Africa*. Manchester, U.K.: Manchester University Press.

Gonzalez, R., L. Nader, and J. Ou. 1995. Between Two Poles—Bronislaw Malinowski, Ludwik Fleck, and the Anthropology of Science. *Current Anthropology*. December.

Goodall, Jane. 1986. *The Chimpanzees of Gombe: Patterns of Behavior*. Cambridge, Massachusetts: Belknap Press.

———. 1990. *Through a Window: My Thirty Years with the Chimpanzees of Gombe*. Boston: Houghton Mifflin.

Goodenough, Ward H. 1953. *Native Astronomy in the Central Carolines*. Philadelphia, Pennsylvania: University Museum, University of Pennsylvania.

Goodenough, Ward H., and Stephen D. Thomas. 1987. Traditional Navigation in the Western Pacific. *Expedition* 29 (3):3–14.

Goody, J. 1977. *The Domestication of the Savage Mind*. Cambridge, U.K.: Cambridge University Press.

Gould, Stephen Jay. 1981. *The Mismeasure of Man*. New York: W. W. Norton & Co.

Government of Canada. House of Commons. 1990. The Standing Committee on Aboriginal Affairs. *Minutes of Proceedings and Evidence*, March 1990.

Government of the Northwest Territories. Department of Culture and Communications. 1991. *Report of the Traditional Knowledge Working Group*.

Graham, David. 1990. When U.S. Has a Science Question, It Asks JASON. *San Diego Union* (July 15): B1, B4.

Gross, Paul R., and Norman Levitt. 1994. *Higher Superstition: The Academic Left and Its Quarrels with Science.* Baltimore, Maryland: Johns Hopkins University Press.

Guerrant, R. L. 1983. Pathophysiology of the Enterotoxic and Viral Diarrheas. In *Diarrhea and Malnutrition,* edited by L. C. Chen and N. S. Scrimshaw. New York: Plenum Press.

Guerrant, R. L., J. A. Lohr, and E. K. Williams. 1986. Acute Infectious Diarrheal Epidemiology, Etiology and Pathogenesis. *Pediatric Infectious Diseases* 5:353–359.

Guiteras-Holmes, C. 1961. *Perils of the Soul.* New York: The Free Press of Glencoe, Inc.

Gundel, Max. 1926. Einige Beobachtungen bei der rassenbiologischen Durchforschung Schleswig-Holsteins. *Klinische Wochenschrift* 5:1186.

Gusterson, Hugh. 1996. *Nuclear Rites: An Anthropologist Among Weapons Scientists.* Berkeley: University of California Press.

Gutmann, M. 1992. Cross-Cultural Conceits—Science in China and the West. *Science as Culture* 3(2):208–239.

Haberer, Joseph. 1969. *Politics and the Community of Science.* New York: Van Nostrand Reinhold.

Hagos, M., et al. 1987. Isolation of Smooth Muscle Relaxing 1,3, Diarylpropan-2-ol Derivatives from *Acacia tortilis. Planta Medica* 1:21–31.

Hall, K. R. L., and I. DeVore. 1965. Baboon Social Behavior. In *Primate Behavior. Field Studies of Monkeys and Apes,* edited by I. DeVore. New York: Holt, Rinehart, and Winston.

Haller, Mark H. 1963. *Eugenics: Hereditarian Attitudes in American Thought.* New Brunswick, New Jersey: Rutgers University Press.

Hallpike, C. R. 1979. *The Foundations of Primitive Thought.* Oxford: Clarendon Press.

Hamer, Dean, S. Hu, V. L. Magnuson, N. Hu, A. M. L. Pattatucci. 1993. A Linkage Between DNA Markers on the X Chromosome and Male Sexual Orientation. *Science* 261 (July 16):321–327.

Hammer, R. H., and J. R. Cole. 1965. Phytochemical Investigation of *Acacia angustissima. Journal of Pharmacological Science* 54:235–237.

Handlin, Oscar. 1972. Ambivalence in the Popular Response to Science. In *Sociology of Science,* edited by Barry Barnes. Harmondsworth, U.K.: Penguin Books.

Haraway, D. 1989. *Primate Visions: Gender, Race, and Nature in the World of Modern Science.* New York: Routledge.

———. 1990. A Manifesto for Cyborgs: Science, Technology, and Socialist Feminism in the 1980s. In *Feminism/Postmodernism,* edited by Linda J. Nicholson. New York: Routledge. First published in *Socialist Review* 80 (1985).

———. 1994. A Game of Cat's Cradle: Science Studies, Feminist Theory, Cultural Studies. *Configurations* 1:59–71.

Harborne, J. B. 1984. Use of High-Performance Liquid Chromatography in the Separation of Flavonol Glycosides and Flavonol Sulphates. *Journal of Chromatography* 299 (2):377–385.

Harding, Sandra. 1994. Is Science Multicultural? Challenges, Resources, Opportunities, Uncertainties. *Configurations* 2:301–330. vol. 2. Baltimore: Johns Hopkins University Press and the Society for Literature and Science.

Harman, R. C. 1974. *Cambios Medicos y Sociales en una Comunidad Maya-Tzeltal.* Mexico D. F.: Instituto Nacional Indigenista.

Harootunian, H.D. 1988. *Things Seen and Unseen: Discourse and Ideology in Tokugawa Nativism.* Chicago, Illinois: University of Chicago Press.

Harris-Warrick, Ronald M., and Eve Marder. 1991. Modulation of Neural Networks for Behavior. *Annual Review of Neuroscience* 14:39–57.

Harwood, Jonathan. 1993. *Styles of Scientific Thought: The German Genetics Community, 1900–1933.* Chicago, Illinois: University of Chicago Press.

Hass, M. 1986. *Cognition-in-Context: The Social Nature of the Transformation of Mathematical Knowledge in a Third Grade Classroom.* Irvine, California: School of Social Sciences, University of California at Irvine.

Hayakawa, Satio, and Morris F. Low. 1991. Science Policy and Politics in Post-War Japan: The Establishment of the KEK High Energy Physics Laboratory. *Annals of Science* 48:207–229.

Hazen, Robert M., and James Trefil. 1991a. *Science Matters: Achieving Scientific Literacy.* New York: Doubleday.

———. 1991b. Quick? What's a Quark? *The New York Times Magazine* (January 1):26.

Herken, Gregg. 1982. *The Winning Weapon: The Atomic Bomb in the Cold War, 1945–1950.* New York: Vintage Books.

Herrnstein, Richard J., and Charles Murray. 1994. *The Bell Curve: Intelligence and Class Structure in American Life.* New York: The Free Press.

Hess, David, and Linda Layne, eds. 1992. *Knowledge and Society: The Anthropology of Science and Technology.* Greenwich, Connecticut: JAI Press, Inc.

Hesse, Mary B. 1961. *Forces and Fields: The Concept of Action at a Distance in the History of Physics.* London: Thomas Nelson and Sons.

———. 1980. *Revolutions and Reconstructions in the Philosophy of Science.* Bloomington: Indiana University Press.

Hessen, Boris. 1971. *The Social and Economic Roots of Newton's Principia.* New York: Howard Fertig.

Hoddeson, Lillian. 1983. Establishing KEK in Japan and Fermilab in the United States: Internationalism, Nationalism, and High Energy Accelerators. *Social Studies Science* (April): Vol 13 pp. 1–48.

Holbrook, Bruce. 1981. *The Stone Monkey: An Alternative, Chinese-Scientific, Reality.* New York: William Morrow.

Holland, W. R. 1963. *Medicina Maya en Los Altos de Chiapas.* Mexico City: Instituto Nacional Indigenista.

Holland, W. R., and R. G. Tharp. 1964. Highland Maya Psychotherapy. *American Anthropologist* 66:41–60.

Holtzman, Neil A. 1989. *Proceed with Caution: Predicting Genetic Risks in the Recombinant DNA Era.* Baltimore, Maryland: Johns Hopkins University Press.

Homans, George. 1941. Anxiety and Ritual: The Theories of Malinowski and Radcliffe-Brown. *American Anthropologist* 43:164–172.

Horton, Robin. 1967. African Traditional Thought and Western Science. *Africa* 37:50–71, 155–187.

Hubbard, Ruth. 1990. *The Politics of Women's Biology.* New Brunswick, New Jersey: Rutgers University Press.

Hunt, S. E., and P. J. McCosker. 1970. Serum Adenosine Diaminase Activity in Experimentally Produced Liver Diseases of Cattle and Sheep. *British Veterinary Journal* 126:74–81.

Huq, M., A. Rahman, A. Al-Sadiq, et al. 1987. Rotavirus as an Important Cause of Diarrhoea in a Hospital for Children in Dammam, Saudi Arabia. *Annals of Tropical Pediatrics* 7:173–177.

Hutchins, E. 1992. Learning to Navigate. In *Understanding Practice,* edited by S. Chaiklin and J. Lave. New York: Cambridge University Press.

Hyndman, David. 1979. Wopkaimin Subsistence: Cultural Ecology in the New Guinea Highland Fringe. Ph. D. diss., University of California at San Diego.

———. 1994. *Ancestral Rain Forests and the Mountain of Gold: Indigenous Peoples and Mining in New Guinea.* Boulder, Colorado: Westview Press.

Ikawa-Smith, Fumiko. 1990. Protohistoric Yamato-Archaeology of the First Japanese State. *Journal of Asian Studies* 49 (1):151–152.

Iliev, S., and G. Papanov. 1983. Antibiotic Activity of Terpenoid Compounds. *Nauchni Tr. Plovdiski University* 21 (3):105–112.

Imanishi, K. 1957. Conservation of Japanese Monkeys. Proceedings and Papers of Sixth Technical Meeting, International Union for the Conservation of Nature and Natural Resources.

———. 1960. Social Organization of Subhuman Primates in Their Natural Habitat. *Current Anthropology* 1 (5–6):393–407.

Imanishi, K., and S. A Altmann, eds. 1965. *Japanese Monkeys: A Collection of Translations.* Edmonton, Canada: S. A. Altmann.

Immele, John, and Paul Brown. 1988. Correspondence. *International Security* 13 (1):196–210.

Itani, J. 1959. Paternal Care in the Wild Japanese Monkey, Macaca fuscata. *Primates* 2:61–93.

———. 1961. The Society of Japanese Monkeys. *Japan Quarterly* 8 (4):421–430.

———. 1963. Vocal Communication of the Wild Japanese Monkey. *Primates* 4 (2):11–66.

———. 1967. Editorial. *Primates* 8 (1):89–90; (2):187–188.

Iyaituk, Markusi. 1985. Interview at Ivujivik. Avataq Cultural Institute, Oral History Project, Montreal.

Jarvie, Ian. 1986. *Thinking About Society.* Boston Studies in the Philosophy of Science, vol. 93. Boston: D. Reidel Publishing Company.

Jasanoff, Shiela, G. E. Markle, J. C. Petersen, and T. Pinch. 1995. *Handbook of Science and Technology Studies.* Thousand Oaks, California: Sage Publications.

Jensen, Arthur R. 1969. How Much Can We Boost IQ and Scholastic Achievement? *Harvard Educational Review* (Winter) vol. 39:1, 123.

Johannes, Robert E. 1981. *Words of the Lagoon—Fishing and Marine Lore in the Palau District of Micronesia.* Berkeley, California: University of California Press.

———. 1989. Fishing and Traditional Knowledge. In *Traditional Ecological Knowledge: A Collection of Essays,* edited by R. E. Johannes. Gland, Switzerland: International Union for the Conservation of Nature.

Johnson, Martha. 1987. Inuit Folk Ornithology in the Povungnituk Region of Northern Quebec. M.A. thesis, University of Toronto.

———, ed. 1992. *Lore: Capturing Traditional Environmental Knowledge.* Dene Cultural Institute, Hay River. Ottawa, Canada: International Development Research Centre.

Jordan, Kathleen, and Michael Lynch. 1992. The Sociology of a Genetic Engineering Technique: Ritual and Rationality in the Performance of the Plasmid Prep. In *The Right Tools for the Job: At Work in Twentieth Century Life Sciences,* edited by A. E. Clarke and J. H. Fujimura. Princeton: Princeton University Press.

Kaneseki, Yoshinori. 1974. The Elementary Particle Theory Group. In *Science and Society in Modern Japan: Selected Historical Sources,* edited by Shigeru Nakayama, David L. Swain, and Eri Yagi. Boston: Massachusetts Institute of Technology.

Katz, S. H. 1995. Is Race a Legitimate Concept for Science? In *The AAPA Revised Statement on Race: A Brief Analysis and Commentary.* Philadelphia: University of Pennsylvania.

Kaufman, Felix. [1944] 1958. *Methodology of the Social Sciences* New York: Humanities Press.

Kaufmann, J. H. 1967. Social Relations of Adult Males in a Free-Ranging Band of Rhesus Monkeys. In *Social Communication among Primates,* edited by S. A. Altmann. Chicago, Illinois: University of Chicago Press.

Kawai, M. 1958. On the System of Social Ranks in a Natural Troop of Japanese Monkeys. *Primates* 1 (2):111–148. Republished in *Japanese Monkeys: A Collection of Translations,* edited by K. Imanishi and S. A. Altmann. Edmonton, Canada: S. A. Altman.

———. 1969. *Nihonzaru no seitai* (Life of Japanese Monkeys). Tokyo: Kawade-shoho-shinsha (J).

Kawamura, S. 1958. Matriarchal Social Ranks in the Minoo-B Troop: A Study of the Rank System of Japanese Monkeys. *Primates* 1:149–156.

———. 1959. The Process of Sub-Cultural Propagation among Japanese Macaques. *Primates* 2:43–60.

Keller, Evelyn Fox. 1982. *A Feeling for the Organism: The Life and Work of Barbara McClintock.* New York: W. H. Freeman.

———. 1991. Genetics, Reductionism, and the Normative Uses of Biological Information. Paper presented at the conference on "Genes-R-Us, But Who Is That?" University of California Humanities Research Institute, Irvine, California, May 2–3, 1991.

Kety, Seymour S. 1976. Genetic Aspects of Schizophrenia. *Psychiatric Annals* 6:6, 15.

Kety, S., D. Rosenthal, P. Wender, and F. Schulsinger. 1968. An Epidemiological-Clinical Twin Study on Schizophrenia. In *The Transmission of Schizophrenia,* edited by D. Rosenthal and S. Kety. Oxford: Pergamon Press.

Keusch, G. T. 1983. The Epidemiology and Pathology of Invasive Bacterial Diarrheas with a Note on Biological Considerations in Control Strategies. In *Diarrhea and Malnutrition*, edited by L. C. Chen and N. S. Scrimshaw. New York: Plenum Press.

Kevles, Daniel J. 1985. *In the Name of Eugenics: Genetics and the Uses of Human Heredity*, New York: Alfred A. Knopf.

Kevles, Daniel J., and Leroy Hood. 1992. *The Code of Codes: Scientific and Social Issues in the Human Genome Project*. Cambridge, Massachusetts: Harvard University Press.

Khadem, H. E., and Y. S. Mohammed. 1958. Constituents of the Leaves of *Psidium guaiava* L. Part II. Quercetin, Evicularin, and Guajaverin. *Journal of the Chemical Society* 60:3320–3323.

Khan, M. R. 1978. Studies on the Rationale of African Traditional Medicine. *Pakistan Journal of Scientific Industrial Research* 27:189–192.

Kidder, Ray. 1987. Maintaining the U. S. Stockpile of Nuclear Weapons During a Low-Threshold or Comprehensive Test Ban. Document UCRL-53820, Lawrence Livermore National Laboratory.

Kikuchi, M. 1981. Creativity and Ways of Thinking: The Japanese Style. *Physics Today* 34:42–51.

Klag, Michael, P. K. Whelton, J. Coresh, C. E. Grim, and L. H. Kuller. 1991. The Association of Skin Color with Blood Pressure in US Blacks with Low Socioeconomic Status. *Journal of the American Medical Association* 265 (5):599–602.

Knorr-Cetina, Karin. 1981. *The Manufacture of Knowledge: An Essay on the Constructivist and Contextual Nature of Science*. Oxford: Pergamon Press.

Knorr-Cetina, Karin, and Michael Mulkay, eds. 1983. *Science Observed*. Beverly Hills, California: Sage Publications.

Knox, Richard. 1991. New Study of Twins Finds Genetic Basis for Homosexuality. *The Boston Globe* (December 5):20.

Koford, P. 1963. Group Relations in an Island Colony of Rhesus Monkeys. In *Primate Social Behavior*, edited by C. H. Southwick. Princeton: von Nostrand.

Kolata, Gina. 1993. Cystic Fibrosis Surprise: Genetic Screening Falters. *The New York Times* (November 16):C1.

Koshland, D. E., Jr. 1991. War and Science. *Science* 251 (4993):497.

Kringlen, Einar. 1968. An Epidemiological Twin Study of Schizophrenia. In *The Transmission of Schizophrenia*, edited by D. Rosenthal and S. Key. Oxford, U.K.: Pergamon Press.

Kroeber, A. L. 1944. *Configurations of Cultural Growth*. Berkeley: University of California Press.

Kuhn, T. S. 1970. *The Structure of Scientific Revolutions*. 2nd ed. Chicago, Illinois: University of Chicago Press.

Kupchan, S. M., A. C. Patel, and E. Fujita. 1965. Tumor Inhibitors VI Cissampereine, New Cytotoxic Alkaloid from *Cissampelos pareira*. *Journal of Pharmacological Sciences* 54:580–582.

Kuroda, S. 1987. *Rigakuteki hôhô toshite no (tsutawaru). Saru to no tsukiai o toshite* (A Scientific Methodology—Through the Fellowship with Monkeys). *Nurse Station* 17(2):31–35.

Kwagley, A. Oscar. 1995. *Yupiaq Worldview—A Pathway to Ecology and Spirit.* Prospect Heights, Illinois: Waveland Press.

Lancaster, J. 1973. In Praise of the Achieving Female Monkey. *Psychology Today,* September:30ff.

Landers, Peter. 1991. Apathy, Court Vacillation Buttress Vote Disparity and Small Parties Fight to Block Reform. *Japan Times Weekly International Edition* 31 (35):1, 6–7.

Langmuir, Irving. 1960–1962. *The Collected Works of Irving Langmuir.* New York: Pergamon Press.

Latour, Bruno. 1987. *Science in Action: How to Follow Scientists and Engineers through Society.* Cambridge, Massachusetts: Harvard University Press.

Latour, Bruno, and S. Woolgar. 1979. *Laboratory Life: The Social Construction of Scientific Facts.* Beverly Hills, California: Sage.

Lave, Jean. 1989. The Culture of Acquisition and the Practice of Understanding. in *Cultural Psychology: Essays on Comparative Human Development,* edited by J. W. Stigler, R. A. Shweder, and G. Herdt. Chicago, Illinois: University of Chicago Press.

———. 1988. *Cognition in Practice: Mind, Mathematics, and Culture in Everyday Life.* Cambridge, U.K. Cambridge University Press.

Le Grande, H. E. 1988. *Drifting Continents and Shifting Theories.* Cambridge, U.K.: Cambridge University Press.

Leach, Edmund R. 1957. The Epistemological Background to Malinowski's Empiricism. In *Man and Culture,* edited by R. Firth. London: Routledge and Kegan Paul.

Leggett, Anthony J. 1966. Notes on the Writing of Scientific English for Japanese Physicists. *Nippon Butsuri Gakkaishi* 21:790–805.

Leifertova, I. 1979. The Antifungal Properties of Higher Plants Affecting some Species of the Genus *Asgergillus. Folia Pharmacologia* 2:29–54.

Levi-Strauss, Claude. 1966. *The Savage Mind.* Chicago, Illinois: University of Chicago Press.

———. 1973. Structuralism and Ecology. *Social Science Information* 12 (1):7–23.

Levy-Bruhl, Lucien. [1910] 1979. *How Natives Think.* New York: Arno Press.

Lewis, David. 1973. *We, the Navigators.* Honolulu: University Press of Hawaii.

Lifton, Robert Jay. 1982. Imagining the Real. In *Indefensible Weapons: The Political and Psychological case Against Nuclearism,* edited by Robert Lifton and Richard Falk. New York: Basic Books/Harper Colophone.

Lippman, Abby. 1991. Prenatal Genetic Testing and Screening: Constructing Need and Reinforcing Inequities. *American Journal of Law and Medicine* 17:15–50.

Lloyd, G. E. R. 1987. *The Revolutions of Wisdom: Studies in the Claims and Practice of Ancient Greek Science.* Berkeley, California: University of California Press.

Lock, Margaret. 1993. *Encounters with Aging—Mythologies of Menopause in Japan and North America.* Berkeley and Los Angeles: University of California Press.

Lopez, Barry. 1986. *Arctic Dreams: Imagination and Desire in a Northern Landscape.* Toronto, Canada: Bantam Books.

Lorenz, Edward. 1979. Predictability: Does the Flap of a Butterfly's Wings in Brazil Set Off a Tornado in Texas? Paper presented at the annual meeting of the American Association for the Advancement of Science, Washington, D. C.

Lorenz, Konrad. 1979. *The Year of the Greylag Goose.* New York: Harcourt Brace Jovanovich.

Löwry, I. 1988. Ludwick Fleck on the Social Construction of Medical Knowledge. *Sociology of Health and Illness* 10 (2):133–155.

Lozoya, X., G. Becerril, and M. Martínez. 1990. Intraluminal Perfusion Model of In Vitro Guinea Pig Ileum as a Model of Study of the Antidiarreic Properties of the Guava (*Psidium Guajava*). *Archivos de Investigaciones Médicas* 21:155–162.

Lozoya, X., M. Meckes-Lozoya, M. Abu-Zaid, J. Tortoriello, C. Nozzolillo, and J. T. Arnason. 1992. Identification of Quercetin Glycosides in Extracts of *Psidium guajava.* Determination of the Active Components with Spasmolytic Activity. *Archives of Medical Research* 21:155–162.

Ludmerer, Kenneth M. 1972. *Genetics and American Society.* Baltimore and London: John Hopkins University Press.

Lutterdot, G. D. 1989. Inhibition of Gastrointestinal Release and Acetylcholine by Quercetin as a Possible Mode of Action of *Psidium Guajava* Leaf Extracts in Treatment of Acute Diarrhoeal Disease. *Journal of Ethnopharmacology* 25:235–247.

Lutterdot, G. D., and A. Maleque. 1988. Effects on Mice Locomotor Activity of a Narcotic-like Principle from *Psidium Guajava* Leaves. *Journal of Ethnopharmacology* 24:219–231.

Lynch, Michael. 1993. *Scientific Practice and Ordinary Action: Ethnomethodology and Social Studies of Science.* New York: Cambridge University Press.

Lynch, Michael, and Steve Woolgar, eds. 1990. *Representation in Scientific Practice.* Cambridge, Massachusetts: Massachusetts Institute of Technology Press.

Mackenzie, Marguerite. 1982. *Cree Dictionary.* Val d'Or, Quebec: Cree School Board.

Malcom, S. A., and E. A. Sofowora. 1969. Antimicrobial Activity of Selected Nigerian Folk Remedies and Their Constituent Plants. *Lloydia* 32:512–517.

Malinowski, Bronislaw. [1925] 1948. *Magic, Science and Religion and Other Essays.* Garden City, New York: Doubleday Anchor.

Mannheim, Karl. 1936. *Ideology and Utopia.* London: Routledge and Kegan Paul.

Maracle, Lee. 1992a. *Sundogs.* Penticton, Canada: Theytus Books.

———. 1992b. A New Sensibility Beyond Survival: Bridging Cultures. Presented at First Nations: A Current Event, Institute for the Humanities at Salado, Salado, Texas, October 23–25, 1992.

Marcus, A. 1991. Out in the Cold: Canada's Experimental Inuit Relocation to Grise Fiord and Resolute Bay. *Polar Record* 27 (163):285–296.

Margolis, Joseph. 1987. *Science Without Unity: Reconciling the Human and Natural Sciences.* The Persistence of Reality, vol. 2. Oxford, U.K.: Basil Blackwell.

Markey, Edward. 1982. *Nuclear Peril: The Politics of Proliferation.* Cambridge, Massachusetts: Ballinger.

Marshak, Robert E., Eldred C. Nelson, and Leonard I. Schiff. 1946. *Our Atomic World*. Albuquerque: University of New Mexico Press.

Martin, Emily. 1987. *The Woman in the Body: A Cultural Analysis of Reproduction*. Boston: Beacon Press.

———. 1994. *Flexible Bodies: Tracking Immunity in American Culture from the Days of Polio to the Age of AIDS*. Boston: Beacon Press.

Martínez, A. 1891. Estudios Sobre la Yerba del Carbonero: *Baccharis conferta*. *El Estudio* 2:91–96.

Marx, Karl. [1867] 1972. *Capital, Book I*. London: Everyman Library.

Mascolo, N., and G. Autore. 1987. Biological Screening of Italian Medicinal Plants for Anti-inflammatory Activity. *Phytotherapy Research* 1 (1):28–31.

Massey, Douglas S., and Nancy Denton. 1993. *American Apartheid: Segregation and the Making of the Underclass*. Cambridge, Massachusetts: Harvard University Press.

Mausner, Bernard, and Judith Mausner. 1955. A Study of the Anti-Scientific Attitude. *Scientific American* 192(2):35–39.

Meckes-Lozoya, M., and I. Gaspar. In press. Photoxic Effect of Methanolic Extract from *Porophyllum macrocephallum* and *Tagetes erecta*. *Fitoterapia* SXIV(1):35–42.

Mednick, Sarnoff A. 1985. Biosocial Factors and Primary Prevention of Antisocial Behavior. In *Biology, Crime, and Ethics*, edited by Frank H. Marsh and Janel Katz. Cincinnati, Ohio: Anderson Publishing.

Mednick, Sarnoff A., W. F. Gabrelli, Jr. , and B. Hutchins. 1984. Genetic Influences in Criminal Convictions: Evidence from an Adoption Cohort. *Science* 224:891–893.

Merculieff, Ilarion. 1990. Western Society's Linear Systems and Aboriginal Cultures: The Need for Two-Way Exchanges for the Sake of Survival. Address presented at the Sixth International Conference on Hunting and Gathering Societies, Fairbanks, Alaska, May 30. Available from Commissioner's Office, Department of Commerce and Economic Development, State of Alaska.

Metzger, D., and G. Williams. 1963. Tenejapa Medicine: The Curer. *Southwestern Journal of Anthropology* 19:216–234.

Midgley, M. 1992a. Can Science Save Its Soul? *New Scientist* (August): vol 135 N1832:24–27.

———. 1992b. *Science as Salvation—A Modern Myth and Its Meaning*. New York: Routledge.

Miller, George, Paul Brown, and Carol Alonso. 1987. Report to Congress on Stockpile Reliability, Weapon Remanufacture, and the Role of Nuclear Testing. Document UCRL-53822, Lawrence Livermore National Laboratory.

Mills, C. Wright. 1956. *The Power Elite*. Oxford, U.K.: Oxford University Press.

Mirowski, Philip. 1989. *More Heat than Light: Economics as Social Physics, Physics as Nature's Economics*. Cambridge, U.K.: Cambridge University Press.

Miyadi, D. 1959. On Some New Habits and Their Propagation in Japanese Monkey Groups. In *Proceedings of the Fifteen International Congress, Zoological Society of London*, edited by H. R. Hewer and N. D. Rilet. London: Linnaean Society.

Mohamed, S., and H. Ayoub. 1983. Molluscicidal Properties of *Acacia* species. *Fitoterapia* 54:183–187.

Mohandas, V., J. Unni, and M. Mathew. 1987. Aetiology and Clinical Features of Acute Childhood Diarrhoea in an Outpatient Clinic in Vellore, India. *Annals of Tropical Pediatrics* 7:167–172.

Mokkhasmit, M., and W. Ngarmwathana. 1971. Pharmacological Evaluation of Thai Medicinal Plants. *Journal of the Medical Association of Thailand* 54 (7):490–504.

Molella, Arthur, and Carlene Stephens. 1995. Science and Its Stakeholders: The Making of Science in American Life. In *Exploring Science in Museums*, edited by Susan Pearce. London: The Athlone Press.

Moore, Sally F., and Barbara Myerhoff, eds. 1977. *Secular Rituals*. Amsterdam: Assen: Van Gorcum.

Morowitz, Harold J., and Temple Smith. 1987. Report of the Matrix of Biological Knowledge Workshop, July 13–August 14. Santa Fe, New Mexico: Santa Fe Institute.

Morris, B. 1976. Wither the Savage Mind? Notes on the Natural Taxonomies of a Hunting and Gathering People. *Man* 2:542–557.

Morse, E. S. 1936. *Japan Day by Day*. 2 vols. Tokyo: Kobunsha.

Morton, J. F. 1981. *Atlas of Medicinal Plants of Middle America*. Springfield, Illinois: Charles C. Thomas.

Motokawa, Tatsuo. 1989. Sushi Science and Hamburger Science. *Perspectives in Biology and Medicine* 32 (4): 489–504.

Mukerji, Chandra. 1990. *A Fragile Power: Scientists and the State*. Princeton, New Jersey: Princeton University Press.

Murtaugh, M. 1985a. A Hierarchical Decision Process Model of American Grocery Shopping. Ph. D. diss., University of California at Irvine.

———. 1985b. The Practice of Arithmetic by American Grocery Shoppers. *Anthropology and Education Quarterly* 16(3):186–192.

Nadel, S. F. 1957. Malinowski on Magic and Religion. In *Man and Culture: An Evaluation of the Work of Malinowski*, edited by R. Firth. London: Routledge and Kegan Paul.

Nader, Laura. 1981. Barriers to Thinking New About Energy. *Physics Today* 34 (3):9, 99–102.

Nader, Laura, with N. Milleron, J. Palacio, and C. Rich. 1977. Belief, Behavior, and Technologies as Driving Forces in Transitional Stages—The People Problem in Dispersed Energy Futures. *Distributed Energy Systems in California's Future: A Preliminary Report* vol. 2, (September):177–238.

Nader, Laura, with S. Beckerman. 1978. Energy as It Relates to the Quality and Style of Life. *Annual Review of Energy* 3:1–28.

Nader, Laura and N. Milleron. 1979. Dimensions of the "People Problem," in Energy Research and the Factual Basis of Dispersed Energy Futures. *Energy* 4 (5):953–967.

Nader, Laura, et al. 1980. *Energy Choices in a Democratic Society*. A Resource Group Study for the Synthesis Panel of the Committee on Nuclear and Alternative Energy Systems for the National Academy of Sciences.

Nadler, R. 1976. Rann vs Calabar: A Study in Gorilla Behavior. *Yerkes Newsletter* 13 (2):11–14.

National Academy of Sciences. 1975. *Genetic Screening: Programs, Principles, and Research.* Washington, D.C.: National Academy of Sciences.

National Center for Human Genome Research. 1990. Annual Report I—FY 1990. Washington, D.C.: Department of Health and Human Services, Public Health Service, National Institutes of Health.

National Research Council. 1988. *Mapping and Sequencing the Human Genome.* Washington, D. C. : National Academy Press.

Needham, Joseph. 1956. *Science and Civilisation in China.* Cambridge, U. K. : Cambridge University Press.

Nelkin, Dorothy, and Lawrence Tancredi. 1989. *Dangerous Diagnostics: The Social Power of Biological Information.* New York: Basic Books.

Nelson, Richard. 1983. *Make Prayers to the Raven: A Koyukon View of the Northern Forest.* Chicago, Illinois: University of Chicago Press.

Nickel, L. G. 1959. Antimicrobal Activity of Vascular Plants. *Economic Botany* 13:281–318.

Nietschmann, Bernard. 1989. Traditional Sea Territories, Resources and Rights in Torres Strait. In *A Sea of Small Boats. Cultural Survival Report 26,* edited by John Cordell. Cambridge, Massachusetts: Cultural Survival.

Nishida, T., ed. 1990. *The Chimpanzees of the Mahale Mountains: Sexual and Life History Strategies.* Tokyo: University of Tokyo Press.

Northrup, Robert. 1987. Editorial. *PRITECH Technical Literature Update* 2 (5):4.

Office of the Secretary of Defense. 1987. Memorandum for Executive Secretariat, Office of the Secretary of Defense, 12 January 1987; Subject: Superconducting Super Collider (SSC).

Okasaki, A. 1958. Antimicrobial Effect of Essential Oils and Ascaridol. *Archives of Pharmacology* 291:66–70.

Ortner, Sherry. 1973. On Key Symbols. *American Anthropologist* 75 (5):1338–1346.

Ortony, A., ed. 1979. *Metaphor and Thought.* Cambridge, U. K.: Cambridge University Press.

Overing, Joanna, ed. 1985. *Reason and Morality.* London: Tavistock Publications.

Oyama, Susan. 1985. *The Ontogeny of Information: Developmental Systems and Evolution.* Cambridge, U.K.: Cambridge University Press.

Palincsar, A. S. 1989. Less Chartered Waters. *The Educational Researcher* 18 (2):5–7.

Paprotte, W., and R. Dirven, eds. 1985. *The Ubiquity of Metaphor: Metaphor in Language and Thought.* Amsterdam: John Benjamins Publ.

Pass, M. A., and A. Seawright. 1979. Lantadine Toxicity in Sheep. *Pathology* 11:89–92.

Patton, Cindy. 1990. *Inventing AIDS.* New York: Routledge, Chapman and Hall.

Penley, Constance, and Andrew Ross. 1990. Cyborgs at Large: Interview with Donna Haraway. *Social Text* 25–26:8–23.

Pepper, Stephen. 1942. *World Hypotheses.* Berkeley: University of California Press.

Petitto, A. 1979. Knowledge of Arithmetic among Schooled and Unschooled African Tailors and Cloth-Merchants. Ph. D. diss., Cornell University.

Pfaffenberger, Brian. 1992. Social Anthropology of Technology. *Annual Review of Anthropology* 21:491–516.

Pickering, Andrew. 1984. *Constructing Quarks: A Sociological History of Particle Physics.* Chicago, Illinois: University of Chicago Press.

———. 1992. *Science as Practice and Culture.* Chicago, Illinois: University of Chicago Press.

Pinch, Trevor. 1993. Testing—One, Two, Three ... Testing: Towards a Sociology of Testing. *Science, Technology, and Human Values* 18 (1):25–41.

Polednak, Anthony P. 1989. *Racial and Ethnic Differences in Disease.* New York: Oxford University Press.

Posner, J. 1978. The Development of Mathematical Knowledge among Baoule and Dioula Children in Ivory Coast. Ph.D. diss., Cornell University.

Preston, Richard. 1975. *Cree Narrative: Expressing the Personal Meanings of Events.* Canadian Ethnology Service Paper No. 30, Mercury Series. Ottawa, Canada: National Museum of Man.

———. 1978. La Relation Sacree Entre les Cris et les Oies. *Recherches amerindiennes au Quebec* 8(2):147–152.

Quiros, C., S. Brush, D. Douches, K. Zimmerer, and G. Huestes. 1990. Biochemical and Folk Assessment of Variability of Andean Cultivated Potatoes. *Economic Botany* 44:256–266.

Rabinow, Paul. 1992. Studies in the Anthropology of Reason. *Anthropology Today* 8 (5).

Rajbahandari, A., and M. F. Roberts. 1985. The Flavonoids of *Stevia Cuscoensis, S. Galeopsidifolia, S. Serrata,* and *S. Soratensis. Journal of Natural Products* 48 (5):858–859.

Rappaport, Roy. 1968. *Pigs for the Ancestors.* New Haven, Connecticut: Yale University Press.

———. 1979. *Ecology, Meaning and Religion.* Richmond, Virginia: North Atlantic Books.

Rather, L. J. 1969. *The New Physician* (February):

Raven, P. H., B. Berlin, and D. E. Breedlove. 1971. The Origins of Taxonomy. *Science* 174:1210–1213.

Reice, Seth R. 1994. Nonequilibrium Determinants of Biological Community Structure. *American Scientist* 82 (5):424–35.

Reichel-Dolmatoff, Gerardo. 1976. Cosmology as Ecological Analysis: A View from the Rain Forest. *Man* 11 (3):307–318.

Reilly, Philip R. 1991. *The Surgical Solution: A History of Involuntary Sterilization in the United States,* Baltimore, Maryland: Johns Hopkins University Press.

Reiser, Stanley J. 1966 Smoking and Health: The Congress and Causality. In *Knowledge and Power,* edited by Sanford A. Lakoff. New York: Free Press.

Restivo, Sal, and Michael Zenzen. 1982. The Mysterious Morphology of Immescible Liquids. *Social Science Information.*

Reynolds, Peter. 1991. *Stealing Fire: The Atomic Bomb as Symbolic Body.* Palo Alto, California: Iconic Anthropology Books.

Reynolds, V. 1976. The Origins of a Behavioural Vocabulary: The Case of the Rhesus Monkey. *Journal for the Theory of Social Behaviour* 6:105, 142.

———. 1992. The Discovery of Primate Kinship Systems by Japanese Anthropologists. In *Topics in Primatology 2*, edited by N. Itoigawa, Y. Sugiyama, G. Sackett, and R. Thompson. Tokyo: University of Tokyo Press.

Richards, Paul. 1985. *Indigenous Agricultural Revolution.* London: Hutchinson and Co.

———. 1986. Coping with Hunger—Hazard and Experiment in an African Rice Farming System. *The London Research Series in Geography* 11: London; Boston: Allen and Unwin.

Ricoeur, Paul. 1977. *The Rule of Metaphor: Multidisciplinary Studies in the Creation of Meaning in Language.* Toronto, Canada: University of Toronto Press.

Riesenberg, Saul H. 1972. The Organization of Navigational Knowledge on Puluwat. *The Journal of the Polynesian Society* 81:19–56.

Rios, J. L., M. C. Recio, and A. Villar. 1988. Screening Methods for Natural Products with Antimicrobial Activity: A Review of the Literature. *Journal of Ethnopharmacology* 23:127–150.

Roberts, Leslie. 1990. The Worm Project. *Science* 248 (June 15):1310–1313.

Roediger, David R. 1994. *Towards the Abolition of Whiteness.* New York: Verso.

Rommetveit, R. n.d. Negative Rationality.

Rose, Hilary, and Steven Rose, eds. 1977. *The Political Economy of/in Science.* London: Macmillan.

Ross, S. A., and N. E. El-Keltani. 1980. Antimicrobial Activity of Some Egyptian Aromatic Plants. *Fitoterapia* 51:201–205.

Rouse, Joseph. 1992. What Are Cultural Studies of Scientific Knowledge? *Configurations* 1:1–22.

Rowell, T. 1966. Forest Living Baboons in Uganda. *Journal of Zoology* (London) 149:344–364.

Roy, P. K., and A. T. Dutta. 1952. A Preliminary Note on the Pharmacological Action of the Total Alkaloids Isolated from *Cissampelos pareira*. *Indian Journal of Medical Research* 40:95–97.

Rumelhart, David E., and James L. McClelland. 1986. *Parallel Distributed Processing: Explorations in the Microstructure of Cognition Volume 1: Foundations.* Cambridge, Massachusetts: Massachusetts Institute of Technology Press.

Ryan, Joan, and Michael P. Robinson. 1990. Implementing Participatory Action Research in the Canadian North: A Case Study of the Gwich'in Language and Cultural Project. *Culture* 10 (2):57–71.

Sachs, W., ed. 1992. *The Development Dictionary.* London: Zed Books.

Sade, D. S. 1972. A Longitudinal Study of Social Behavior of Rhesus Monkeys. In *The Functional and Evolutionary Biology of Primates,* edited by R. Tuttle. Chicago, Illinois: Aldine.

Salmon, M., E. Díaz, and A. Ortega. 1973. Chistenine, A New Epoxguianolid from *Stevia serrata* Cav. *Journal of Organic Chemistry* 38:1759–1761.

———. 1977. Epoxylactones from *Stevia serrata*. *Revista Latino Americana de Química* 8:172–174.

Salmond, Anne. 1985. Maori Epistemologies. In *Reason and Morality*, edited by Joanna Overing. London: Tavistock Publications.

Saunders, A., ed. 1992. Indigenous Knowledge. *Northern Perspectives* 20 (1):1, 16.

Saxe, Geoffrey. 1990. *Culture and Cognitive Development: Studies in Mathematical Understanding*. Hillsdale, New Jersey: Erlbaum Associates.

Sayer, Andrew. 1984. *Method in Social Science: A Realist Approach*. London: Hutchinson.

Schatz, Bruce R. 1991. Building an Electronic Scientific Community. In *Proceedings of the Twenty-Fourth Annual Hawaii International Conference on Systems Science*, edited by Jay Nunamaker, vol. 3. Los Almaritos, California: IEEE Society Press.

Scheler, Max. 1926. *Die Wissensformen und die Gesellschaft*. Bern: Francke.

Schultes, R. E. 1984. Ethnopharmacological Conservation: A Key to Progress in Medicine. Paper presented at VIII Simposio de Plantas Medícinais do Brasil. Manaus.

Schusterov, G. A. 1927. Isohaemoagglutinierenden Eigenschaften des menschlichen Blutes nach den Ergebnissen einer Untersuchung an Straflingen des Reformatoriums (Arbeitshauses) zu Omsk. *Moskovskii Meditsinksii Jurnal* 1:1–6

Schutz, Alfred. 1973. Common Sense and Scientific Interpretation of Human Action. In *Collected Papers I: The Problem of Social Reality*, edited and with an introduction by Maurice Natanson. The Hague: Martinus Nijhoff.

Schwartz, Charles. 1975. The Corporate Connection. *Bulletin of the Atomic Scientists* (October):15.

Scott, Colin. 1983. The Semiotics of Material Life among Wemindji Cree Hunters. Ph. D. diss. McGill University.

———. 1989. Knowledge Construction among Cree Hunters: Metaphors and Literal Understanding. *Journal de la Societe des Americanistes* 75:193–208.

Scribner, S., and E. Farmeier. 1982. Practical and Theoretical Arithmetic: Some Preliminary Findings. Industrial Literacy Project, Working Paper No. 3. Graduate Center, City University of New York.

Sebeok, Thomas. 1975. Zoosemiotics: At the Intersection of Nature and Culture. In *The Tell-Tale Sign: A Survey of Semiotics*, edited by Thomas Sebeok. Lisse/Netherlands: Peter de Ridder Press.

Selin, Helaine. 1992. *Science across Cultures. An Annotated Bibliography of Books on Non-Western Science, Technology, and Medicine*. New York: Garland.

Selvin, Paul, and Charles Schwartz. 1988. Publish and Perish. *Science for the People* 20 (January, February):6–10, 48.

Sharaf, A., and M. Naguib. 1959. A Pharmacological Study of the Egyptian Plant *Lantana camara*. *Egyptian Pharmacology Bulletin* 41:93–97.

Sharma, O. P., and H. P. S. Makkar. 1982. Changes in Blood Constituents in Guinea Pigs in *Lantana* Toxicity. *Toxicology Letters* 11:73–76.

———. 1983. Effects of Lantana Toxicity on Lysosomal and Cytosol Enzymes in Guinea Pig Liver. *Toxicology Letters* 16:41–45.

Sharma, O. P., H. P. Makkar, R. K. Dawra, S. S. Negi. 1981. Hepatic and Renal Toxicity of Lantana in the Guinea Pig. *Toxicology Letters* 347–351.

Sharma, O. P., J. P. Makkar, R. N. Pal, et al. 1980. Lantadene Content and Toxicity of the *Lantana camara. Toxicology* 18:485–488.

Shipman, Pat. 1994. *The Evolution of Racism: Human Differences and the Use and Abuse of Science.* New York: Simon and Schuster.

Shipochliev, T. 1968. Pharmacological Investigation into Several Essential Oils. *Veterinary Medicine Nauki* 5 (6):63–66.

Shiva, Vandana. 1991. The Green Revolution in the Punjab. *The Ecologist* 21 (2):57–60.

Siegal, Harvey. 1987. *Relativism Refuted: A Critique of Contemporary Epistemological Relativism.* Dordrecht: D. Reidel.

Silver, D. B. 1966. *Zinacanteco Shamanism.* Unpublished PhD. diss., Harvard University.

Singh, K. V. 1984. Effects of Leaves Extracts of Some Higher Plants on Spore Germination of *Ustillago maides. Fitoterapia* 55 (5):189–192.

Sioui, Georges E. 1992. *For an Amerindian Autohistory.* Montreal and Kingston, Canada: McGill-Queen's University Press.

Skolnick, Jerome. 1966. *Justice without Trial: Law Enforcement in a Democratic Society.* New York: Wiley.

Smith, Jeffrey. 1990. America's Arsenal of Nuclear Time Bombs. *Washington Post National Weekly Edition* (May 28–June 3).

Smith, John Maynard. 1968. *Mathematical Ideas in Biology.* Cambridge, U.K.: Cambridge University Press.

Smith, M. Estellie. 1982. Fisheries Management: Intended Results and Unintended Consequences. In *Modernization and Marine Fisheries Policy,* edited by J. Maiolo and M. Orbach. Ann Arbor, Michigan: Ann Arbor Science Publishers.

———. 1988. Fisheries Risk in the Modern Context. *MAST: Maritime Anthropological Studies* 1 (1):29–48.

Smuts, B. 1985. *Sex and Friendship in Baboons.* New York: Aldine.

Sofowara, A. 1982. *Medicinal Plants and Traditional Medicine in Africa.* New York: John Wiley and Sons.

Stallybrass, P., and A. White. 1986. *The Politics and Poetics of Transgression.* Ithaca, New York: Cornell University Press.

Star, Susan Leigh. 1989. *Regions of the Mind: Brain Research and the Quest for Scientific Certainty.* Stanford, California: Stanford University Press.

Stevens, A. L., and D. Gentner. 1983. Introduction. In *Mental Models,* edited by D. Gentner and A. L. Stevens. Hillsdale, New Jersey: Erlbaum.

Stober, Dan. 1990. Lawrence Livermore Braces for Change. *San Jose Mercury News* (September 9):1B.

Stokes, Terry D. 1985. The Role of Molecular Biology in an Immunological Institute. Paper presented at the International Congress of History of Science, University of California at Berkeley, 31 July–8 August.

Strauss, Lewis L. 1962. *Men and Decisions.* New York: Doubleday.

Strum, S. 1987. *Almost Human: A Journey into the World of Baboons.* New York: Random House.

Suzuki, David T., Anthony J. F. Griffiths, Jeffrey H. Miller, and Richard C. Lewontin. 1989. *An Introduction to Genetic Analysis.* 4th ed. New York: W. H. Freeman and Co.

Tambiah, Stanley. 1990. *Magic, Science, Religion and the Scope of Rationality.* Cambridge, U.K.: Cambridge University Press.

Tanner, Adrian. 1979. *Bringing Home Animals: Religious Ideology and Mode of Production of the Mistassini Cree Hunters.* St. John's: Institute of Social and Economic Research, Memorial University.

Tavris, Carol. 1992. *The Mismeasure of Woman.* New York: Simon and Schuster.

Taylor, D. N., P. Echeverria, and T. Pal. 1986. The Role of *Shigella* spp., Enteroinvasive *Escherechia Coli,* and Other Enteropathogens as Causes of Childhood Dysentery in Thailand. *Journal of Infectious Diseases* 153 (4):1132–1138.

Tetu, Hiroshige. 1974. Social Conditions for Prewar Japanese Research in Nuclear Physics. In *Science and Society in Modern Japan: Selected Historical Sources,* edited by Shigeru Nakayama, David L. Swain, and Eri Yagi. Boston, Massachusetts: MIT Press,.

Thomas, Stephen D. 1987. *The Last Navigator.* New York: Henry Holt.

Tokuda, K. 1961–1962. A Study of the Sexual Behavior in the Japanese Monkey. *Primates* 3 (2):1–40.

Toledo, V. M. 1988. La Diversidad Biologica de México. *Ciencia y Desarrollo* 81:17–30.

Traweek, Sharon. 1988a. *Beamtimes and Lifetimes: The World of High Energy Physicists.* Cambridge, Massachusetts: Harvard University Press.

———. 1988b. *Things Seen and Unseen: Discourse and Ideology in Tokugawa Nativism.* Chicago, Illinois: University of Chicago Press.

———. 1992a. Big Science as Colonialist Discourse: Regional Differences in Japanese High Energy Physics. In *Big Science,* edited by Peter Galison. Stanford, California: Stanford University Press.

———. 1992b. Border Crossings: Narrative Strategies in Science Studies and Among Physicists in Tsukuba Science City. In *Science as Practice and Culture,* edited by Andy Pickering. Chicago, Illinois: University of Chicago Press.

———. 1993. An Introduction to Cultural, Gender, and Social Studies of Science and Technology. *Journal of Culture, Medicine, and Psychiatry* 17:3–25.

Trivedi, C. P., and N. T. Modi. 1978. Preliminary Pharmacological Studies of *Acacia farnesiana.* *Indian Journal of Physiology and Pharmacology* 22:234–235.

Tsui, Lap-Chee. 1992. The Spectrum of Cystic Fibrosis Mutations. *Trends in Genetics* 8 (11):392–398.

Tsunoda, T. 1975. Functional differences between Right- and Left-Cerebral Hemispheres Detected by Key-Tapping Method. *Brain and Language* 2:152–170.

———. 1978. *Nihonjin no no (The Japanese Brain).* Tokyo: Taishukan Shoten.

Tuniq, Martha. 1985. Interview at Wakeham Bay. Avataq Cultural Institute, Oral History Project, Montreal.

Turnbull, David. 1991. *Mapping the World in the Mind: An Investigation of the Unwritten Knowledge of the Micronesian Navigators.* Geelong, Victoria: Deakin University Press.

Turner, Victor. 1969. *The Ritual Process; Structure and Anti-Structure.* Chicago, Illinois: Aldine.

Tylor, E. B. 1871. *Primitive Culture: Researches into the Development of Mythology, Philosophy, Religion, Language, Art and Customs.* London: J. Murray.

Umesao, Tadao. 1986. Keynote Address: The Methodology of the Comparative Study of Civilization. *Senri Ethnological Studies* 19:1–8.

United Nations University. 1990. *An Archive of Traditional Knowledge.* Unpublished Proposal. Tokyo, Japan.

Uppal, R. P., and B. S. Paul. 1971. Preliminary Studies with Crude Lantadine, Toxic Principle of *Lantana Camara*, in Albino Rats. *Haryana Agricultural University Journal of Research* 1:98–102.

Vackimes, Sophia. 1995. Science and Anti-Science at the Smithsonian Institution. B. A. honors thesis, University of California at Berkeley.

Van Gennep, Arnold. 1909. *The Rites of Passage.* London: Routledge and Kegan Paul.

Vogt, E. Z. 1970. *The Zinacantecos of Mexico.* New York: Holt Rinehart and Winston.

———. 1976 *Tortillas for the Gods.* Cambridge, Massachusetts: Harvard University Press.

Von Heijne, Gunnar. 1987. *Sequence Analysis in Molecular Biology: Treasure Trove or Trivial Pursuit.* San Diego, California: Academic Press Inc.

Wagner, Roy. 1977. Scientific and Indigenous Papuan Conceptualizations of the Innate: A Semiotic Critique of the Ecological Perspective. In *Subsistence and Survival: Rural Ecology in the Pacific,* edited by Timothy P. Bayliss-Smith and Richard G. Feacham. New York: Academic Press.

———. 1981. *The Invention of Culture.* Chicago, Illinois: University of Chicago Press.

Waldram, James. 1986. Traditional Knowledge Systems: The Recognition of Indigenous History and Science. *Saskatchewan Indian Federated College Journal* 2 (2):115–124.

Wanjari, D. G. 1983. Antihaemorrhagic Activity of *Lantana camara. Nagarjun* 27:40–41.

Warren, D. M. 1991. *Indigenous Knowledge Systems: The Cultural Dimension of Development.* New York: Kegan Paul International.

Washburn, Sherwood, and Irven DeVore. 1961. The Social Life of Baboons. *Scientific American* 204 (6):62–71.

Watanabe, M. 1974. The Conception of Nature in Japanese Culture. *Science* 183:279–282.

Waterman, Michael S. 1990. Genomic Sequence Databases. *Genomics* 6:700–701.

Watson, James D. 1987. *Molecular Biology of the Gene.* vol. 2. Menlo Park, California: W. A. Benjamin, Inc.

———. 1990. The Human Genome Project: Past, Present, and Future. *Science* 249 (April 6):44–49.

Weatherford, J. 1988. *Indian Givers: How the Indians of the Americas Transformed the World.* New York: Crown Publishers.

Webster, Andrew. 1991. *Science, Technology and Society.* New Brunswick, New Jersey: Rutgers University Press.

West, Bruce J., and Michael Shlesinger. 1990. The Noise in Natural Phenomena. *American Scientist* 78 (January/February):40–45.

White, Leslie A. 1938. Science is Sciencing. *Philosophy of Science* 5:369–89. Reprinted in *The Science of Culture*. New York: Grove Press, 1949.

White, Lynn, Jr., 1967. Historical Roots of Our Ecologic Crisis. *Science* 155 (3767): 1203–1207.

———. 1979. The Ecology of Our Science. *Science* 80:72–76.

White, Merry. 1988. *The Japanese Overseas: Can They Go Home Again?* New York: The Free Press.

Wiley, H. W. 1902. The Dignity of Chemistry. An unnumbered volume Supplement to the *Journal of the American Chemical Society* 148–164.

Williams, Raymond. 1980. *Problems in Materialism and Culture*. London: Verso.

———. 1985. *Keywords: A Vocabulary of Culture and Society*. Rev. New York: Oxford University Press.

Wilson, Andrew. 1983. *The Disarmer's Handbook of Military Technology and Organisation*. London: Penguin Books.

Wilson, E. O. 1975. *Sociobiology: The New Synthesis*. Cambridge, Massachusetts: Belknap.

Wilson, James Q., and Richard Herrnstein. 1985. *Crime and Human Nature*. New York: Simon and Schuster.

Yearley, Steven. 1988. *Science, Technology and Social Change*. London: Unwin Hyman.

Yoffee, Norman, and George L. Cowgill, eds. 1988. *The Collapse of Ancient States and Civilization*. Tucson: University of Arizona Press.

York, Herbert. 1987. *Making Weapons, Talking Peace: A Physicist's Odyssey From Hiroshima to Geneva*. New York: Basic Books.

Young, Robert. 1972. The Anthropology of Science. *New Humanist* 88 (3):102–105.

Yoxen, E. J. 1983. *The Gene Business: Who Should Control Biotechnology?* New York: Oxford University Press.

Ziegler, Charles A. 1988. Waiting for Joe-1: Decisions Leading to the Detection of Russia's First Atomic Bomb Test. *Social Studies of Science* 18:197–229.

Ziegler, Charles A., and David Jacobson, 1995. *Spying Without Spies: Origins of America's Secret Nuclear Surveillance System*. Westport, Connecticut: Praeger.

Zuckerman, Sir S. 1932. *The Social Life of Monkeys and Apes*. London: Kegan Paul, Trench and Trubner.

CONTRIBUTORS

PAMELA J. ASQUITH is Associate Professor of Anthropology at the University of Calgary in Alberta, Canada. Her areas of research include anthropomorphism in animal behavior studies, Japanese and Western approaches to the natural sciences, and modern Japanese and Korean society. Books include *The Monkeys of Arashiyama, Thirty-five Years of Research in Japan and the West* (coedited with Linda Fedigan, 1991), *The Culture of Japanese Nature* (coedited with Arne Kalland, forthcoming, Curzon Press), and a translation of Imanishi Kinji, *The World of Living Things* (with Heita Kawakatsu, Hiroyuki Takasaki, and Shusuke Yagi, forthcoming, SUNY Press).

B. BERLIN is Professor of Anthropology at the University of Georgia in Athens. He is recognized for his work in ethnobiology and cognitive anthropology, especially in the area of general principles of human classification of plants and animals. He has worked and published on the ethnobotany of the Highland Maya for several decades.

E. A. BERLIN is Professor of Anthropology at the University of Georgia. She holds degrees in public health and medical anthropology and has conducted extensive field research on epidemiology, nutrition, and health among the Jivaro of the Peruvian Amazon. For the last six years she has worked on the ethnomedicine of the Highland Maya of Chiapas, Mexico.

ELLEN BIELAWSKI is a scholar affiliated with the Arctic Institute of North America, a leading institute in research conducted cooperatively with northern communities at the University of Calgary in Alberta, Canada. She is an anthropologist who has worked with Inuit, scientists, and Dene in the Canadian North since 1975.

TROY DUSTER is Professor of Sociology and Director of the Institute for the Study of Social Change at the University of California, Berkeley. He currently serves on the Committee on Social and Ethical Impact of Advances in Biomedicine, Institute of Medicine, National Academy of Sciences. Duster is the author of a number of articles on theory and methods. His books and monographs include *Cultural Perspectives on Biological Knowledge* (coedited with Karen Garrett, 1984). His most recent book is on the social implications of the new technologies in molecular biology, *Backdoor to Eugenics* (1990), released as *Retour a l'Eugenisme* in French (1992).

MICHAEL FORTUN is Executive Director of the Institute for Science and Interdisciplinary Studies at Hampshire College, where he is Visiting Assistant Professor of Science Studies. He received his Ph.D. in the history of science at Harvard University. His current writing explores the political, cultural, and literary dimensions of speed and time in genomics. Fortun is also conducting fieldwork and writing on quantum cryptography and quantum teleportation, and on the theory and practice of dialogue as a tool for changing how science is done.

JOAN H. FUJIMURA is Henry Luce Professor of Biotechnology and Society at Stanford University. She is a member of the anthropology department and the program in history and philosophy of science. Fujimura received her Ph.D. from the University of California, Berkeley, and has taught at Harvard University. Her current research is of genomics and bioinformatics in Japan and the United States and on the human genome diversity project. Fujimura also writes on gender and science, tools and technologies in the biosciences, and AIDS research, oncogene research, and theories and methods in science studies. As part of her Luce Professorship, Fujimura also organizes workshops on the biosciences. Last year's workshop was on "Vital Signs: Cultural Perspectives on Coding Life and Vitalizing Code."

WARD H. GOODENOUGH is University Professor Emeritus at the University of Pennsylvania, where he joined the faculty in 1949. His ethnographic research has been in Chuuk (Truk) and Kiribati in Micronesia and in New Britain in Melanesia. He has served as editor of the *American Anthropologist*, has been president of the American Ethnological Society and the Society for Applied Anthropology, and has published extensively in anthropology.

HUGH GUSTERSON is Assistant Professor of Anthropology and Science Studies at the Massachusetts Institute of Technology. His book on the cultural world of nuclear weapons scientists at the Lawrence Livermore Laboratory is titled *Nuclear Rites: An Anthropologist Among Weapons Scientists* (1996). He has been published in *The Sciences, New Scientist, Tikkun,* the *Journal of Contemporary Ethnography,* and the *Journal of Urban and Cultural Studies.*

DAVID JACOBSON is Associate Professor of Anthropology at Brandeis University. With Charles Ziegler, he has done research on the cultural systems of scientists and nonscientists, focusing on the structure of social networks and on the relationships between beliefs and behavior. He has published, with Ziegler, a study of scientific methods of nuclear surveillance, *Spying Without Spies* (1995). He has examined similar issues in quite different ethnographic contexts, reporting the results in *Itinerant Townsmen* (1973) and *Reading Ethnography* (1991).

JEAN LAVE is Professor of Anthropology in the School of Education at the University of California, Berkeley. She is interested in theories of learning and apprenticeship, activity theory, and everyday practices. Recent publications include *Cognition and Practice: Mind, Mathematics and Culture in Everyday Life* (1988), *Situated Learning: Legitimate Peripheral Participation* (with E. Wenger, 1991), and *Understanding Practice* (with S. Chaiklin, 1993).

X. LOZOYA founded the Mexican Institute for the Study of Medicinal Plants (IMEPLAM) and served as its Director until 1980, when he became Director of the Research Center for Medicinal Plants and Traditional Medicine of the Mexican Social Security Institute in Xochitepec, Morelos. He has been involved in the study of the medicinal plants of Mexico for the last twenty years.

EMILY MARTIN is Professor of Anthropology at Princeton University. She is the author of *The Woman in the Body: A Cultural Analysis of Reproduction* (1987), which won the Eileen Basker Memorial Prize of the Association of Medical Anthropology, and *Flexible Bodies: Tracking Immunity in American Culture from the Days of Polio to the Age of AIDS* (1994). Bjorn Claeson, Wendy Richardson, Monica Schoch-Spana, and Karen-Sue Taussig are Ph.D. candidates in the Johns Hopkins Department of Anthropology.

M. MECKES is currently Chief of the Division of Pharmacology of Medicinal Plants, Centra Biomedical Research Unit of the Mexican Social Security Institute in Mexico City. She is a cofounder of the Mexican Institute for the Study of Medicinal Plants and has worked on the pharmacology of Mexican medicinal plants for many years.

LAURA NADER is Professor of Anthropology at the University of California, Berkeley. Her current work focuses on how central dogmas are made and how they work. *Energy Choices in a Democratic Society* (1980), is a multidisciplinary collaborative effort of the National Academy of Sciences. *Harmony Ideology—Justice and Control in a Zapotec Mountain Village* (1991), and *Essays in Controlling Processes* (1994), emerge from a theoretical perspective that crosses disciplinary boundaries. She is a member of the American Academy of Arts and Sciences. In 1995 the Law and Society Association awarded her the Kalven prize.

CHARLES SCHWARTZ, is a Professor of Physics at the University of California at Berkeley. A theoretical physicist, he has been on the faculty since 1960. For many years he has focused his attention on questions of social responsibility and the politics of science—as public critic and activist, as well as scholar and teacher. Schwartz was a founder (in 1969) of the activist organization, Science for the People.

COLIN SCOTT is Associate Professor in the Department of Anthropology at McGill University. His research concentrates on indigenous ecological knowledge, community-based resource management among aboriginal people in northern

Canada, and the cultural politics of aboriginal self-determination. He directs a multidisciplinary, interuniversity research program called AGREE (Aboriginal Government, Resources, Economy, and Environment), headquartered at McGill.

M. ESTELLIE SMITH is Professor of Anthropology at the State University of New York in Oswego. She has published on the development of state systems, politicking, and bureaucratic structures; political and informal economies; the European Community; fisheries management policy; urban anthropology; and other topics.

SHARON TRAWEEK is Associate Professor in the History Department and is the Director of the Center for Cultural Studies of Science, Technology, and Medicine at UCLA. Her research spans cultural, feminist, and rhetorical studies and is focused on the transnational high energy physics community. She is the author of *Beamtimes and Lifetimes: The World of High Energy Physicists* (1988), in addition to many articles in scholarly journals.

J. TORTORIELLO is chief of Pharmaceutical Division of Medicinal Plants, Center for Biomedical Investigations of the South, Mexican Institute of Social Security, Xocoitepèc, Morelos.

M. L. VILLARREAL is Director of Cytotoxicity Laboratory of the Center for Biomedical Investigations of the South, Mexican Institute of Social Security, Xochitepec, Morelos.

CHARLES A. ZIEGLER is Lecturer in Cultural Anthropology at Brandeis University. He holds advanced degrees in anthropology and in physics and has been active professionally in both fields. With Jacobson, he has recently published a study of the origins of America's secret nuclear surveillance system, *Spying Without Spies* (1995). He is currently working on a study of the relationship between organizational culture and decision making in science-based institutions.

Index

AAAS xiii
Acacia angustissima 52, 53, 54, 55, 59, 63
Academic senates 151, 154
Adams, Richard 24
Adams, W. P. 218, 225
Adas, Michael 12
Ageratina 53, 54, 55, 56, 57, 63, 66
Agnew, Harold 154
Agriculture: 12, 203; principles 11; technique 7
AIDS 102–115
Akuliaq 216
Alayco, Salmonie 222
Allium sativum 53, 54, 60, 63, 66
Altair 30, 33
Altieri 12
Altmann, Stuart 244
Alvares 12
Amazon 8, 11 (Amazonia)
American Physical Society 157–158
Anderson, Eugene 6
Anthropology: 183; and cultures 274; and primitive knowledge 259; and realist inquiry 219; and religion 4; cultural 40; ethnoscience and ethnoscientists 6, 7; nineteenth century 4; of science 87, 89, 239–240, 261; role of xii, 3 ff; scarcity of 41
Anthropomorphic 85, 254, 255
Archimedes 7
Arctic: 7, 20, 21; science 21, 222–227; and integration with Inuit knowledge 225–227; ethnography of 222–225
Aspin, Les 133
Asquith, Pamela J. 22, 304
Atomic monopoly 237
Australian aborigines 5, 274

Baccharis 52, 53, 54, 55, 56, 57, 63, 66

Barnes, Barry 95–96, 230
Barro Colorado Island 241
Bartlett, Frederick 94, 95
Bateson, Gregory 72, 86
Behaviorism 244
Belau (formerly Palau) 29
Beliefs: about fallout 230; about the bomb 230; and lay beliefs 235–236; authoritative nature 237; confirmation of 235–236; hegemonic effect of 240; implication for science and policy 236–237; influence on decisions 228; influence on policy 231; laypersons' 15, 21, 148, 229–231; scientists' 229–235; systems 206; (also see Nature, Energy)
Berkes, Fikret 71
Berlin, B. 6, 14, 43, 304
Berlin, E. A. 304
Beta (and Gamma Aquilae) 30
Beta-Thalassemia 126, 127
Bielawski, Ellen 7, 21, 304
Biology: and representation of nature 166; as explanation of crime 128–129; as technological science 19; biochemical experiments 165; of AIDS 101; of cancer 101; of sex and HIV 105; paradigm shift in 166–167; representations of and politics 19; standardization problems 168
Biotechnology (see Human Genome Project)
Black, M. 72
Boas, Franz xiii
Borreria laevis 53, 54, 56, 59, 63
Boundaries: 114; and anthropology 24; and diseases 105; and power 2; blurring of 171, 201; conflict between ways of knowing 217; disciplinary 121; ecological 82; in metaphor 106; in risky

sexual practices 107; in science debates xiv; integrating indigenous knowledge and science 226; in Western science xii; of science 3–4; preventative health and national priorities 111; safety 107; self and non-self 115; social significance of 108
Bradbury, Norris 133
Brecht, Bertolt 158
Broad, William 135, 137
Brokensha, David 11
Brown, Joseph 165
Bureaucracy 224, 267–268, 270–271
Bush, George xiii, 132, 186
Business 190, 201, 274; corporations 151; industrial firms 229
Byrsonima crassifolia 53, 54, 56, 59, 63

Callan, Curtis 189, 190, 196
Calliandra 52, 53, 54, 55, 56, 60, 63
Canada: 7, 21; Canadians 8, 15; Dept. of Energy, Mines and Resources 224; government 21, 216, 219; Government of N.W. Territories 217–218, 219; Minister of Environment 217
Capra, Fritjof 239
Career paths: (see Japanese scientists)
Caribou: Beverly Kaminuriak Management Board (BKCMB) 227
Caroline Islands 29–42
Carpenter, Clarence 241–242, 244–245
Carraher, T. 90
Carter, Jimmy 131, 133, 153
Cayo Santiago colony 241, 242, 246
Censorship: 267, 272 (see Group-think)
CERN 184, 185, 195
Chambliss, W xiv
Chenopodium ambrosioides 53, 54, 55, 56, 61, 63
Chiapas 14, 43–68
CIA 188, 235
Cissampelos pareira 53, 54, 55, 60, 63
Cizek, Petr 227
Claeson, Bjorn 306
Class: and gender 179, and science/culture 16
Cognitive framework: 69–70, 72, 85, 134; dual categories 99; generalizing mind 98; intuition vs abstract thought 95; mental models 97; naturalistic vs personalistic 43–44; negative rationality 97, 114, 115; "primitive" vs "civilized" mind 93–95, 97 (and see Comparison); rationality of science 4; social evolution 5; the inferior "other" 88, 95, 98
Cognitive processes: authoritative 134; concrete thought 94; judgments 134; logical mind 98; normative models 98; representational mind 98
Cognitive studies 88
Colson, Elisabeth 6, 263
Comanagement: 227; consensual 205
Committees: Advisory 149; Atomic Energy 233; history of Federal Science 150–151; selection to 151; Military Liaison 153
Communication: as universal network 81; concepts vs sign 95; signs and culture 73; human and goose 80
Community: control 226–227
Comparison: xii, 1, 2, 7, 10, 15, 20, 44, 218, 222, 269; and positional superiority 2; as boundary 4; false 6; Inuit with West 222–225; Japanese with Western science 11, 19, 254; of epistemologies 239; of Japanese and Western primatology 22, 244; science/superstition 1–2
Committee on Nuclear and Alternative Energy Systems (CONAES) 261–262, 271
Conklin, Harold 6
Controversy: 9, 201 ff; Gulf War 188–189; IQ and genetics 122; public xi
Crataegus pubescens 52, 53, 54, 55, 56, 59, 63
Cree: 15; animal and human reciprocators 81–84; knowledge construction 76–81; metaphors of eating and sexuality 75–76; modeling of experience 77; relating and differentiating 73–75; resource management 7; science 69; signs in epistemology 72–73; Wemindji community 78
Crime genetic basis 119, 128–129
Cultural survival xiv

Culture: 40–41, 73; and nature 74; and politics 99; and science 229–230, 274; and society 212
Cystic fibrosis: 126, 162; gene for 162

Databases: 19, 48, 49, 160; and ethnography 172; consequences of 170; definition of 163; HGP sequence 164, 170
De La Rocha, O. 91
De Waal, Frans 253
Defense Science Board 150, 155
Department of Defense 135, 150, 155, 157, 159, 192, 197
Department of Energy xii, 135, 150, 153, 155, 157, 163, 182, 271
Deutsches Elektronen-Synchrokron (DESY) 195
DeWitt, Hugh 132
Discourse: analysis and study of science 240; symbolic 72
Disease classes 67
Divination: 44, 263; and numbers 263–264; and risk 6
Division of labor 191
Domhoff, G. William 149
Doolittle, Russell 165, 165
Douglas, Mary 3
Downey, Gary L. 8
Dubinkas, Frank 6
Durkheim, Emile 138
Duster, Troy 17, 304

Easlea, Brian 144
Ecology: 12; adaptation 71; cultural 15, 71; environmental anxieties 69; principles 11
Elizabetsky, E. 43
Energy: 261; barriers 264–266, 271–272; ideology of 262; Ireland 263; New Zealand 263; nuclear 264; solar 262–264
Escobar, Arturo 6
Ethnocentrism 11, 69
Ethnography: 40–41; advantages of 71; and sequence databases 172; categories of 70; limitations of 71; magic 71; of Arctic 222–225; scarcity of 41
Ethnomedicine: bioactivity & potential pharmacological effect 57; consensus of naming 51; curers 44; data sets 47, 48; folk diagnoses 67; gastrointestinal conditions 44, 45–47, 49, 67; healing rituals 44; healing, alternative 103; multidisciplinary research 45–46; naturalistic/personalistic cognitive framework 44; portable herbarium 47; selection of treatment 52–57; self-reports 46; signs & symptoms 44, 49, 51, 62, 64; specificity 51, 62–67; "symptoms"/bacteria 65; tests 49; treatment of abdominal pain 60–61; treatment of diarrhea 57–60; treatment of worms 61–62
Ethology 244
Etkin, Nina 8, 14
European expansion: 12; and "civilizing mission" 7, 12, 93–97
Evans-Pritchard, E. E.: 6, 94, 146; and myth 70
Experts: local 71; Plato on 273; scientists 16

Farnsworth, N. R. (and Pezzuto) 43
Federal Housing Authority 126, 127
Feit, H. 71, 86
Feyerabend, Paul 5, 10, 273
Fienup-Riordan, Ann 222, 223
Firth, R. 260
Fisheries Research Institute (U of W) 204
Fishermen 10, 20, 210–211
Fishing: 20; as science 202–204; capital intensive 202; catch limits 203; commercial 203; family firms 203; in Palau 10; global 203; labor intensive 202; management of stocks 201; management problems 205; managers 20; marine resources 205; nets 202; research 204–205; salmon industry 204; supertrawler 203; technology 202, 210; total allowable catch 206; trawlers 202; traps 204
Fleck, Ludwik 10
Fleising, U. xiv
Foeniculum vulgare 53, 56, 60, 63, 66
Forrestal (Navy Secretary) 231
Fortes, Meyer 6
Fortun, Michael 6, 19, 305
Foster, George 44

Foucault, Michel 12, 69, 144, 147
Fox, William 210
Frazer, James 4, 5
Freedom of Information Act 155
Frisch, Jean 244
Fuchsia 53, 54, 55, 62
Fujimura, Joan 19, 305
Funding: agencies 155; and business 90; and Japanese physics 181–183, 185–190, 190–192; and military power 175; military 187, 190; research 202

Gaiatsu 185, 176, 183
Gamble, Clive 7
Garwin, Richard 133
Gastrointestinal conditions: 14; diarrhea 49; epigastric pain 51; gallbladder disease 49; worms 49
Gay gene 171
Geertz, Clifford 102
GenBank 163, 165
Gender, and science/culture 16
Gene: markers 162; myth 102
Genetics: and Human Genome Project 171; and sociology of knowledge 122–123; explanations 17, 119–122, 160; population 120; Mendelian 120
Gentner, D.(and Stevens) 97
Geographers 7
Gieryn, Thomas 2, 24
Gilbert Islands (see Kiribati)
Gilbert, Walter 166–168
Gladwin, Thomas 14, 40, 41, 42
Gleick, James 208, 214
Gonzalez, R. 10
Goodall, Jane 244
Goodenough, Ward 14, 41, 305
Goody, Jack 95
Gorbachev, Mikhail 131
Government: 201; agencies 229
Graham, David 196
Great Basin tribes 274
Green Revolution 11
Grise Fiord 219
Gross, Paul (and Levitt) xi, xii
Group-think 264–265, 269, 271
Guam 35, 36
Guerrant, R. L. 62,64

Gulf War xiii, xiv, 188
Gusterson, Hugh 6, 18, 305
Gutmann, M. xii, 2
Gwembe Tonga 6

Hallpike, C. R. 95, 261
Hamer, Dean 171
Handlin, Oscar 229
Haraway, Donna 22–23, 170–171, 172, 256
Harding, Sandra 23, 24
Harwood, Jonathan 123
Hayakawa, Satiol 95
Hazen, Robert (and Trefil) 101, 115, 116
Hegemony: 160, 169–170, 239–255; effect on beliefs 240; (see also Ideology, Worldview, Beliefs)
Helianthemum glomeratum 53, 54, 55, 56, 59, 63
Herken, Gregg 234
Herrnstein, Richard J. (and Murray) 127
Hess, David (and Linda Layne)
Hesse, Mary 72
High energy physics: community 19–20; surplus scientific labor 192
Hiroshima 12
Hoddeson, Lillian 195
Homology: definition of 164
Horton, Robin 6, 146
Hubbard, Ruth 102
Human Genome Project (HGP): 19, 160; and genetic determinism 171; and homo-sexuality 171; databases 163–164; description of 161; dry vs wet lab 165; homologies 164–165; mapping and sequencing genes 162–163; technologies 161; linkage maps of 161; sequencing 161
Human: -plant interaction 14; reproduction 76
Hunting: restraint in 82; settlement-based 79
Huntington's disease 127
Hutchins, E. 90
Hyndman, David 6

Ideology: 3, 262, 264; and nonWestern peoples 12; development and science 11; of knowledge 69; of progress xi, 146; (see also Beliefs)

Ikawa-Smith, Fumiko 25
Imanishi, Kinji 244–245
Immune system: 16, 103–115; body-country metaphor 109; bounded and independent 109; compared to social system 109
Immunology, understanding of: 16; at university level 102–105; in gay circles 105–108; and war metaphor 108–111, 112, 114
Interlocking directorships 149
Inuit: elders 217; epistemology 219–222; integration with Arctic science 225–227; knowledge 21; significance of 224
Inukjuak 219, 220
Irvine, D. xiv
Itani, J. 249–251, 252, 256
Iyaituk, Markusi 221,222

Jacobson, David 21, 305
Japan Liberal Democratic Party 182
Japan Monkey Center 244
Japan: and Soviet Union 187; assistance to poorer countries 186; science cities 185, 186, 193–195, 255; U.S. occupation 180
Japanese scientists: career strategies 175–179; collaboration with foreigners 177; generations 179–180; physicists 19
Japanese: 11, 22; character 181; culture terms: Gaiatsu 185, 176, 183, Kokusaika: 175–194, and nihonjinron 175; Physics community: and power 193–194, funding 181–183, 185–190, 190–192, language and generations 179–181, and TRISTAN 184–185; worldview 175 ff, 249–252, 254
Jasanoff, Sheila 23, 25
Jason 189, 190, 196
Jenson, Arthur 127
Johannes, Robert 7, 10
Joint Chiefs of Staff 231

Katayama, Tsuneo 214
Katz, S. H. 124
Kaufman, J. H. 252
Kaufmann, Felix 125
Kawai, M. 252

KEK Laboratory 182–185, 187, 188, 191, 194–195
Keller, Evelyn Fox 170, 171, 240
Keusch, G. T. 64
Khadem, H. E. (and Mohamed) 58
Kidder, Ray 133–134, 140
Kiribati 29
Klag, Michael 123
Knowledge production: xiii, 8, 13, 76–81, 239; sociology of 17
Knowledge systems, traditional: and renewed interest in 8, 14, 69, 71, 218; and similarity with western science/thought 14, 41, 43, 84, 100; ethnomedicine 7, 43–68; everyday math 87–100; hunting 69–86; immunology 101–116; Inuit 219–222; lay beliefs 230–238; lost 7; Maori epistemology 218; navigation 7, 29–42; Traditional Environmental Knowledge (TEK) field 8, 70; achievements of TEK 84; Western epistemology 222
Knowledge: and memory 39; authorization of 140; consumption of 13; control over 8; empirical 67, 70; hierarchy 140; integration of 216; localized 23; ritual framing of 15; sharing of 71; sociology of 17, 122–123; subjugation of 15; transmission 174; types 221–222
Kobe earthquake 214
Koford, P. 252
Kokusaika: 175–194; and *nihonjinron* 175
Koshland, D. E. xiv
Kroeber, A. 274
Kwagley, Oscar 1, 2, 8

La Jolla, California 189, 196
Laboratories: KEK 182–185, 187, 188, 191, 194–195; Livermore xii, 131–147, 149, 153–155; Los Alamos xii, 143–144, 146, 149, 153–155, 163; Nuclear Weapons: history of 153, hierarchy in 153–154, and University of California 154–155; research 225
Laboratory discourse 145;"dry" /"wet" labs 164–165
Lancaster, J. 256
Langmuir, Irving 25

Language: and generations 179–181; figurative, literal, metaphoric 72–74; strategies 181
Lantana 53, 54, 55, 56, 58, 63, 66
Latour, Bruno 102, 239
Laurence, William 231
Lave, Jean 15–16, 306
Layne, Linda (see Hess)
Le Grande, H. 23
Leach, Edmund R. 260, 273
Leggett, Anthony 256
Legitimacy, and exclusion 13
Lepechinia schiedeana 53, 54, 61
Lepidium virginicum 53, 54, 55, 56, 60
Levi-Strauss, Claude: 6, 70, 95, 146; and structuralism 70, 71
Levy-Bruhl, Lucien 94, 99, 269, 271–272
Lewis , David 42
Lieberman, L. xii
Lifton, Robert 141
Linnaeus 130
Lippman, Abby 170
Livermore Laboratory xii, 131–147, 149, 153–155
Lloyd, G. E. R. 116
Lobelia laxiflora 53, 54, 55, 62, 63, 66
Lock, Margaret 6
London Zoo 242
Long Range Detection System 228, 229
Lopez, Barry 216
Lorenz, Edward 209
Lorenz, Konrad 76–77, 244
Los Alamos xii, 143–144, 146, 149, 153–155, 163
Lozoya, X. 306
Lucknow, India 241

MacArthur, Douglas 180
Maeda, Yoshiaki 244
Magic: 259; and science 4, 260; and structuralism 70
Magnuson Act 202, 205
Malinowski, Bronislaw xiii, 4, 5, 6, 7, 8, 70, 259, 260, 261, 273,
Mannheim, Karl 122
Mark, Carson 133
Marshall Islands 29
Martin, Emily 6, 16, 306

Marx, Karl 145
Mathematics practices: research on 90–91, 98–99; idealization of 16, transformation in problem solving 91–93, the "cottage cheese" problem 96–97,
Mathematics practitioners: tailors 90, blue collar workers 90, farmers 90, cloth merchants 90, bookies 90, street venders 90, in school 90, dairy workers 90
Mathematics: and schooling 88, 90; calculation 90; cognitive studies 93; Adult Math Project 98–99; everyday 15, 89–91; relation of quantities 91–93
May, Michael 131
May, Robert 214
Maya: 14, 43–68, 274; and science 7
Meckes-Lozoya, M. 306
Media reports 230
Meiji Restoration (1868) 180, 181
Melanesians 14
Metaphor: 15, 16, 69–70, 72, 75–76, 85, 103; and worldview 85, 114; of birth 143, 145; of war 108–111, 112, 114; in nuclear weapons testing 143–145; of Nation states 108
Method: description 40; discourse analysis 72; fieldwork 70; genealogical methods 247; group elicitation 48; in ethnomedicine 47–49; interpretative 229; interviews 48, 222; limitations 68; long and short term observation 252; medical ethnobotany 47–48; multidisciplinary 45–46, 229; naturalistic studies 244; of primatology 247; oral histories 220; participant-observation 174, 224; participatory action research 226; power structure research 149; questionnaires 224; situation specific 91; survey 46, 48, 51
Mexico 11
Micronesia 10, 14, 29, 40
Midgley, Mary xv, 13
Militarization: xiii, 156–157; of science 266
Military: 150; and funding 175; and jobs 159; and science xiii, 155, 156–157; (see Nuclear)
Mills, C. Wright 149

Mizuhara, Hiroki 244
Molecular biology: biologists 17, 19, 176; as technology 166
Molecular genetics 119
Molella, Arthur 9
Motokawa, Tatsuo 255
Murray Hill Area 232, 235
Myth: and Levi-Strauss 70; and native cosmologies 71; and ritual in scientific experiment 146; and science 74

Nadel, Sigfried F. 6, 260
Nader, Laura 9, 306
Nadler, Ron 253
Nagasaki 12
Nakasone (Prime Minister) 175
National Academy of Sciences xii, 121
National Cancer Institute 171
National Center for Human Genome Research 167, 169
National High Energy Physics Laboratory 182
National Marine Fisheries Service (NMFS) 203, 204, 210
National Science Foundation 150, 151, 155, 157, 166, 190, 191, 197
Nature: and science 235–237; beliefs about 237; concept of 160; environmental philosophy 240; human place in 245; linear and non-linear 207; the nature of 207; two paradigms of 207, 214; unpredictable universe 208
Navigation (Micronesian):"aimers" 37; as practical science 39–40, 42; conceptual tools 14; constructing mental equivalents 35–36; drills 37; "ghost islands" 35; keeping track ("drags") 33–35; living seamarks 33; magnetic compass 14; memory storage 39; navigators as ritual specialists 38; predicting the weather 36–37; putting the system to work 37–38; sailing directions 31–33, 37; schematic mapping 35–36; sidereal calendar 36–37; star structure 29–31
Needham, Joseph 239
New England: 20; council 206; and cult belief systems 206

Nicotiana tabacum 53, 54, 55, 56, 60, 63, 66
Nietschmann, Bernard 71
Nixon, Richard 151
Nuclear accidents 141
Nuclear weapons banning: 131–132, 135; Arms Control Treaties: 131–132; Freeze movement 131; Limited Test Ban Treaty (1963) 131
Nuclear weapons testing: and metaphor 143–145; and power/knowledge 140; and scientists 231–235; as cultural process 138; as culture 18; as ritual 138–139; as scientific experiment 138; as status 139; as system of deterrence 143; deconstructing of 132–135; explosion 233; fallout 230, 233; high tech rituals 132; initiation into 139–141; lay beliefs 230, 235–236; method 135–139; number of 132; purpose of 135; Sandstone 235–236; significance 138
Nuclear Weapons: careers 153; cliques 153; conversion of xiii; design of 135–138; mastery of technology 141–143; monitoring 231; Pincher 234; Polaris 133; reliability of 133, 134–135, 139, 140, 147
Nuclear: mafia 154; scientists 17; surveillance system 234; low grade ores 232
Nunavik 220

Ocimum selloi 53, 54, 55, 61, 63
Oppenheimer, J. Robert 144, 153
Ou, Jay 10

Palau 10, 29 (see Belau)
Palincsar, A. S. 96–97
Patton, Cindy 6
Pauling, Linus 131
Petitto, A. 90
Pfaffenberger, Brian 3, 25
Philosophy of science 95
Photon factory 184, 188
Physics: and power structure research 149; ascendance of 149; community 19–20; high energy physicists 17; high energy 174–197; peer review 150; surplus scientific labor 192; (see also Japanese Physics)

Plato on experts 273
Polar Continental Shelf Project 224
Polaris 29 (star)
Polynesia 7, 11
Posner, J. 90
Power: 2; and big science 193–194; and "geneticization" 160; and information 170–172; and Japanese physicists 193–194; and juxtaposition 7; and knowledge 140–141; and national elites 18; and paradigms 10; and peer review 150; and politics 9; and public discourse 23; and scientific hegemonies 255; and society 3, 265–266; and status 158–159; in cultural authority 16; in genetic explanation 17; in physicists' strategies 19; in science xi, xii, 12; of knowledge 148; of scientists 228; of the group 3; problems of science and power 150; shifts 237; social organization 18; (see also Military, Boundaries)
President's Science Advisory Committee 150, 151
Preston, Richard 86
Primatology: 22; at Kyoto 243–244; background to 240–242; comparison 240; definition 241; dominance relationships 253; EuroAmerican 241–242; initiation of 241; individual and group 247–249; Japanese 242–244; long-term studies 243, 246–247; methodology 246–249; theoretical perspectives 245–246; Takasakiyama colony 244, 249–251, 253
Prism of heritability 17, 119
Protagoras 273
Psidium 53, 54, 55, 56, 58, 63
Public policy : 201, 228, 229, 231; a triage policy 201; and science 201, 213, 236; comanagement theory of 205–206

Rabi, I. I. 265
Race: 123–129; and second-order constructs 126–128; science & genetic explanations 123–124; "science" of race 124–126
Rappaport, Roy 71
Rationality: 15, 87, 216, 239, rational mind 98, 7, tradition of xii, of science 4, 6, 236, and relativism 218, and cultural knowledge 6, competency 100, utilitarian 99, non rational xii, 94, "negative" 97
Reagan, Ronald 110, 132, 154, 155
Reciprocity: 15, 75, 81–84; and asymmetry 74; human-animal 75–76; social and ecological 84
Reice, Seth 215
Reichel-Dolmatoff, Gerardo 71
Research: and politics 273; and public 27; multidisciplined 45, 229
Resolute Bay 219
Reynolds, V. 253
Richards, Paul 12
Richardson, Wendy 306
Riesenberg, Saul 41, 42
Ritual: and human-animal reciprocity 82, framing 15; of nuclear weapons testing (and initiation) 138–139, 143–145, 147; theory of 138–139
Rommetveit, R. 97
Rouse, Joseph 256
Rowell, Thelma 247
Rubus coriifolius 53, 55, 60, 63
Ryan, Joan (and Robinson) 226

Sachs, W. 11
Sade, Donald 252
Saipan 35, 36
Salk Institute 240
Salmond, Anne 216, 218
SALT I, II 132
Satawal 42
Saxe, Geoffrey 90
Scheler, Max 122
Schoch-Spana, Monica 306
Schultes, R. E. 43
Schwartz, Charles 18, 306
Science, types of: "big" and "little" 177, "cosmopolitan" 23; Cree 69; education 13; expert science 21; neolithic science 95; of domestic life 99; of mind 9; of schooling 99; practical 39–40, 42; protoscience 69; pseudoscience 4, 69; pure and applied 202

Index

Science: and bureaucracy 267–268; and consumers 13; and culture 22, 229–230, 274; and exclusion 13; and hierarchy 4, 20, 85, 268; and image 9, 144–145; and magic 70, 260; and the military 156–157; and myth 74, 95; and Nature 235–237; and non-science 87–88; and politics 9, 148–149, 158–159, 273; and power 85; and public controversy xi; and secrecy 232, 267; and society 8–12, 148, 201, 212–213; anthropology of 87, 239–240, 261; autonomy of 3, 23; broadened 101; commodification of xiii; conflict 217; cosmopolitan 23; cultural studies of 239; definition of 1, 69, 87, 101–102, 115, 213, 239, 269, 272; decolonization of 254–255; decontextualization of 3; history of term 3, 150; idealized 9; images 16; juxtaposed 20; literacy xiii, 16, 101; limitations of 6; militarization of xiii, 266; neutrality of 148–149; omnicompetence of 24; philosophy of 218; rationales 157; social concerns 123; superstition and 1; technology and science studies xiv; universality 7
Scientific: activity, exceptional nature of 99; beliefs 229–210, 231–236; discourse 99–100; experiment as ritual 18; literacy xiii, 15, 16, 111, defined 101; laboratories 153; technocratic nature of 115; national establishment 149; reports 255
Scientist : 95, 201; and power 228; and shaman 70; as company engineer 191; as expert 16, 273; bureaucrats 190; career strategies of 175; community of 17; graduates 191; non-scientists 190; post-doctoral 191; recruitment of 155, 157; role of 17; technicians 191
Scott, Colin 15, 306
Scribner, Sylvia 90
Shinkolobwe mine, S. Africa 232
Shiva, Vandana 11
Sickle-cell anemia 126
Skin color and hypertension 124
SLC 184, 185, 195
Smith, J. Maynard 208

Smith, M. Estellie 20, 307
Smithsonian xi, 9
Social responsibility and advocacy 18
Sofowara, A. 43
Soviet Union: and Japan 186–187; bomb project 234, 235, 236; technological infrastructure 232
Stevia ovata 53, 56, 61, 63, 66
Strategic Defense Initiative (SDI) 154, 156, 187–188
Strauss, Lewis (Atomic Energy Commissioner) 231
Strauss, Monica 195
Sugiyama, Yukimaru 254
Superconducting Super Collider (SSC) 155, 186
Surveillance: covert systems 21; Long Range Detection System 228
Syphilis 10

Tagetes lucida 53, 54, 55, 56, 61, 63
Tallensi 6
Tambiah, Stanley 4, 6, 70, 86
Taussig, Karen-Sue 306
Technology: attributes of humans and machines 145
Teller, Edward 144, 153
Test-ban cause 133
Theory: and ideal scientific practice 96; and religion 246; Chaos 20, 207–210, 212, 214; cognitive 88; evolutionary 5; of public policy 205–206; of ritual 141; realist 219–227; relativism 230, 237; sociobiological 248
Thomas, Stephen 40, 41
Three Mile Island 209
Tithonia 53, 56, 61, 63
Toledo 43
Tortoriello, J. 307
Traweek, Sharon 6, 8, 11, 19–20, 195, 255, 307
TRISTAN 183–185
Trobriand Islanders 5, 23, 260, 273, 274
Truk 33
Tuktoyuktak 225
Tuniq, Martha 220, 221
Turnbull, David 42
Turner, Victor 138

Tylor, Edward B. 4, 5
Tzeltal 6, 19, 43–68
Tzotzil 43–68

U. S. Air Force 231, 233
Umesao, Tadao 195
United Nations: xii, university 20, UNESCO 124, 125, 126
Universities: 150, 155–158, 229; and DoD 155–157; and secret research 156; and SDI 156; and Superconducting Super Collider 155; Hachioji University Center 176; Japanese 176; Kyoto University 176, 193, 244; MIT 188; Osaka 176, 183, 193, 244; University of California: board of regents 154, managers of 154, weapons laboratories 154, San Diego 153, 196; University of Washington, fisheries program 204; Yale 241

Vackimes, Sophia xi
Verbena 53, 54, 55, 56, 57, 63
Villarreal, M. L. 307
Von Heijne, Gunnar 164, 169

Wagner, Roy 73, 74, 75
Wald, Eliza 102
War on Drugs 129
Warren, D. M. 11
Weatherford, J. 7

Weber, Max 210
Weight Watchers 87, 91, 93, 96
Western xi, xiv, science 1, 3, 6, 10, 17 and rationalist tradition xii, epistemology 222
White, Lynn 3, 12
Whitehead, Alfred North 11
Williams, Raymond 3, 13
Wilson, Edward 248
Woolgar, S. 239
Worldviews: 1; and scientific image 16; and root metaphor 85, 114, 115; Christian 246; Cree 72–72; EuroAmerican 252–254; growth as religion 266; identity of "others" 98; inferior other 97; in scientific images 115; Japanese 240–252

Yap 29
Yerkes Primate Center 253
Yerkes, Robert 245
York, Herbert 133, 153
Young, A. xiv
Young, R. 261, 262, 273
Yup'ik Eskimo 223
Yupiak 1, 2, 23

Zapotec xiii
Ziegler, Charles A. 21, 305
Zuckerman, Solly 242

For Product Safety Concerns and Information please contact our EU
representative GPSR@taylorandfrancis.com
Taylor & Francis Verlag GmbH, Kaufingerstraße 24, 80331 München, Germany